T0139046

Green Chemistry
for Beginners

Green Chemistry for Beginners

With a Foreword by Paul Anastas

edited by

Rakesh K. Sharma | Anju Srivastava

JENNY STANFORD
PUBLISHING

Published by

Jenny Stanford Publishing Pte. Ltd.
Level 34, Centennial Tower
3 Temasek Avenue
Singapore 039190

Email: editorial@jennystanford.com
Web: www.jennystanford.com

British Library Cataloguing-in-Publication Data
A catalogue record for this book is available from the British Library.

Green Chemistry for Beginners

Copyright © 2021 by Jenny Stanford Publishing Pte. Ltd.
All rights reserved. This book, or parts thereof, may not be reproduced in any form or by any means, electronic or mechanical, including photocopying, recording or any information storage and retrieval system now known or to be invented, without written permission from the publisher.

For photocopying of material in this volume, please pay a copying fee through the Copyright Clearance Center, Inc., 222 Rosewood Drive, Danvers, MA 01923, USA. In this case permission to photocopy is not required from the publisher.

ISBN 978-981-4316-96-5 (Hardcover)
ISBN 978-1-003-18042-5 (eBook)

Contents

Foreword xv
Preface xvii

1. Genesis of Green Chemistry **1**
Anju Srivastava, Reena Jain, Manavi Yadav, and Rakesh K. Sharma
1.1 Introduction 1
1.2 Early History 3
1.3 Need for Green Chemistry: The Whys and Wherefores 8
1.4 Designing of the 12 Principles of Green Chemistry 11
1.5 Green Chemistry and Sustainable Development 20
1.6 Parameters to Evaluate Chemical Processes: *E*-Factor and LCA 22
1.7 Atom Economy 25
1.7.1 Atom-Economical Reactions 27
1.7.2 Atom-Uneconomical Reactions 27
1.8 Hazards and Risks in Chemistry 28
1.9 Learning Outcomes 29
1.10 Problems 29

2. Waste: A Misplaced Resource **33**
Anju Srivastava, Sriparna Dutta, and Rakesh K. Sharma
2.1 Introduction 33
2.2 Sources of Waste Generation 37
2.2.1 Chemical Wastes Generated from Industrial and Academic Sectors 37
2.2.1.1 Pharmaceutical wastes 37
2.2.1.2 Wastes from the academic research sector 43
2.2.2 Plastic Wastes 44
2.2.3 Electronic Wastes 47

2.2.4 Paper Wastes 50

2.3 Problems Associated with the Generation and Mismanagement of Waste 51

2.3.1 Global Case Studies Reflecting Mismanagement of Waste 53

2.3.1.1 Minamata mercury poisoning incident 53

2.4 Waste as a Resource 53

2.4.1 Biomass: A Renewable Feedstock 57

2.4.2 Biodiesel 58

2.4.3 Polymers from Renewable Raw Materials: Thinking Green 59

2.4.3.1 Bioplastics 60

2.4.3.2 Bioadhesives 61

2.5 Waste Minimization Techniques 61

2.5.1 Minimizing the Use of Derivatives in Chemical Processes: A Way toward Improving the Environmental Credentials of Chemical Synthesis 61

2.5.1.1 Sitagliptin 63

2.5.2 Recycling Reagents 64

2.5.2.1 Recycling reagents in chemical industries and laboratories 65

2.5.3 Miniaturization 68

2.5.4 Reduce, Reuse, and Recycle 70

2.5.4.1 Reduce 71

2.5.4.2 Reuse 71

2.5.4.3 Recycle 72

2.6 Design for Degradation 72

2.7 Conclusion 74

2.8 Learning Outcomes 74

2.9 Problems 75

3. Catalysis: A Promising Green Technology **79**

Manavi Yadav, Radhika Gupta, Gunjan Arora, and Rakesh K. Sharma

3.1 Introduction 79

3.1.1 What Is a Catalyst? 80

3.1.2 History of Catalysis 81

3.1.3 Catalytic Route vs. Stoichiometric Route: The Greener Aspect 82
3.1.4 Nobel Prize Awards in the Development of Catalysis 88
3.2 Role of Catalysis 89
3.3 Next-Generation Catalysts 90
3.4 Classification of Catalysts 91
3.5 Homogeneous Catalysis 91
3.5.1 Hydroformylation Reaction 91
3.5.2 Olefin Hydrogenation Using Wilkinson's Catalyst 92
3.5.3 Monsanto and Cativa Process 92
3.5.4 Reppe Carbonylation Process 93
3.5.5 Koch Reaction 94
3.6 Heterogeneous Catalysis 94
3.6.1 Haber–Bosch Process 95
3.6.2 Ziegler–Natta Polymerization 96
3.6.3 Ostwald Process 96
3.6.4 Contact Process 97
3.6.5 Catalytic Converters 97
3.7 Phase Transfer Catalysts 100
3.8 Asymmetric Catalysis 103
3.9 Nanocatalysis: Emerging Hybrid Catalysis 104
3.9.1 What Is Nanocatalysis? 104
3.9.2 Synthetic Approaches 105
3.9.3 Catalytic Applications 106
3.9.3.1 Metal nanoparticles 106
3.9.3.2 Metal oxide nanoparticles 109
3.9.3.3 Magnetic nanoparticles 110
3.10 Biocatalysis 112
3.11 Current Challenges and Future Development in Catalysis 113
3.12 Learning Outcomes 113
3.13 Problems 114

4. Alternative Reaction Media **119**

Radhika Gupta, Reena Jain, and Rakesh K. Sharma

4.1 Introduction 119
4.2 Need for Solvents 120
4.3 Problems Related to Traditional Solvent Use 120

4.4 Criteria for the Selection of Green Solvents 122
4.5 Green Solvents for Organic Synthesis 124
4.5.1 Water 124
4.5.2 Supercritical Fluids 129
4.5.2.1 Introduction to supercritical fluids 129
4.5.2.2 Properties of supercritical fluids 131
4.5.2.3 Supercritical CO_2 ($T_c = 31.1°C$, $P_c = 73.8$ bar) 132
4.5.2.4 Supercritical H_2O ($T_c = 374.2°C$, $P_c = 220.5$ bar) 134
4.5.3 Ionic Liquids 135
4.5.3.1 Introduction to ionic liquids 135
4.5.3.2 Properties of ionic liquids 136
4.5.3.3 Ionic liquids as solvents 137
4.5.4 Polyethylene Glycols 139
4.5.5 Organic Carbonates 141
4.5.6 Solvents Obtained from Renewable Resources 142
4.5.6.1 Glycerol 143
4.5.6.2 2-Methyltetrahydrofuran 145
4.5.6.3 Ethyl lactate 146
4.5.6.4 γ-Valerolactone 147
4.5.7 Fluorous Biphasic Solvents 148
4.5.7.1 Introduction to fluorous biphasic solvents 148
4.5.7.2 Advantages of using fluorous solvents 148
4.5.7.3 Fluorous biphasic system as a reaction media 149
4.6 Solvent-Free Synthesis 151
4.6.1 When At Least One of the Reactants Is a Liquid 151
4.6.2 Gas-Phase Catalytic Reactions 152
4.6.3 Solid–Solid Reaction 152
4.6.4 Benefits of Solvent-Free Synthesis 153
4.7 Immobilized Solvents 154
4.8 Learning Outcomes 155
4.9 Problems 157

5. Greening Energy Sources 161

Gunjan Arora, Pooja Rana, and Rakesh K. Sharma

5.1 Introduction 161
5.2 Microwave as a Greener Energy Source 164
5.2.1 Mechanism of Microwave Heating 164
5.2.1.1 What makes microwave technology superior to conventional heating? 165
5.2.2 Microwave-Assisted Chemical Reactions 167
5.2.2.1 Non-solid-state reactions 167
5.2.2.2 Solid-state reactions 171
5.2.3 Challenges Faced by Microwave Technology 173
5.3 Chemistry Using Ultrasonic Energy 173
5.3.1 How Sonochemistry Works 175
5.3.2 Factors Affecting the Cavitation Effect 176
5.3.3 Sonochemistry for Efficient Organic Synthesis 177
5.3.4 Applications in Wastewater Treatment 179
5.3.5 Challenges Faced by Sonochemical Processes 180
5.4 Visible Light–Driven Processes: Photochemistry 180
5.4.1 Classification of Photocatalysts 181
5.4.2 Basic Principle of Photochemistry 181
5.4.3 Photocatalytic Organic Transformations 182
5.4.3.1 Photochemical cycloaddition reactions 183
5.4.3.2 Photoinduced isomerization 184
5.4.3.3 Photodimerization effect in water 184
5.4.4 Industrial Applications of Photochemistry 185
5.4.4.1 Photonitrosation 185
5.4.4.2 Photo-oxygenation 185
5.4.5 Advantages of Photochemistry 186
5.4.6 Photocatalytic Degradation of Organic Pollutants 187

5.4.6.1 Mechanism of photocatalytic degradation of organic pollutants 188
5.4.7 Factors Affecting Photocatalytic Degradation 189
5.4.7.1 Effect of the concentration of organic pollutants 189
5.4.7.2 Effect of the catalyst amount 190
5.4.7.3 Effect of pH 190
5.4.7.4 Size, structure, and surface area of the photocatalyst 190
5.4.7.5 Effect of the reaction temperature 190
5.4.7.6 Effect of the light intensity and wavelength of irradiation 191
5.4.8 Challenges Faced by Photochemical Synthesis 191
5.5 Electrochemistry for Clean Synthesis 192
5.5.1 Basis of Electrochemical Synthesis 193
5.5.2 Types of Organic Electrochemical Synthesis 194
5.5.2.1 Anodic oxidative processes 194
5.5.2.2 Cathodic reductive processes 194
5.5.2.3 Paired organic electrosynthesis 194
5.5.3 Examples of Electrochemical Synthesis 195
5.5.3.1 Anodic oxidations 195
5.5.3.2 Cathodic reductions 196
5.5.3.3 Paired organic electrosynthesis 197
5.5.4 Advantages of Electrochemical Synthesis 198
5.5.5 Challenges Faced by Electrochemical Synthesis 199
5.6 Future Outlook 199
5.7 Learning Outcomes 199
5.8 Problems 200

6. Implementation of Green Chemistry: Real-World Case Studies **205**

Sriparna Dutta, Manavi Yadav, and Rakesh K. Sharma

6.1 Introduction 205
6.2 Synthesis of Valuable Compounds: Greener Protocols 206
6.2.1 Synthesis of Adipic Acid and Catechol 207
6.2.2 Synthesis of Styrene 208
6.2.3 Synthesis of Citral 209
6.2.4 Synthesis of Disodium Iminodiacetate 210
6.2.5 Synthesis of Acetaldehyde 211
6.2.6 Synthesis of Urethane 211
6.2.7 Selective Methylation of Active Methylene Using Dimethyl Carbonate 212
6.3 Some Real-World Cases: Green Chemistry Efforts Honored 213
6.3.1 Development of NatureWorks™ PLA: An Efficient, Green Synthesis of a Biodegradable and Widely Applicable Plastic Made from Corn (A Renewable Resource) 214
6.3.1.1 Greener alternative: Polylactic acid 215
6.3.1.2 NatureWorks LLC: Synthesizing PLA from corn 215
6.3.2 Healthier Fats and Oils via a Greener Route: Enzymatic Interesterification for the Production of Trans-Free Fats and Oils 219
6.3.2.1 Green chemistry: Interesterification of oils 222
6.3.2.2 Enzymatic interesterification of oils 223
6.3.3 Development of Fully Recyclable Carpet: Cradle-to-Cradle Carpeting 224
6.3.3.1 Green chemistry: Production of a cradle-to-cradle carpet 225
6.3.3.2 Recycling of PO-backed nylon-faced carpeting 226

6.3.4 Design and Development of Environmentally Safe Marine Antifoulant 226
6.3.4.1 Development of SeaNine™ 211: Environmentally safe antifoulant 228
6.3.5 Designing Rightfit™ Pigments to Replace Toxic Organic and Inorganic Pigments 230
6.3.5.1 Green chemistry innovation 231
6.3.5.2 Issues underlying and resolved 231
6.3.5.3 How the Rightfit™ pigments can be synthesized 233
6.3.6 Design and Application of Surfactants for CO_2 Replacing Smog-Producing and Ozone-Depleting Solvents for Precision Cleaning and Service Industry 235
6.3.6.1 CO_2 as a greener alternative 236
6.3.6.2 Benefits of scCO_2 237
6.3.6.3 How does a surfactant work? 238
6.3.6.4 Green chemistry innovation 238
6.3.6.5 Mechanism of action 239
6.3.7 Green Synthesis of Ibuprofen by BHC 239
6.3.7.1 Green chemistry innovation 240
6.3.8 TAML Oxidant Activators: General Activation of Hydrogen Peroxide for Green Oxidation Technologies 241
6.3.8.1 Green chemistry innovation 243
6.3.9 Simple and Efficient Recycling of Rare Earth Elements from Consumer Materials Using Tailored Metal Complexes 244
6.3.9.1 Green chemistry innovation 245
6.3.10 Using Naturally Occurring Protein to Stimulate Plant Growth, Improve Crop Quality, Increase Yields, and Suppress Disease 245
6.3.10.1 Green chemistry innovation 247
6.3.10.2 Mechanism of action 248

6.3.10.3 Advantages of harpin 248
6.3.11 Environmentally Advanced Wood Preservatives: Replacing Toxic Chromium and Arsenic with Copper and Quaternary Ammonium Compounds 249
6.3.11.1 Green chemistry: Removing arsenic and chromium from PTW 250
6.4 Need for Industry-Academia Collaboration 251
6.4.1 An Efficient Biocatalytic Process to Manufacture Simvastatin 253
6.4.2 Green Route for the Manufacture of Ranitidine 255
6.5 Conclusion 255
6.6 Learning Outcomes 257
6.7 Problems 257

7. Green Chemistry in Education, Practice, and Teaching 263

Reena Jain, Anju Srivastava, Manavi Yadav, and Rakesh K. Sharma

7.1 Introduction 263
7.2 Green Chemistry in Classroom 265
7.3 Green Chemistry in a Teaching Laboratory 270
7.4 Green Chemistry Institutes and Network Centers 274
7.5 Important Journals and Websites 277
7.6 Career Prospects 280
7.7 Learning Outcomes 281
7.8 Problems 281

8. Green Chemistry: Vision for the Future 283

Pooja Rana, Sriparna Dutta, Anju Srivastava, and Rakesh K. Sharma

8.1 Introduction 283
8.2 Challenges Lying Ahead of Green Chemistry 284
8.3 Future Directions: Focus of the Future Researchers 289
8.3.1 Nondepleting Nature 289
8.3.2 Nontoxic Nature 289

8.3.3 Nonpersistent Nature 290
8.4 Selective Reagents in Organic Transformations 290
8.4.1 Green Solvents 292
8.4.2 Dry Media Synthesis 292
8.4.3 Green Catalysis in Organic Synthesis 293
8.4.4 Catalyst-Free Reactions in Organic Synthesis 295
8.4.5 Energy-Efficient Synthesis 295
8.5 Miniaturization 297
8.5.1 Generic Goals of Miniaturization 297
8.5.2 Miniaturization in Pharmaceutical Industries 298
8.5.3 Miniaturization in Undergraduate Laboratories 299
8.6 Biomimetic: Green Chemistry Solution 300
8.7 Continuous Flow Technology 305
8.8 Combinatorial Chemical Technology 308
8.9 Green Chemistry and Sustainability 311
8.10 Conclusion 313
8.11 Learning Outcomes 314
8.12 Problems 315

Index 321

Foreword

The word green has many definitions and connotations. Often green has been the color of the environment conjuring up images of lush undisturbed forests or grasslands. To others in countries such as the United States, green is the color of money and is the symbol of positive economic growth and profitability. Still for others, green is the metaphorical traffic signal that means Go! and that all impediments are removed and you can proceed. Green chemistry embraces all of these simultaneously. Through an international community of brilliant scientists and engineers, green chemistry has designed, developed and implemented scientific and technological advances that promotes progress and enhance profits that benefit human health and the environment.

And yet there is another definition of green. Young, fresh, and new. Green chemistry embraces the new ideas, the new concepts, the new materials, the new transformations, the new products, the new processes that will create a sustainable society. This is most important for one over-riding reason. In order to keep this pipeline of creativity pumping out innovation, we need motivated, knowledgeable, ingenious students as the essence of the future. Those students need to see the vision of the future and enhance it. They need to know the foundation of today and build on it. They need to see the *status quo* and disrupt it. This is what we need to realize a sustainable tomorrow.

This book facilitates exactly that. Within the pages of this volume is the inspiration and the compass needed to guide students that are seeking to make a positive difference in the world. Students who are looking to distinguish themselves by how much good they can bring to society. Students who understand that success is not about how much money you make but rather how much value you can give.

My compliments to Professor R. K. Sharma for his long-term

dedication to advancing green chemistry in service to a sustainable society and in service to students. His relentless pursuit of green chemistry distinguishes his work and sets a model for his students to follow.

Paul T. Anastas
Yale University
U.S.A.

Preface

This book has been completed at a time when the novel coronavirus has engulfed the entire planet, killing hundreds of thousands of people and stalling economic activity across the world. This unprecedented outbreak can be traced back to the tenuous relationship we humans have had with the biodiversity. This generation-defining pandemic crisis compels us to stop and revisit our practices of exploiting nature and disrupting the integrity of the ecosystem to satiate our rapidly increasing appetite for goods and services. The planet has been showing signs of distress for a while: extreme weather events, extinction of life forms, fast-depleting fossil fuel resources, and other epidemic-level diseases.

If there is anything to learn from this moment, it is that mere concern for our environment is no longer sufficient. The questions that have been raised for decades now need to be answered urgently: how can we transition to low-waste, low-in-toxicity, low-energy practices in order to have tangible positive effects on the environment and society?

In the chemical sciences, green chemistry has developed as a philosophy with a tremendous potential to provide answers to these concerns. The principles of green chemistry promise to allow us to achieve sustainable development goals. Thus, it has never been more crucial for up-and-coming chemists (and even nonchemists) to have a grasp of green chemistry concepts and how they can be used to sustain human life on our fragile planet.

The 12 principles of green chemistry are the basic tenets that promote sustainable synthesis and innovations in the form of greener technologies that are broadly applicable to our work as chemists and the wider sciences.

This book has been designed to introduce high school and undergraduate students to the beautiful and captivating world of green chemistry. Beginning with what motivates its need and a discussion of its origins, the book takes the reader through the popular principles that are most commonly used to modify or replace reckless polluting methods, eventually discussing the future challenges and the future of this subject. The book also provides insights into efforts made by the pioneers in this field in encouraging real-world practitioners to embrace this science.

The book is divided into eight chapters, is written in a simple language for easy understanding, and is relevant for courses in clean technology and green chemistry, environment chemistry, and others. Each chapter has been concluded with the expected learning outcomes.

Chapter 1 takes us back several decades, to the emergence of this field, with the vision to curb pollution and reduce the environmental impact of chemical industries. The 12 principles of green chemistry, postulated by the father of green chemistry, Professor Paul T. Anastas, and Dr. John C. Warner, have brought a paradigm shift in the development of new products and processes that are compatible with human health and environment.

Chapter 2 focuses on the serious concerns pertaining to waste generation and attempts toward waste minimization through green chemistry. The chapter illustrates how green chemistry can resolve the dual problems of resource deficits and waste surpluses by turning waste into a resource itself. Effective projection of the various innovative waste management techniques—such as minimization of derivatives, recycling of reagents, miniaturization, the "three R" concept of the waste management hierarchy (reduce, recycle, and reuse), and the design for degradation—could drive us toward a zero-waste planet, one which is cleaner, greener, and healthier.

Catalysis, a key tool of green chemistry that is virtually behind every chemical we use today, is the focus of Chapter 3. Catalysts embody several advantages, with the main aim being moving away from stoichiometric reagents. This chapter covers various important transformations with the aid of chemical and biological catalysts. Chapter 4 provides a brief overview of some of the most promising alternative solvents, such as water, supercritical fluids, ionic liquids, polyethylene glycols, organic carbonates, biosolvents, and fluorous biphasic solvents. Given the lethal effects of conventionally practiced solvents, this chapter presents an effort to educate beginners about the potential benefits that these alternative green solvents offer in the context of sustainability. This chapter also discusses the ways in which organic reactions can be performed under solvent-free conditions.

The burning of fossil fuels is one of the key reasons for greenhouse gas emissions, which continues to be a serious threat to our environment. Clean, safe, energy-efficient, and convenient sources of energy are thus being explored vigorously. In Chapter 5, some commonly used nonconventional sources of energy are discussed. Specific attention has been paid to the introduction, advantages, applications, and limitations of microwave technology, ultrasonic energy, photochemistry, and electrochemistry. Shorter reaction times, by-product elimination, better yields, improved selectivity, and homogeneous heating are the major advantages offered by these alternative energy sources.

We are beginning to realize the indispensable need to empower our succeeding generations with the right skills as well as the knowledge to practice chemistry in a benign manner. This can be accomplished through the effective projection of the implementation of green chemistry in the real world. Chapter 6 showcases green chemistry in action by illustrating some of the most stimulating real-world case studies, broadly based on the US Presidential Green Chemistry Challenge Awards. This chapter also sheds light on some of the most innovative green processes adopted by industries for the synthesis of valuable compounds. Each of the examples in this chapter illustrates how the various principles of green chemistry could be put to practice and what would be the underlying advantages of the redesigned greener protocols. The need for academic-industrial collaborations in the current global scenario has also been highlighted with the aid of interesting examples.

There is a big mismatch between the way chemistry is taught at secondary and senior secondary levels and the global initiatives that guide scientific and public sustainability discourse. It is, therefore, pertinent that the worldwide status of green chemistry education be highlighted to chemists, educators, and policy makers.

This is the discussion done in Chapter 7, with efforts made by the pioneers in pulling out green chemistry from the side columns and boxes in chapters and integrating it with the mainstream text. Ultimately, the mission is to remove the word "green," that being the only way chemistry is taught and understood. It is crucial to understand how to introduce to stakeholders the concept of green chemistry in the classrooms, laboratories, and research so that it captures their attention and they start thinking of providing solutions to real-world problems.

Chapter 8 summarizes the various future directions of research that require an extensive focus while designing and developing any new chemical process or product. The significance of this chapter is to communicate to the readers a brief sense of the various promising innovative techniques, such as biomimetic, continuous flow, combinatorial chemical technology, and miniaturization, that reflect the need of employing traditional science coupled with emerging systems thinking and systems redesigning for accomplishing high performance in terms of both primary functionality as well as sustainability.

All the chapters are explained with ample examples and real-world cases as far as possible. We are truly privileged that the father of green chemistry, Professor Anastas, agreed to write the foreword for this book, which is a huge honor but at the same time has raised the bar in terms of our responsibility and public expectations.

We hope the readers find the book worth their time and interest.

Rakesh K. Sharma

Anju Srivastava

January 2021

Chapter 1

Genesis of Green Chemistry

Anju Srivastava,[a] Reena Jain,[a] Manavi Yadav,[a,b] and Rakesh K. Sharma[b]

[a]*Department of Chemistry, Hindu College, University of Delhi, Delhi 110007, India*

[b]*Green Chemistry Network Centre, Department of Chemistry, University of Delhi, Delhi 110007, India*

dr.anjusrivastava@gmail.com

> *We do not inherit the earth from our fathers, we are borrowing it from our children.*
>
> —David Brower

1.1 Introduction

The past few decades of the twentieth century saw chemistry contributing significantly to the advancement of human civilization. Chemists, with their magical powers to play with the chemical molecule, have been looked at as the problem solvers of the society. They have synthesized crop-enhancing chemical fertilizers and pesticides that enhance crop yields to ensure a constant and

Green Chemistry for Beginners
Edited by Rakesh K. Sharma and Anju Srivastava
Copyright © 2021 Jenny Stanford Publishing Pte. Ltd.
ISBN 978-981-4316-96-5 (Hardcover), 978-1-003-18042-5 (eBook)
www.jennystanford.com

sustainable food supply. Modern medicines, health-care products, advanced diagnostic chemical tests, and analysis have played a key role in the eradication of hard-hitting diseases and infections, increasing the average life of humans from 45 to 75 years [1]. The innovative products arising out of research in chemistry and its interplay with other disciplines have led to cutting-edge advancements and commercially viable applications in numerous fields—transportation, computing, aerospace, electronics, and more. This contribution in specialty applications is over and above the role chemistry plays in our everyday lives by providing consumer products, such as our local grocery, hardware, detergents, clothes, plastics, adhesives, lubricants, paints, coatings, and even fire retardants.

Undeniably, therefore, without the continuous efforts of chemists and the enormous productivity of the chemical industries, the high standards of modern society could not have been attained. And in these enhancements, the environment has been used as a source of natural resources that can be exploited for the society by the scientific tool called chemistry. However, the rapid pace of development and application of chemical sciences has come at a heavy price, where the misuse of the otherwise fascinating science has led to the release of pollutants and toxic substances into land, air, and water and the production of nonbiodegradable materials, resulting in a harmful impact on the environment and living beings.

The scientific development of and research on molecules and products, until quite recently, was being done with a focus only on the desired molecules or products, with a very superficial understanding of the fate and toxicity of the chemicals in the environment. Obviously, there was no awareness and intent among scientists and policy makers to study any deleterious large-scale or global impacts of the development of new molecules and chemicals on the environment and health. In most cases, therefore, the negative and destructive impacts of chemicals were realized only when they had reached alarmingly high levels. Numerous examples can be discussed to support this fact.

In fact, even the environment problems for which chemicals have a negative public image can be mitigated with long-term sustainable solutions using the knowledge of chemistry. As is often said, chemical problems require chemical solutions.

In the subsequent sections, the reader will be taken through the various infamous disasters affecting the environment and life that have over time led to the emergence of green chemistry and its philosophy.

1.2 Early History

In the 1940s, the chlorine-based hydrocarbon dichlorodiphenyltrichloroethane (DDT) was developed as a pesticide and was found to be very effective in eradicating not only pests but also the vectors causing diseases like malaria and yellow fever. This proved to be an elixir for both the agriculture and health sectors and distinguished itself during World War II, clearing malaria-causing insects for US troops. The indiscriminate use of DDT proliferated, until in 1962, Rachel Carson, a marine biologist and writer, courageously published her book *Silent Spring*, which exposed the hazards of the pesticides manufactured by chemical industries. The book described how these chlorine-containing compounds entered the food chain and bioaccumulated to pose as a threat to the survival and existence of several species of birds and other living beings. *Silent Spring* helped in the launch of a new public awareness that nature was vulnerable to human intervention. Ten years after the book's publication, the government banned the use of this pesticide [2, 3].

Around the same time, in Europe, pharmaceutical companies were selling a newly synthesized thalidomide-based drug for treating morning sickness and nausea in pregnant women. While the medicine proved to be very effective and brought a wave of joy among the expecting mothers, the children born to these mothers were found to have acute birth defects with deformed or missing limbs. These thalidomide babies were born because the drug had its enantiomers acting as a teratogen, which was being synthesized inadvertently along with the actual drug. The infamous incident caused the birth of 10,000 such deformed babies worldwide, with as many as 5000 such babies in Germany alone. The drug was immediately banned from use and was removed from all the pharmacists' outlets [4].

Invented in the 1930s, chlorofluorocarbons (CFCs), popularly called freons, were considered a class of wonder hydrocarbons that, due to their low toxicity and nonreactive nature, found numerous industrial and consumer applications, including as refrigerants, propellants, hospital sterilant, industrial solvents, and foam-blowing agents. Their worldwide market increased multifold in a very short span of time. After about 40 years of indiscriminate use of these so-called benign compounds, it was found in 1974 that these compounds, when released into the atmosphere, diffuse to the stratosphere, where their chlorine component depletes and destroys the ozone layer, which acts as a protective shield against the harmful ultraviolet radiations, which otherwise are dangerous for life on this planet. When a definite link was established for this observed relation, CFCs were banned in 1987 under the Montreal Protocol [5].

In the early 1970s, an oil slick and debris within the Cuyahoga River caught fire in Cleveland, Ohio, US, drawing national attention to the environmental problems in the United States. This was not the first time that the river had caught fire. Fires had occurred on the Cuyahoga River several times earlier also. In fact, one such fire in 1952 caused damage worth over US$1.5 million. Another incident, which in particular brought this problem into spotlight, was the Love Canal site, Niagara Falls, New York, US. It was a chemical waste dumpsite constructed in an old abandoned canal where approximately 82 waste chemicals, including a number of suspected carcinogens, such as benzene, chlorinated hydrocarbons, and dioxin, had been dumped. Later, the site was sealed and a residential locality was constructed over it. However, the chemicals began to ooze out over time and affected people of that area by leading to severe birth defects and miscarriages. The entire town had to be evacuated, which led to an expenditure of many millions of US dollars to relocate nearly one thousand households and to remediate the site [6].

Times Beach, a small town in in Missouri, USA, used to spray used motor oil on the unpaved streets to settle the rampant dust in the area. The hired person started spraying waste oily chemicals along with the oil. These chemicals, generated as by-products of the production of chemical disinfectants, were toxic (dioxins). Oblivious to the toxic nature of the oil spray, the town continued this practice for four years, when widespread unexplained deaths of birds and animals and poisonous effects of dioxin manifesting as acute headaches, diarrhea, and nose bleed among residents were observed. In 1982, soil tests showed the alarmingly high levels of dioxin present. As if this was not enough, a flash flood in the area worsened the situation, with more people coming in contact with the hazardous chemicals. The site was at once declared uninhabitable and evacuation of the town was ordered immediately [7].

In 1984, one of the worst industrial catastrophes occurred when an accidental release of highly toxic methyl isocyanate (MIC) gas, in the pesticide manufacturing plant at Union Carbide India in Bhopal, Madhya Pradesh, India, on the night of December 2–3, killed nearly 3600 people in their sleep, 8000 in the following week, and another 8000 over a period of time. Over 200,000 people suffered from long-term effects due to the disastrous exposure to this poisonous gas, including birth defects in children born to affected women. The plant was shut down after the incident [8].

On April 26, 1986, the worst nuclear disaster in history unfolded at the Chernobyl nuclear power station in Ukraine, in the former Soviet Union. This accident was followed by the release of huge amounts of radioactive materials into the environment that necessitated the evacuation of tens of thousands of people and farm animals from the surroundings. This accident was a result of a flawed reactor design and inadequately trained plant operators [9].

Amongst disastrous oil spills, the March 24, 1989, Exxon Valdez oil spill left a deep impact. It occurred when an oil tanker owned by Exxon Shipping Company spilled millions of gallons of crude oil into Prince William Sound in Alaska, one of the most ecologically sensitive areas in the world. Within a few days, this oil slick covered nearby beaches and killed hundreds of thousands of seabirds, seals, otters, and whales [10].

Since then, the world has seen massive oil spills, nuclear plant meltdowns, explosions at chemical plants, and waves of toxic sludge flooding towns and rivers. Figure 1.1 illustrates a few such environmental disasters.

Major Environmental Disasters in India

- Bombay docks explosion, 1944
- Chasnala mining accident, 1975
- Bhopal gas tragedy, 1984
- Jaipur oil depot fire, 2009
- Korba chimney collapse, 2009
- Styrene gas release, 2020

Figure 1.1 (a) Bhopal gas tragedy (1984, India), (b) Chernobyl nuclear accident (1986, Ukraine, erstwhile USSR), (c) Cuyahoga River fire (1969, USA), (d) Los Angeles smog (1943, USA), (e) Love Canal disaster (1984, USA).

These incidents threw light on the dangerous extent to which human activities were contaminating the various segments of the environment, which compelled several countries to deliberate and pass regulatory laws that would reduce the release of pollutants and help clean up waste chemical sites. On June 5, 1972, the first ever UN conference marked a turning point in the management of natural resources and maintenance of a clean environment. It was thereafter decided to celebrate this day as the World Environment Day. Additionally, the Environmental Protection Agency (EPA) was constituted in 1970 in the United States.

In 1987, the Brundtland Commission published a report to reconcile environmental issues and defined "sustainable development," for the first time, as development that meets the needs of the current generation without compromising the future generation. This report also highlighted the dangers of ozone depletion and its effects on global warming. Figure 1.2 provides a time line of the evolution of green chemistry [11–13].

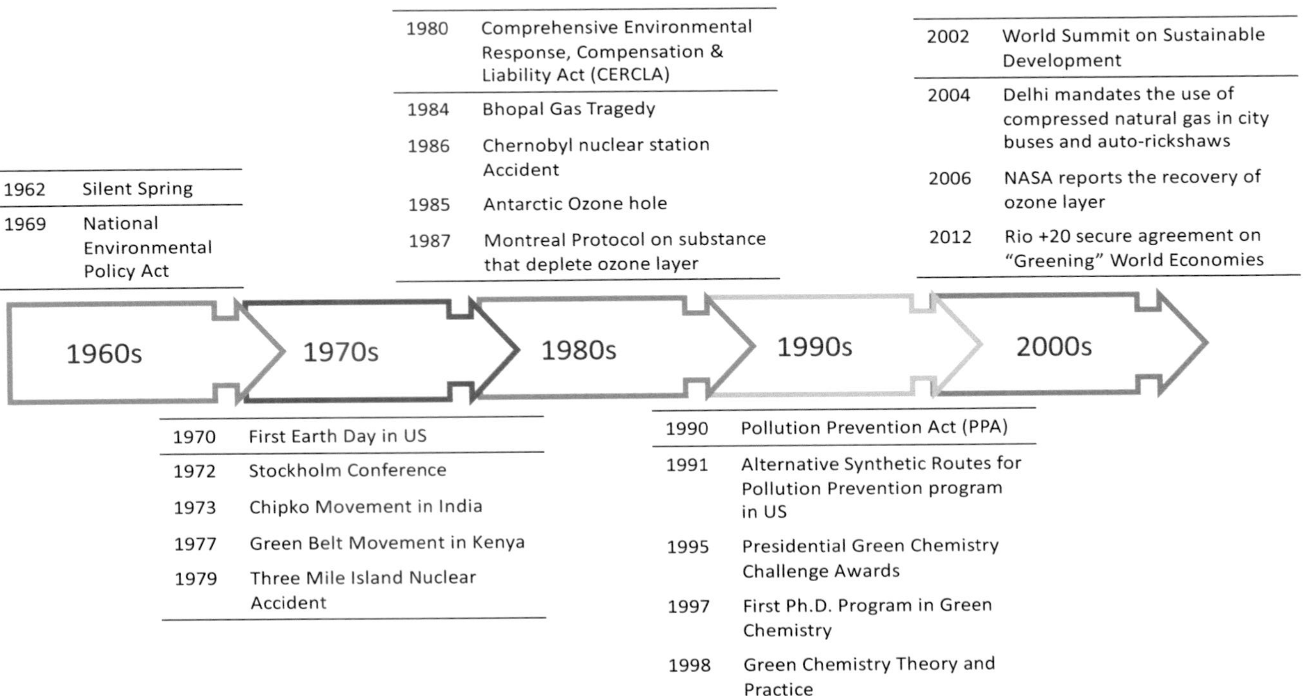

Figure 1.2 Time line representing the origin of green chemistry—from chemical accidents to environmental movements.

Regulatory Agencies in India

With the rapid pace of industrialization, India has its fair share of environmental issues, such as industrial accidents, air and water pollution, water scarcity, soil erosion, and deforestation. Following the Stockholm Declaration, the Department of Science and Technology established the National Council for Environmental Policy and Planning to protect the environment. Later, this council became the Ministry of Environment and Forest, one of the most important governmental agencies in India. The Central Pollution Control Board was founded in 1974, which came up with the National Air Quality Monitoring Program and in 1986 the Environment Protection Act. Till date, India has framed many legislative rules and acts for environmental protection and preservation.

Recently, a styrene gas leak, reminiscent of the Bhopal disaster, occurred in Visakhapatnam, Andhra Pradesh, India, that claimed 11 lives and affected thousands of residents in nearby villages [14]. All these examples reflect how chemicals can act as double-edged weapons.

1.3 Need for Green Chemistry: The Whys and Wherefores

Prior to the advent of laws that control the release of chemical wastes into the environment, it was believed that mere dilution in the concentration of a chemical substance in an appropriate medium would be sufficient to alleviate its ultimate impact. However, this philosophy was soon found to be absurd when factors such as chronic toxicity, bioaccumulation, and carcinogenicity emerged and started affecting the lives of human beings. Another approach was identified by the environmental law, namely the "command and control" method, which specifies the maximum concentration level of a substance that could be released into any particular receiving stream (air, water, land). It uses standards or guidelines such as maximum concentration guideline levels (MCGLs) to strictly control the amount of a particular chemical, thereby limiting the discharge of pollutants. This approach has also resulted in the implementation of several technologies that deal with pollutants and transform waste into innocuous form and are often termed as "end-of-pipe measures." Examples include neutralization of acidic pollutants, chemical precipitation of water pollutants, stack gas scrubbing, and waste immobilization. However, this method does not consider the synergistic effects of the released substances with the other substances already present in the environment and, thus, cannot be considered as an environmentally benign solution.

In the 1990s, the United States passed the Pollution Prevention Act, stating that wherever possible, in the first place, waste should not be generated and if generated, it must be minimized [15]. This could be achieved via several means, ranging from very simple approaches, such as careful inventory control and reduction of unnecessary solvent evaporation, to much more sophisticated pathways, such as complete redesigning of manufacturing processes keeping waste minimization as a first priority. Shortly after this, in 1991, the EPA launched the Alternative Synthetic Routes for Pollution Prevention program in the United States, which brought about a new outlook on controlling the risks associated with toxic chemicals, giving priority to nonproduction of these materials. This was followed by the expansion of new approaches, such as the use of environmentally friendly solvents and safer chemicals. This new approach was later named as "green chemistry."

Even though there was a progression in environmental regulations, what was still imperative was the desire to keep searching for sustainable solutions to address these increasing global issues that are resulting in the continued deterioration of the environment. With a mission to find strategic solutions to environmental protection and conservation, green chemistry was introduced. This term was first coined in the early 1990s by Paul T. Anastas, who was then working at the EPA Office of Pollution Prevention and Toxics. Since then, he is widely known as the father of green chemistry. Green chemistry is defined as the design of chemical products and processes that reduce or eliminate the use and production of toxic or hazardous substances. It is not a separate branch of chemistry but is rather considered as a code of conduct to minimize reagents, steps, costs, and energy, thereby reducing the environmental impact of chemical processes. More specifically, it focuses on environmental protection, not by cleaning waste but by designing new chemical pathways that do not generate pollutants. Thus, green processes are benign by design and are all about protecting health and environment while promoting innovation and increasing profits [1, 16]. It is a rapidly evolving and important area of chemical sciences that will eventually bring back the perspective that chemicals are not bad but are needed for the development of mankind [17–19].

Soon after the research in green chemistry gained momentum, leaders of innovation from both academia and industry working in the field of green chemistry were recognized by an annual awards program called the Presidential Green Chemistry Challenge (PGCC), launched by the EPA in 1995 [20]. This is the only award that is presented by the president of the United States explicitly for work in chemistry. Shortly after this, in 1997, the world's first PhD program began in green chemistry at the University of Massachusetts, Boston, US, founded by John C. Warner with the goal to educate the new generation about cleaner and safer chemistry. In the same year, the Green Chemistry Institute (GCI), a nonprofit institute, was created for advancement in the green chemistry domain. Later, the GCI joined the American Chemical Society to address global issues in the field of chemistry and environment.

The groundbreaking book on green chemistry entitled *Green Chemistry: Theory and Practice* by Anastas, coauthored by Warner, was published in 1998 [21]. This book is another important landmark in the field of green chemistry. It gives a precise definition of green chemistry and enumerates its 12 principles, which have become the generally accepted guidelines for green chemistry. The book soon became so popular that it was translated into numerous languages. Since the 1990s, various scientific journals have dedicated their research to green chemistry and published plenty of articles on this field. As of now, all the major publishing houses have at least one journal exclusively on green and sustainable chemistry and more than 40 nations and networks from all over the world are involved in promoting green chemistry. With time, the green chemistry community realized the needs and gaps and has thrived gradually.

1.4 Designing of the 12 Principles of Green Chemistry

In the book *Green Chemistry: Theory and Practice*, the following 12 principles of green chemistry have been stated (Fig. 1.3):

Principle 1: It is better to prevent waste than to treat or clean up waste after it is formed.

Principle 2: Synthetic methods should be designed to maximize the incorporation of all materials used in the process into the final product.

Principle 3: Wherever practicable, synthetic methodologies should be designed to use and generate substances that possess little or no toxicity to human health and the environment.

Principle 4: Chemical products should be designed to preserve efficacy of function while reducing toxicity.

Principle 5: The use of auxiliary substances (e.g., solvents, separation agents, etc.) should be made unnecessary wherever possible and innocuous when used.

Principle 6: Energy requirements should be recognized for their environmental and economic impacts and should be minimized. Synthetic methods should be conducted at ambient temperature and pressure.

Principle 7: A raw material of feedstock should be renewable rather than depleting whenever technically and economically practicable.

Principle 8: Unnecessary derivatization (blocking groups, protection/deprotection, temporary modification of physical/ chemical processes) should be avoided whenever possible.

Principle 9: Catalytic reagents (as selective as possible) are superior to stoichiometric reagents.

Principle 10: Chemical products should be designed so that at the end of their function they do not persist in the environment and break down into innocuous degradation products.

Principle 11: Analytical methodologies need to be further developed to allow for real-time, in-process monitoring and control prior to the formation of hazardous substances.

Principle 12: Substances and the form of a substance used in a chemical process should be chosen so as to minimize the potential for chemical accidents, including releases, explosions, and fires.

Figure 1.3 The 12 principles of green chemistry.

These green chemistry principles have provided a roadmap to researchers, scientists, and technologists in their quest to reduce or eliminate hazardous substances by designing appropriate pathways in a scientifically sound and cost-effective manner [22]. Some examples of their applications to basic and applied research are given next.

Prevent waste: A traditional method to produce phenol involves benzene, sulfuric acid, and sodium hydroxide as reactants in a multistep reaction. Scheme 1.1 depicts the overall reaction.

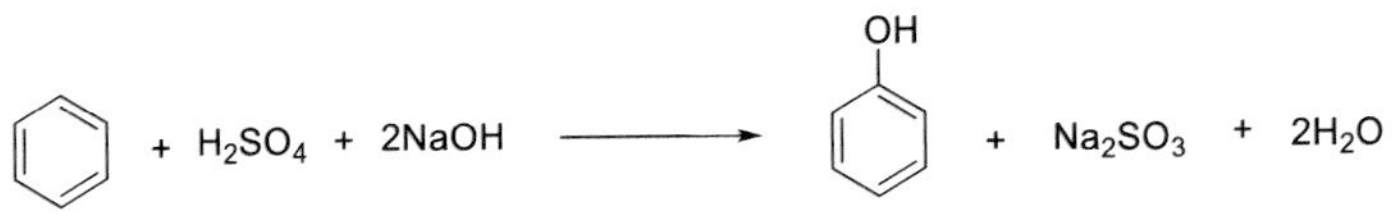

$$C_6H_6 + H_2SO_4 + 2NaOH \longrightarrow C_6H_5OH + Na_2SO_3 + 2H_2O$$

Scheme 1.1 Conventional method of manufacturing phenol.

This chemical equation shows that 1 mole of benzene (78 g) should yield 1 mole of phenol (94 g). However, the actual yield (Eq. 1.1) is about 82% (77 g), which is quite good, but this reaction also generates 1 mole of sodium sulfite (126 g) for each mole of phenol produced. This side-product poses a problem of waste management and adds significantly to the costs. An alternative process involves the reaction of benzene with propene and dioxygen. In this, phenol is produced in multiple steps, but with propanone as a coproduct, which is a valuable chemical. The net reaction is represented in Scheme 1.2.

$$\%\text{yield} = \frac{\text{Actual yield of products}}{\text{Expected yield of products}} \times 100\% \tag{1.1}$$

$$C_6H_6 + CH_2{=}CHCH_3 + O_2 \longrightarrow C_6H_5OH + CH_3COCH_3$$

Scheme 1.2 Alternative method to manufacture phenol.

Maximize atom economy: Atom economy is the concept of evaluating the efficiency of a chemical reaction and is expressed as a ratio of the total mass of atoms in the desired product to the total mass of atoms in all the reactants (Eq. 1.2). The nearer its value is to 100, the lesser will be the waste generated.

$$\text{Atom economy} = \frac{\text{Relative molecular mass of the desired product}}{\text{Relative molecular mass of all reactants}} \times 100\% \tag{1.2}$$

For example, phenol produced from benzene and propene has 100% atom economy since the coproduct has an appreciable value. In comparison, the atom economy is only 37% when phenol is manufactured by the conventional method, where sodium sulfite is obtained as a side-product. Another example of 100% atom economy is the Diels–Alder reaction (Scheme 1.3), where all the atoms of reactants are involved in the final product.

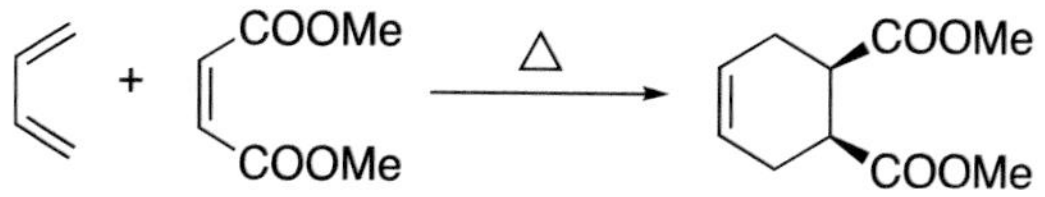

COOMe + COOMe → (Δ) COOMe COOMe

Scheme 1.3 Diels–Alder reaction.

Synthesize less hazardous chemicals: CFCs, chief cause behind ozone depletion, global warming, and smog, were initially used to manufacture polystyrene and have now been replaced by nontoxic carbon dioxide. Another example is the synthesis of polycarbonates that have high strength and optical properties. These are generally obtained by the reaction of bisphenol A (Fig. 1.4) with carbonyl chloride (Scheme 1.4). However, carbonyl chloride is a highly poisonous gas and hence has now been substituted by diphenyl carbonate, which is produced from dimethyl carbonate, which in turn is readily obtained from methanol, carbon monoxide, and oxygen in the presence of copper (II) chloride (Scheme 1.5). Therefore, the overall process of manufacture of polycarbonate that employs diphenyl carbonate is less hazardous than the one that employs carbonyl chloride.

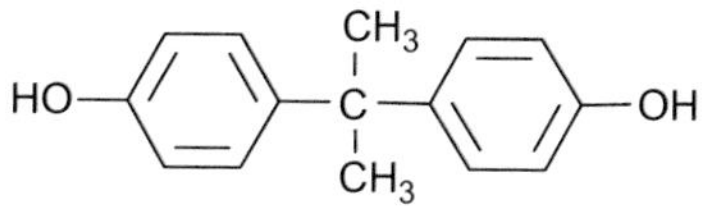

HO–C₆H₄–C(CH₃)₂–C₆H₄–OH

Figure 1.4 Bisphenol A.

$$CO(g) + Cl_2(g) \longrightarrow Cl_2C{=}O$$

Scheme 1.4 Formation of carbonyl chloride.

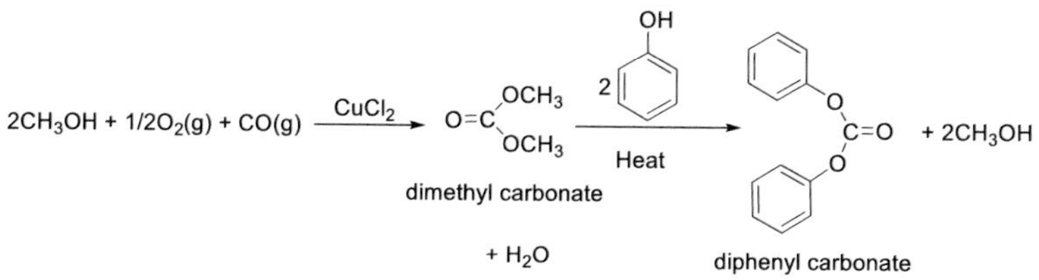

$2CH_3OH + 1/2O_2(g) + CO(g) \xrightarrow{CuCl_2} O{=}C(OCH_3)_2$ (dimethyl carbonate) $+ H_2O$; dimethyl carbonate + 2 PhOH $\xrightarrow{Heat}$ $(PhO)_2C{=}O$ (diphenyl carbonate) $+ 2CH_3OH$

Scheme 1.5 Formation of diphenyl carbonate.

Design safer chemicals: Pharmaceutical products are usually chiral in nature, where one enantiomer can be beneficial and the other can have detrimental effects. For instance, racemic thalidomide taken during pregnancy was found to have curing effects as well as produce birth defects due to different properties of the enantiomers. Hence, it is important to produce two chiral forms separately. This can be done by using catalysts that can enhance the rate of the desired reaction and produce one of the two chiral molecules. Another example is the synthesis of polymers that not only are less flammable but also have high toughness to be able to absorb more severe impacts than the popular ones. Polyphenylsulfone is one such polymer, with the formula depicted in Fig. 1.5.

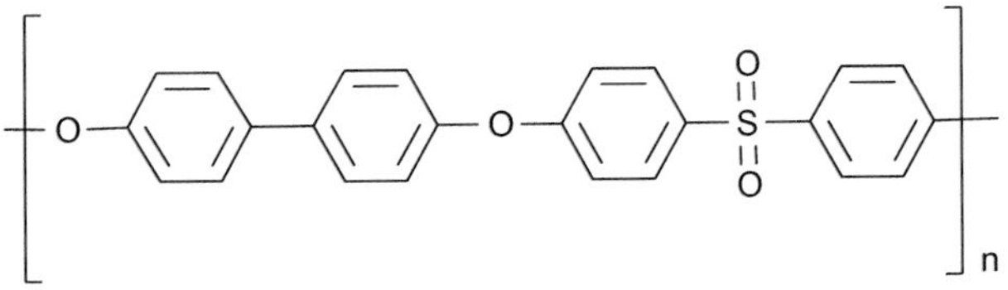

Figure 1.5 Structure of polyphenylsulfone.

Use safer solvents and auxiliaries: Chlorinated solvents such as CCl_4, $CHCl_3$, perchloroethylene, and CH_2Cl_2 are toxic and volatile and are responsible for the depletion of the ozone layer. Other toxic solvents benzene, toluene, etc. Purification techniques such as recrystallization and chromatographic separation use large amounts of solvent and generate wastes. Some alternatives are discussed in Table 1.1 [23].

Table 1.1 Greener alternatives to toxic solvents

Safer green solvents	Properties
Ionic liquids (ILs)	ILs are liquids at room temperature and below. They are nonvolatile, have no vapor pressure, can provide nonaqueous reaction media of varying polarities, do not require any special apparatus or methods to carry out reactions in them, and can be reused. These properties make ILs a suitable green alternative to toxic solvents.

(Continued)

Table 1.1 (*Continued*)

Safer green solvents	Properties
Supercritical fluid (supercritical CO_2 and supercritical water)	• Supercritical CO_2: It has been an excellent alternative to chlorinated solvents due to its low viscosity, nonflammability, zero surface tension, and low toxicity. It has the unique ability to diffuse through solids and dissolve a wide range of organic compounds. Besides, it is stable and most of the compounds can be easily extracted from it. CO_2 is the only by-product and easily evaporates, leaving no residue behind, thereby making it an inexpensive choice. It is widely used as a solvent in caffeine extraction from coffee beans and also in the dry-cleaning equipment. • Supercritical water: Organic substances are insoluble in water, but most of them become soluble in water when it becomes supercritical at 374°C and 218 atm. Hence, this can be used as a green solvent for many organic transformations.
Solvent free	No solvent is the best solvent. Some chemical reactions run under neat conditions.
• Reactions in solid phase	• Numerous reactions take place in the solid state that are simple to operate and economical and prevent solvent-based issues.
• Reactions in gaseous phase	• Examples of reaction in the gas phase include synthesis of ammonia, methanol, and ethene.
Water	Due to limited chemical compatibility, use of water as a solvent was not known until it found success in accelerating rates of some reactions due to its high polarity. Reactions carried out in an aqueous medium include the Diels–Alder reaction, oxidations, reductions, epoxidations, and polymerizations (with or without catalysts). Water-borne paints have substituted volatile organic compounds, such as the hydrocarbons.
Immobilized solvents	Solvents immobilized on polymer beads minimize the volatility associated with organic solvents and offer efficiency in separation through simple filtration, in comparison to tedious procedures like rotary evaporation and distillation.

Design for energy efficiency:

- Microwave (MW) irradiation: Reactions using MW sources have been performed on solid supports that have reduced or eliminated solvent usage, and these reactions occur at a faster rate in comparison to conventional heating. Beckmann rearrangement of oximes is one of the examples in the solid state that gave good product yields without the use of acid catalysts.
- Sonochemistry (ultrasound energy): Another energy source is ultrasound that can be carried out at room temperature. Ullmann's coupling is an excellent example that gives an appreciable yield at low temperatures and in lesser time under ultrasound energy, which otherwise takes place at a high temperature and gives a poor yield when carried out using the traditional method.

Use renewable feedstocks: An example of renewable feedstocks is the replacement of benzene by inexhaustible and nontoxic glucose for synthesizing adipic acid, which is required in the production of nylon, plasticizers, and lubricants, and the reaction is carried out in aqueous media. Some other examples are as follows:

- Surfactants that are readily biodegradable and are derived from plant resources
- Polyols made from soya and used to produce polyurethanes
- Ethene obtained from bioethanol and used to make biobased polyethene.

Avoid chemical derivatives: A typical example of use of a blocking group is the protection of an alcohol by converting it into benzyl ether, which on undergoing oxidation oxidizes the other part of the molecule, leaving the alcohol unaffected. Afterward, the alcohol is regenerated through benzyl ether cleavage. This type of derivatization is extremely important for synthesizing fine chemicals, pharmaceuticals, dyes, and pesticides. However, handling reagents like benzyl chloride (a known hazard) can be dangerous because they are not only being used but are regenerated as waste in the deprotection step. Another problem lies with the temporary modifications in properties like viscosity, vapor pressure, dispersibility, and water solubility to improve the performance of certain compounds. These modifications often result in the generation of a large amount of waste [24].

Use of catalysts: Aluminum chloride has been used to catalyze synthetic reactions, for example, in the production of alkylbenzene sulfonates (the surfactant in many detergents) and phenol. However, it poses the problem of waste disposal due to its nonrecyclable nature. It has now been substituted by alternative solid-supported catalysts, which can be recycled and reused, such as zeolite and silica. These supports are much more benign, and further modifications on their surface can enhance their properties. A solid zeolite with acid groups can catalyze a number of reactions. One example is its use in converting benzene and propene to cumene during the manufacture of phenol (Scheme 1.6). A similar example that involves zeolite acid as a catalyst is the manufacture of nylon 6, a polymer used in fabrics. This has eliminated the use and waste of sulfuric acid in the rearrangement step (Scheme 1.7).

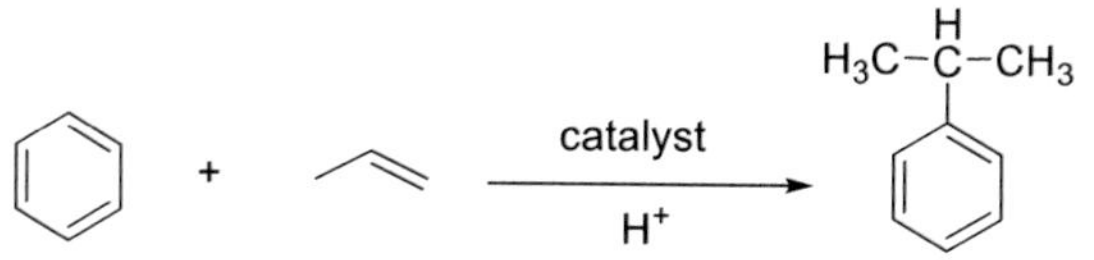

Scheme 1.6 A solid zeolite with acid groups as a catalyst for the conversion of benzene and propene into cumene.

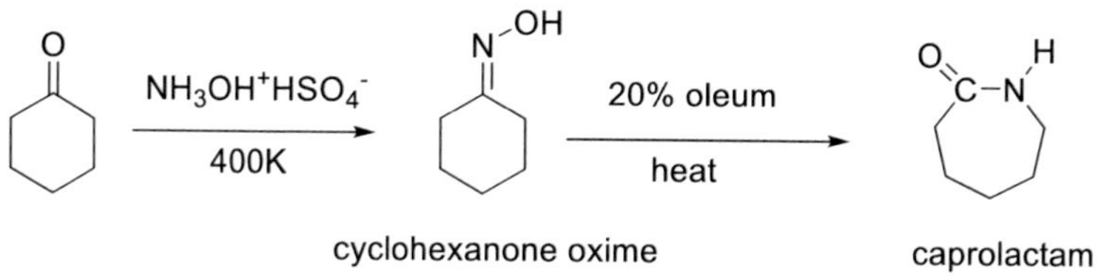

Scheme 1.7 Isomerization of oxime into caprolactum by a solid acid zeolite catalyst.

Another example is a new process called hypochlorite decomposition catalysis (HYDECAT), which was designed to remove hypochlorite as the by-product from effluents in the sewage, generated during chlorination processes (used to disinfect wastewater), which may be carcinogenic and harmful to aquatic species. In this process, finely divided nickel is dispersed on an inert solid through which the effluent passes. The nickel reduces the hypochlorite ion to a chloride ion and oxygen gas. The overall reaction is represented in Scheme 1.8.

$$2OCl^- \longrightarrow 2Cl^- + O_2$$

Scheme 1.8 HYDECAT process.

Design for degradation: A number of researchers have addressed the problems posed by plastics by developing biodegradable plastic products. For example, plastic cups are made with polylactic acid (PLA), which is a biodegradable polymer derived from renewable feedstocks. Biological pesticides help control pests with microorganisms or natural products.

Develop analytical methodologies for real-time monitoring: With the development of both in-process and real-time analysis, the chemical process can be monitored for side-reactions and generation of hazardous waste. And if trace levels of toxic compounds are detected, it is possible to change the parameters so that the process can either reduce or eliminate the generation of these noxious materials. Also, monitoring reaction completion can save huge amounts of reagents, solvents, and energy.

Use inherently safer chemistry for accident prevention: Examples include replacing the highly toxic diborane, which burns in flames at room temperature, and the cancer-causing benzene by environmentally benign sodium borohydride ($NaBH_4$) in the preparation of gold nanoparticles. Another example is the new route to manufacture herbicide glyphosate where sodium salt of 2,2′-iminodiethanoic acid is one of the intermediates that was formerly synthesized from a series of reactions involving ammonia, formaldehyde, and extremely toxic hydrogen cyanide. The new route involves the use of ammonia and epoxyethane as starting materials, which when reacted together form diethanolamine, which is further converted into the sodium salt of 2,2′-iminodiethanoic acid (Scheme 1.9). Thus, in the event of any accident, the repercussion would not be as dangerous as in the event of an accident in the previously adopted route and clean-up would be simpler.

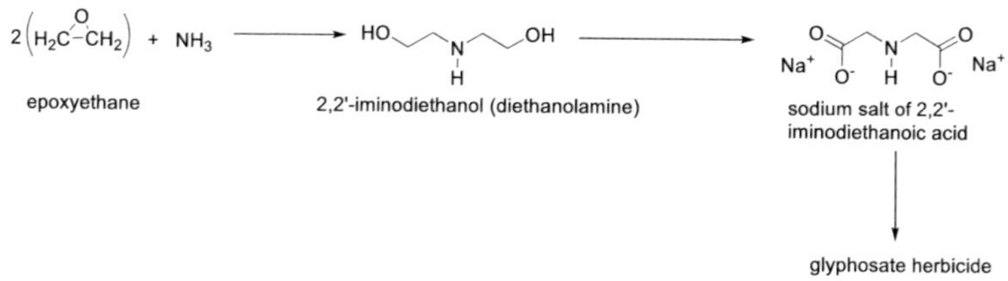

$$2\left(H_2C\overset{O}{-}CH_2\right) + NH_3 \longrightarrow HOCH_2CH_2NHCH_2CH_2OH \longrightarrow Na^+\ {}^-OOCCH_2NHCH_2COO^-\ Na^+ \longrightarrow \text{glyphosate herbicide}$$

epoxyethane — 2,2'-iminodiethanol (diethanolamine) — sodium salt of 2,2'-iminodiethanoic acid — glyphosate herbicide

Scheme 1.9 New route for the of synthesize herbicide glyphosate.

1.5 Green Chemistry and Sustainable Development

As discussed in the previous sections, sustainable development is a concept that meets the needs of the present generation without compromising the needs of future generations. The basic principle behind sustainability is that all activities must be performed in a way that is acceptable by the environment with minimum utilization of resources. With the emergence of green chemistry, an exponential growth was observed in devising ways to reduce pollution through catalysis, procedural efficiency, solventless systems, minimizing energy consumption, use of biobased feedstock, etc. In this way, implementation of green chemistry principles reduces environmental impacts of chemical processes and products, thereby helping humanity move toward the goals of sustainability [25–28].

In 2015, the world came together on a common platform to agree on a set of universal goals, the UN Sustainable Development Goals (SDGs), that carry a vision for the planet as a world of good health, quality education, gender equality, safe drinking water, clean energy, and decent work for all [29]. The SDGs are an urgent call to shift the world onto a sustainable path and hence are centered around attaining a sustainable tomorrow and a sustainable civilization at large, which is the biggest challenge in today's time. These goals, 17 in number, are integrated—they recognize that action in one sphere will affect the outcomes in others and ensure that development balances economic, social, and environmental sustainability. In fact, these goals provide us the greatest chance to make an all-out effort to improve the life of our future generations. These goals can only be achieved when the fundamental chemistry that is the basis of our society is transformed into a sustainable chemistry. Green chemistry can be used as an important tool to address most of these challenges, as it depends on technically feasible, cost-effective, and socially acceptable decisions by academia, regulators, industry, and the wider community. It is imperative to integrate these principles or methods and work in a direction to achieve these goals successfully. A sincere effort has been made in the chapters ahead to highlight how the principles of green chemistry can be helpful in achieving these goals. Some of the examples of sustainable development are discussed in Table 1.2.

Table 1.2 Examples of sustainable development of technologies and processes

Green paints	Archer RC won the Presidential Green Chemistry Challenge Award in 2005 by developing a biobased paint that not only has lower odor but also has good opacity and better scrub resistance.
Green plastics	Nature-Works LLC, a US-based company, developed a biobased polymer PLA by fermenting corn (waste biomass), as a packing material for food and beverages. This product is biodegradable and can be recycled. It has been also reported as a replacement for many petroleum-based polymers, such as carpets, cups, bags, and textile fibers.
Green carpets	Shaw Carpet won the Presidential Green Chemistry Challenge Award in 2003 with its carpet tile backing, EcoWorx. This product contains bitumen, polyvinyl chloride, or polyurethane with polyolefin resins, all of which are less toxic. Furthermore, it ensures their better adhesion, does not shrink, and can be recycled.
Computer chips	Scientists at the Los Alamos National Laboratory developed a process that employs supercritical CO_2 in one of the steps for chip preparation. This process was found to significantly reduce the amounts of chemicals, energy, and water needed to manufacture chips.
Drug synthesis	Zoloft®, an antidepressant drug, is prepared from a redesigned process that eliminates tons of waste.

The best way to achieve sustainability is by integrating science and technology with societal, policy, economic, moral, cultural, and ethical ecosystem. The scientific and technological foundation of the elements is provided by green chemistry, whereas the imperative context is provided by the other elements related to noble goals, humanitarian aims, and enabling system conditions. This daunting task of moving toward sustainability can be achieved by the alternative periodic table, as proposed by Anastas and Zimmerman [30].

1.6 Parameters to Evaluate Chemical Processes: *E*-Factor and LCA

The main aim of green chemistry is to reduce and eliminate the pollution created during chemical processes (Fig. 1.6). However, there is one type of waste material that is common and remains unconverted after the completion of the reaction. These starting materials end up as waste and contribute to pollution, making the overall process ineffective and raising the cost of production. The environmental impact of a reaction or a chemical process can be measured by many parameters. Nevertheless, the first parameter to evaluate the efficiency of a reaction still remains prevalent and is termed as the "environmental factor" or the "*E*-factor." This was first introduced by Sheldon and is defined as the ratio of waste to product [31, 32]. It is calculated by dividing the total mass of the waste produced by the total mass of the product synthesized in a chemical process (Eq. 1.3). This can be calculated with or without considering process water because including water will result in an exceptionally high *E*-factor. Usually water is excluded from the calculation. This implies that when an aqueous waste stream is considered, only the inorganic salts and organic compounds contained in water are counted. The lower the *E* value, the lower is the waste produced and the lesser will be the environmental impact of the process. Table 1.3 highlights the range of *E*-factors in different chemical industries.

$$E\text{-factor} = \frac{\text{Total mass of waste produced}}{\text{Total mass of product synthesized}} \quad (1.3)$$

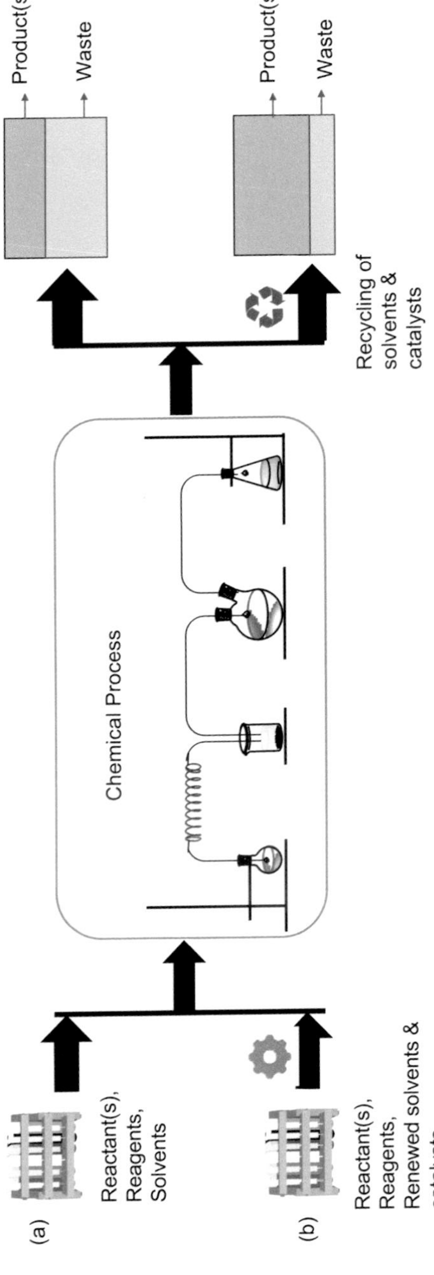

Figure 1.6 (a) General chemical reaction and (b) sustainable chemical process.

Table 1.3 Range of *E*-factors in different chemical industries

Industry segment	Volume/tons year^{-1}	*E*-factor
Bulk chemicals	10^4–10^6	<1–5
Fine chemical industry	10^2–10^4	5–50
Pharmaceutical industry	10–10^3	25–100

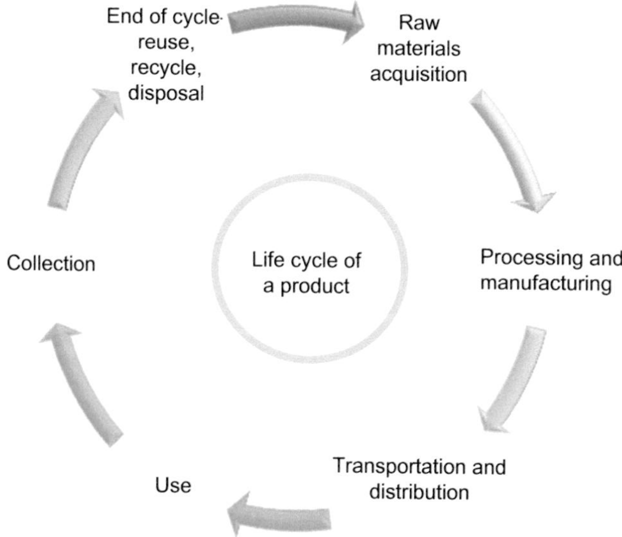

Figure 1.7 Life-cycle stages.

Besides, another useful tool to understand and characterize products, processes, and their impacts on the environment has been life-cycle assessment (LCA). Usually, the life cycle of any product is defined as the time from "cradle to grave," that is, from acquisition of raw materials, to the manufacture of products, to their distribution and use, and finally to their disposal. LCA evaluates the impact of a product over its entire life cycle. Progressively, another term, "cradle to cradle," came into existence, which accounts for the recyclability of a product either in the form of another useful product or its conversion to energy at the end of its life (Fig. 1.7). LCA has been extensively used to compare different products and processes and has helped to make important decisions as to which option will be more environmentally benign. Thus, LCA has aided in quantifying environmental burdens, assessing resource use and emissions, and estimating human health impacts and has recommended opportunities for improvements.

1.7 Atom Economy

The traditional way of evaluating the efficiency of a chemical process relies on yield, which is defined as a percentage of the degree to which a chemical reaction goes to completion. However, it totally neglects the by-products generated that are an intrinsic part of a chemical transformation. It is often observed that a chemical process achieves 100% yield yet generates waste that is far more in mass and volume than the desired product. This is because the yield is calculated according to the mole concept, where moles of reactant versus moles of product are considered. If 1 mole of the starting material gives 1 mole of the product, the yield will be 100%. However, the same transformation can result in a significant amount of waste that remains unseen while evaluating efficiency.

Green chemistry emphasizes more on atom economy, which measures the fraction of the starting material ending up in the final product (Eq. 1.2). Thus, for a reaction to be productive, it is desirable to have 100% atom economy, that is, all the reactants must be incorporated in the product. The concept of atom economy was developed by Trost, a professor of chemistry at Stanford University, who received the PGCC award for his work in 1998 [33, 34]. Since some reactions have 100% atom economy but have poor yields, it becomes necessary to consider both measures, yield and atom economy, to assess efficiency. The concept of improving atom economy gained recognition and today, most industries acknowledge its importance. This has saved money and has helped in improving the environment. Although calculating atom economy is simple and direct, it has a number of limitations as it does not consider the actual chemical yield and the amount of solvents and other reagents employed in the reaction or for workup. Some examples to calculate atom economy are given below.

Example 1: Oxidation of benzene

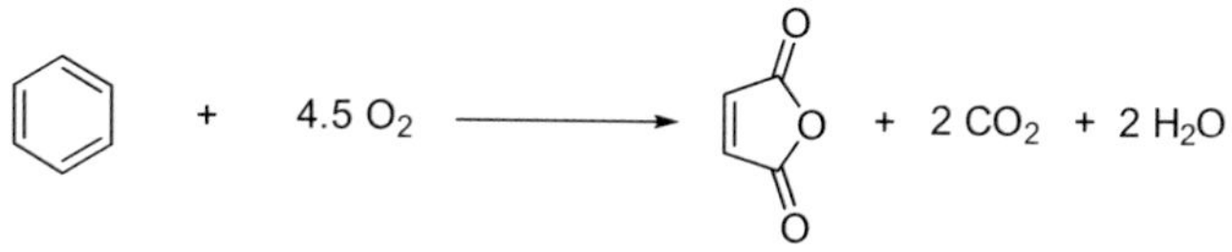

benzene + 4.5 O_2 ⟶ maleic anhydride + 2 CO_2 + 2 H_2O

M.W. = 78.0 4.5×32 (M.W.) = 144   M.W. = 98

$$\text{Atom economy} = \frac{98}{78 + 144} \times 100 = 44.1\%$$

Example 2: Oxidation of 1-phenylethanol

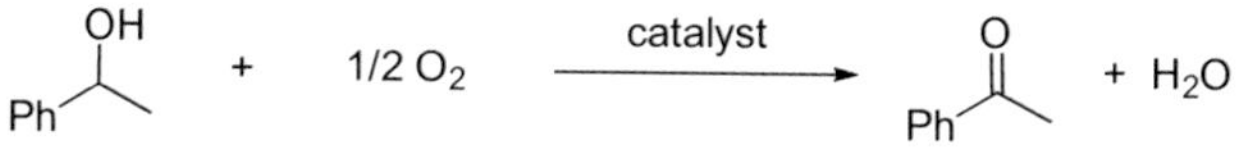

PhCH(OH)CH$_3$ + 1/2 O_2 —catalyst⟶ PhCOCH$_3$ + H_2O

M.W. = 122 0.5×32 (M.W.) =16   M.W. = 120

$$\text{Atom economy} = \frac{120}{122 + 16} \times 100 = 87\%$$

Example 3: Fermentation of glucose

$C_6H_{12}O_6$ (aq.) —enzyme⟶ 2 CH_3CH_2OH(aq.) + 2CO_2 (g)

M.W. = 180 2×46 (M.W.) = 92

$$\text{Atom economy} = \frac{92}{180} \times 100 = 51.1\%$$

The Baylis–Hillman reaction is a perfect example of 100% atom economy, while the Wittig reaction represents a poor atom economy of only 18.5% (Schemes 1.10 and 1.11).

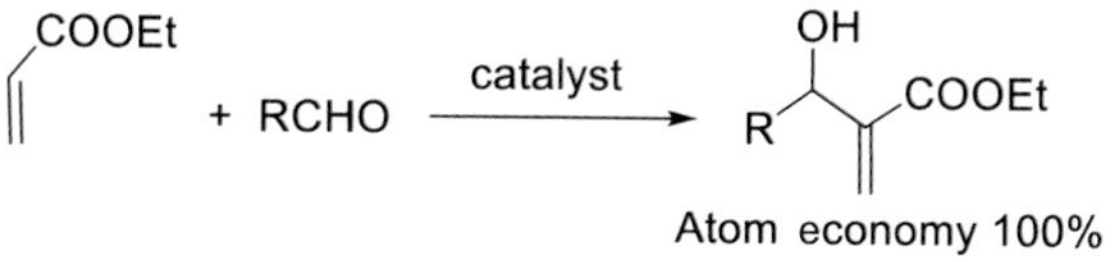

CH_2=CH–COOEt + RCHO —catalyst⟶ R–CH(OH)–C(=CH_2)–COOEt

Atom economy 100%

Scheme 1.10 Baylis–Hillman reaction.

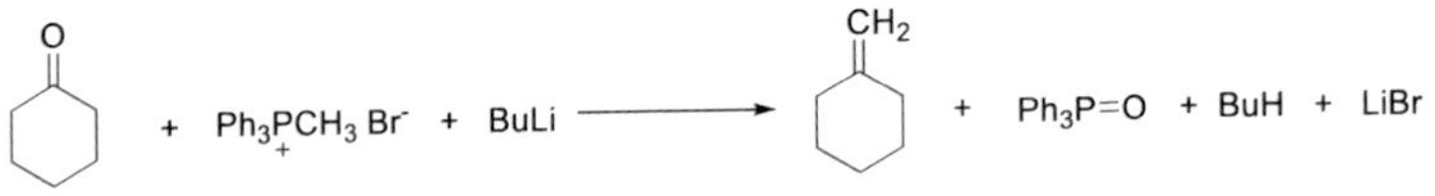

$$\text{cyclohexanone} + Ph_3\overset{+}{P}CH_3\,Br^- + BuLi \longrightarrow \text{methylenecyclohexane} + Ph_3P{=}O + BuH + LiBr$$

Scheme 1.11 Wittig reaction.

1.7.1 Atom-Economical Reactions

Addition reaction: Here, one reactant is added to another, usually across multiple bonds. Hence, it represents another example of 100% atom economy. For example, in the manufacture of poly(chloroethene), chlorine is added to ethene to form 1,2-dichloroethane, which has 100% atom economy (Scheme 1.12).

$$H_2C{=}CH_2 + Cl_2 \longrightarrow ClCH_2CH_2Cl$$

Scheme 1.12 An example of an addition reaction.

Rearrangement reaction: This type of reaction reorganizes the atoms to form a product and since it simply alters the arrangement of atoms in a molecule, it is considered as a 100% atom-economical reaction, for example, the Beckmann rearrangement (Scheme 1.13).

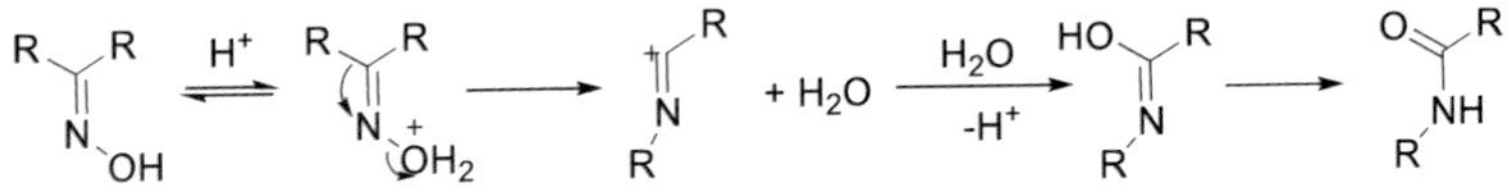

$$R_2C{=}N{-}OH \underset{}{\overset{H^+}{\rightleftharpoons}} R_2C{=}N{-}\overset{+}{O}H_2 \longrightarrow R{-}C{\equiv}\overset{+}{N}{-}R + H_2O \xrightarrow[-H^+]{H_2O} HO(R)C{=}N{-}R \longrightarrow O{=}C(R){-}NH{-}R$$

Scheme 1.13 Beckmann rearrangement.

1.7.2 Atom-Uneconomical Reactions

Substitution reaction: In a substitution reaction, one atom or group of atoms (leaving group) is substituted by another atom or group of atoms (substituting group). Since the leaving group is not included in the final product and is considered as waste, this type of reaction reduces the atom economy, for example, hydrolysis of 1,2-dichloroethane (Scheme 1.14).

$$ClCH_2CH_2Cl + 2H_2O \longrightarrow HOCH_2CH_2OH + 2HCl$$

Scheme 1.14 An example of a substitution reaction.

Elimination reaction: In this type of reaction, one parent molecule breaks down into smaller molecules. Since some of the produced molecules are not desired, they are waste products and reduce atom economy. In fact, an elimination reaction is inherently the least atom economical of all the general chemical reactions, for instance, the production of tetrafluoroethylene (Scheme 1.15).

$$2CHClF_2 \longrightarrow C_2F_4 + 2HCl$$

Scheme 1.15 Production of tetrafluoroethylene.

1.8 Hazards and Risks in Chemistry

The main aim of the manufacturing process and the use of commercial products is to minimize any kind of risk associated with them. This risk has two major aspects—hazard due to product or process and exposure to human or any other potential victim. Their relationship can be represented by Eq. 1.4, which simply states that risk is a function of hazard and exposure. Therefore, in order to reduce risk, one should reduce exposure or minimize hazard.

$$\text{Risk} = f(\text{hazard} \times \text{exposure}) \tag{1.4}$$

The industry and society emphasize on reducing exposure rather than hazard, which can be done through various means, such as engineering controls, protective clothing, and respirators. The most familiar example of reducing exposure is that of wearing safety goggles in a chemistry laboratory so as to protect the eyes. However, goggles would not prevent acid or any other chemical from splashing on the face of the student. Similarly, explosion shields will not prevent explosions. In addition, controlling exposure can reduce risk to the person working in close proximity, but it might affect someone who is not protected by exposure controls. Also, protective equipment depends on how humans are using it. Besides, no face shield, glove, respirator, or goggle is perfect. Therefore, reducing hazard is a much better way of minimizing risk. Here, the chemists play a significant role in limiting hazard through their constant efforts to replace toxic substances by environmentally benign ones. For example, instead of using volatile, toxic, and flammable solvents for cleaning purposes, one can use a water solution containing a nontoxic cleaning agent. Obviously, this requires unceasing efforts and constant vigilance to avoid hazards, but if hazard reduction is done through green chemistry, it would not fail and in that scenario, risk will definitely reduce.

Despite the fact that the accomplishments of green chemistry are astounding, a vast majority of people are unaware of its relevance to human health and environment. Thus, an integral component of its growth is green chemistry education and research. With awareness and understanding, this field will continue to thrive and innovate and make a better tomorrow.

1.9 Learning Outcomes

At the end of this chapter, students will be able to:

- Define green chemistry
- List the 12 principles with examples
- Distinguish between %yield and atom economy
- Calculate atom economy and *E*-factor
- Identify atom-economical and atom-uneconomical reactions
- Describe the need for green chemistry and sustainability concept
- Appreciate the accomplishments in green chemistry
- Relate risk to hazard and exposure

1.10 Problems

1. "Green chemistry is sustainable chemistry." Explain the statement.
2. Write the 12 principles of green chemistry. Explain any two in detail.
3. Define atom economy. How can you improve the atom economy of a reaction?
4. What is the difference between %yield and %atom economy in a chemical reaction?
5. (i) Calculate the atom economy of the following reactions:
    a. Suzuki reaction

$$C_6H_5B(OH)_2 + CH_3C_6H_4OH + 3\,K_2CO_3 \longrightarrow C_6H_5C_6H_4OH + B(OH)_2OCOO^- + I^- + CO_3^{2-} + HCO_3^- + 6K^+$$

b. Hg^{2+} catalyzed hydration of an alkyne

$$\text{1-pentyne} + H_2O \xrightarrow[HgSO_4\ (cat.)]{H_2SO_4\ (cat.)} \text{2-pentanone}$$

c. The fermentation of sugar to make ethanol

$$C_6H_{12}O_6\,(aq.) \xrightarrow{\text{yeast}} 2C_2H_5OH_{(aq.)} + 2CO_{2\,(aq.)}$$

(ii) Which of the above-mentioned reactions is greener and why?

6. Give one example each of an atom-economical reaction and an atom-uneconomical reaction.
7. "Rearrangement reactions are 100% atom economical." Explain giving suitable examples.
8. Briefly write about the goals of green chemistry.
9. Give the relation between risk, hazard, and exposure. Which control—hazard or exposure—leads to the paradigm shift to green chemistry? Justify your answer.
10. Explain briefly which of the following legislations gave birth to green chemistry initiatives.
    (i) The Clean Water Act of 1972
    (ii) The Montreal Protocol of 1989
    (iii) The Pollution Prevention Act of 1990
    (iv) The Superfund Act of 1980

References

1. Anastas, P. T. and Farris, C. A. (1994). *Benign by Design*, ed. Anastas, P. T., Chapter 1, Benign by Design Chemistry (ACS Symposium Series No. 577, American Chemical Society Washington, DC) pp. 2–22.
2. https://www.acs.org/content/acs/en/greenchemistry/what-is-green-chemistry/history-of-green-chemistry.html
3. https://www.acs.org/content/acs/en/greenchemistry/what-is-green-chemistry/history-of-green-chemistry.html

4. https://helix.northwestern.edu/article/thalidomide-tragedy-lessons-drug-safety-and-regulation
5. https://www.acs.org/content/acs/en/education/whatischemistry/landmarks/cfcs-ozone.html
6. https://ohiohistorycentral.org/w/Cuyahoga_River_Fire
7. https://www.atlasobscura.com/places/times-beach-missouri
8. https://www.britannica.com/event/Bhopal-disaster
9. https://www.world-nuclear.org/information-library/safety-and-security/safety-of-plants/chernobyl-accident.aspx
10. https://www.history.com/topics/1980s/exxon-valdez-oil-spill#:~:text=The%20Exxon%20Valdez%20oil%20spill,Horizon%20oil%20spill%20in%202010
11. Anastas, P. T. and Kirchhoff, M. M. (2002). Origins, current status, and future challenges of green chemistry. *Acc. Chem. Res.*, **35**, pp. 686–694.
12. Horváth, I. T. and Malacria, M. (2017). *Advanced Green Chemistry*, eds. Anastas, P. T., Chapter 1, Origins and early history of green chemistry (World Scientific, Singapore) pp. 1–17.
13. Török, B. and Dransfield, T. (2018). *Green Chemistry*, 1st Ed. (Elsevier, USA).
14. https://indianexpress.com/article/explained/vizag-gas-leak-what-is-styrene-gas-6398020/
15. https://www.epa.gov/p2/pollution-prevention-act-1990
16. Kümmerer, K. (2007). Sustainable from the very beginning: Rational design of molecules by life cycle engineering as an important approach for green pharmacy and green chemistry. *Green Chem.*, **9**, pp. 899–907.
17. Anastas, P. T. and Williamson, T. C. (1996). *Green Chemistry: Designing Chemistry for the Environment* (ACS Symposium Series No. 626, American Chemical Society Washington, DC).
18. Clark, J., Sheldon, R., Raston, C., Poliakoff, M. and Leitner, W. (2014). 15 years of green chemistry. *Green Chem.*, **16**, pp. 18–23.
19. Sheldon, R. A. (2017). The E factor 25 years on: the rise of green chemistry and sustainability. *Green Chem.*, **19**, pp. 18–43.
20. https://www.epa.gov/greenchemistry/information-about-green-chemistry-challenge
21. Anastas P. T. and Warner J. C. (1998). *Green Chemistry: Theory and Practice* (Oxford University Press, New York).
22. Anastas, P. T. and Eghbali, N. (2010). Green chemistry: principles and practice. *Chem. Soc. Rev.*, **39**, pp. 301–312.

23. Matlack, A. S. (2010). *Introduction to Green Chemistry*, 2nd Ed. (Taylor & Francis, CRC Press, New York).

24. Lancaster, M. (2016). *Green Chemistry: An Introductory Text*, 3rd Ed. (Royal Society of Chemistry, Cambridge).

25. Dunn, P. J. (2012). The importance of green chemistry in process research and development. *Chem. Soc. Rev.*, **41**, pp. 1452–1461.

26. Linthorst, J. A. (2010). An overview: origins and development of green chemistry. *Found. Chem.*, **12**, pp. 55–68.

27. Newman, S. G. and Jensen, K. F. (2013). The role of flow in green chemistry and engineering. *Green Chem.*, **15**, pp. 1456–1470.

28. Song, J. and Han, B. (2014). Green chemistry: a tool for the sustainable development of the chemical industry. *Natl. Sci. Rev.*, **2**, pp. 255–256.

29. Anastas, P. T. and Zimmerman, J. B. (2018). The United Nations sustainability goals: how can sustainable chemistry contribute? *Curr. Opin. Green Sustainable Chem.*, **13**, pp. 150–153.

30. Anastas, P. T. and Zimmerman, J. B. (2019). The periodic table of the elements of green and sustainable chemistry. *Green Chem.*, **21**, pp. 6545–6566.

31. Calvo-Flores, F. G. (2009). Sustainable chemistry metrics. *ChemSusChem*, **2**, pp. 905–919.

32. Sheldon, R. A. (2007). The E factor: fifteen years on. *Green Chem.*, **9**, pp. 1273–1283.

33. Trost, B. M. (1991). The atom economy: a search for synthetic efficiency, *Science*, **254**, pp. 1471–1477.

34. Trost, B. M. (1995). Atom economy—a challenge for organic synthesis: homogeneous catalysis leads the way. *Angew. Chem., Int. Ed. Engl.*, **34**, pp. 259–281.

Chapter 2

Waste: A Misplaced Resource

Anju Srivastava,[a] Sriparna Dutta,[b] and Rakesh K. Sharma[b]

[a]*Department of Chemistry, Hindu College, University of Delhi, Delhi 110007, India*

[b]*Green Chemistry Network Centre, Department of Chemistry, University of Delhi, Delhi 110007, India*

dr.anjusrivastava@gmail.com

> *A society is defined not only by what it creates, but by what it refuses to destroy.*
>
> —John Sawhill

2.1 Introduction

Chemical products and processes have played an incredible role in creating a myriad of the things we use today. From batteries to painkillers to agricultural chemicals, these innovations have shaped our modern world and lives. What follows is a snapshot of the major discoveries in chemistry. However, the repercussions of many of

Green Chemistry for Beginners
Edited by Rakesh K. Sharma and Anju Srivastava
Copyright © 2021 Jenny Stanford Publishing Pte. Ltd.
ISBN 978-981-4316-96-5 (Hardcover), 978-1-003-18042-5 (eBook)
www.jennystanford.com

these developments have been disastrous. We have already learned about some of the adverse effects of chemicals on human health and environment in Chapter 1. It is evident that the path chemists followed did not consider the consequences of the manufacturing processes. As a result of lack of knowledge pertaining to proper disposal of wastes, the unwanted by-products, unutilized reagents in the form of solvents, and spent catalysts ended up forming toxic wastes. The word "waste" refers to any substance that is undesirable and unwanted.

Basel Convention 1989

According to the Basel Convention, which was designed to reduce the movement of hazardous wastes across various nations, especially from the developed to the developing or less developed countries, waste refers to substances that may be discarded/disposed of or are intended to be discarded under the provisions of national laws. This meaning does not prescribe a hazardous nature to the waste and depends upon national laws to categorize waste substances.

There are numerous sources of waste generation. These can be primarily categorized into (i) municipal solid wastes (MSWs), for example, trash or garbage containing paper, plastic, and food discarded mostly from our houses, (ii) liquid wastes (chemicals and other water pollutants), and (iii) gaseous wastes (pollutants such as CO, NO_x, SO_x, etc.). In fact, solid wastes have been with us since human activities began. During that time, the only option was to dump these wastes. With progressive modernization, we ended up piling more and more wastes. Slowly, we transferred these wastes into environmental burdens. Unfortunately, even the basic waste management techniques, including incineration and landfills, have failed to cope with the quantum of waste produced, impacting the environment severely. Thus, not surprisingly, today waste has emerged as a key developmental and environmental issue. We all know about plastics—in fact, we are the ones who made it and became dependent on it gradually. But we have slowly released plastic wastes into the environment, where it has accumulated. One cannot really imagine the extent of unrecycled plastic waste that reaches the ocean and other water bodies every year that is known to kill an exorbitant number of marine species. In the year 2015, Jenna Jambeck, who is an engineering professor from the

University of Georgia, expressed serious concerns regarding the waste generated from the coastal regions itself each progressive year, which was roughly estimated to lie between 5.3 million and 14 million tons [1]. She along with her colleagues pointed out that "most of it isn't thrown off from ships, but dumped carelessly on land or in rivers, mostly in Asia. It's then blown or washed into the sea." What is ironic here is that "it's unclear how long it will take for that plastic to completely biodegrade into its constituent molecules." If estimates are to be believed, this might take 450 long years or more or the plastic may never degenerate.

The problems that waste may lead to, including the unprofitable use of resources and capital along with the risks to welfare and development, have been recognized by several sectors of society. Perceiving the criticality of the situation, many countries are now considering active programs to reduce the amount of waste disposed of into land, air, and water through various waste minimization techniques or recycling strategies. Among these, "reduction of waste at source" has emerged as one of the most preferred solutions, that is, accepted hierarchy for waste management (Fig. 2.1) [2].

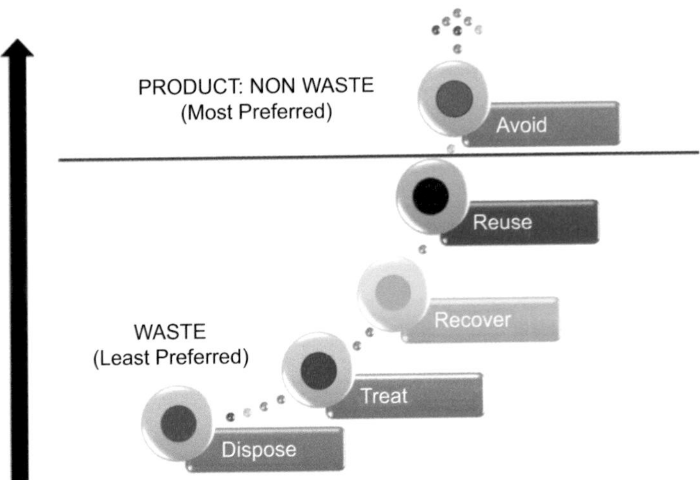

Figure 2.1 Waste prevention hierarchy.

Green chemistry aims to achieve the same [3]. The first principle says, "It is better to prevent waste rather than to treat it after it has been created." If we talk about chemical wastes, from this principle, what we infer is that the chemical processes should be redesigned in a way that minimum amount of waste is generated to the maximum plausibility. In Chapter 1, we learned about the environmental factor (E-factor), a metric helpful in assessing how green a reaction is and developed to estimate the amount of waste a process created. If we compare the E-factors associated with various routes involved in designing/producing a particular chemical or drug, we may infer that the route involving lesser E-factor is the greener and improved one. Not surprisingly, as per the data available, the majority of the drug manufacturing processes had exorbitant E-factors, but green chemistry principles helped in reducing this. For instance, if we talk about ranitidine and ibuprofen synthesis, there was a drastic reduction in the amount of waste generated. In the case of ranitidine drug, a special catalyst was developed that could convert back the toxic methyl sulfide gas, also called methyl mercaptan, to dimethyl sulfoxide (DMSO), the starting material for the reaction. So, practically the waste generated was being converted into a resource (DMSO, the solvent required for the drug synthesis) that helped in cutting down the manufacturing cost largely and also minimized the toxic effects that would have resulted from the release of this gas. This was indeed one of the best findings of green chemistry [4]. Also, in the case of ibuprofen, an improved process was designed in 1991 that eliminated millions of pounds of waste chemical by-products. What was realized soon was that what Paul T. Anastas said, "Responsible chemistry must be as important as creative chemistry," has shown a new direction to several of the practicing researchers and chemists.

Today, waste is being perceived as a precious resource and different types of waste materials are being utilized to derive valuable end-products. Slowly but steadily, the circular economic model has replaced the linear model. Considering this fact, it would not be wrong to state that "waste is a misplaced resource." Green chemistry helps to solve the twin problems of resource and waste by making the latter a solution to the former. The UN Sustainable Development Goals (SDGs) outlined in the UN's 2030 Agenda for Sustainable Development also prioritize waste as a critical issue that must be dealt with immediately. In fact, it is believed that if due attention is given to solid waste management in a sorted manner, then the day is not far when the various global goals of sustainable development would be accomplished readily. In the subsequent sections, we shall discuss the various sources of waste generation, how green chemistry helped in reducing waste, and some of the techniques adopted for waste minimization.

2.2 Sources of Waste Generation

2.2.1 Chemical Wastes Generated from Industrial and Academic Sectors

Every hour of every day, tons of highly dangerous waste materials are disposed of into the various environmental matrices via industrial operations. According to the available data in an article entitled "India Environment-2025," "As compared to Indian chemical and food processing industries' contribution to the total output, which is 25%, the biochemical oxygen demand (BOD) load is exorbitantly high, almost reaching 86%, which is a serious cause of concern" [5]. These data clearly reflect how polluting the industrial sector is. Long back, Sheldon had compared the *E*-factors of various industries, which directly reflect the amount of waste generated, and made interesting observations (Fig. 2.2) [6].

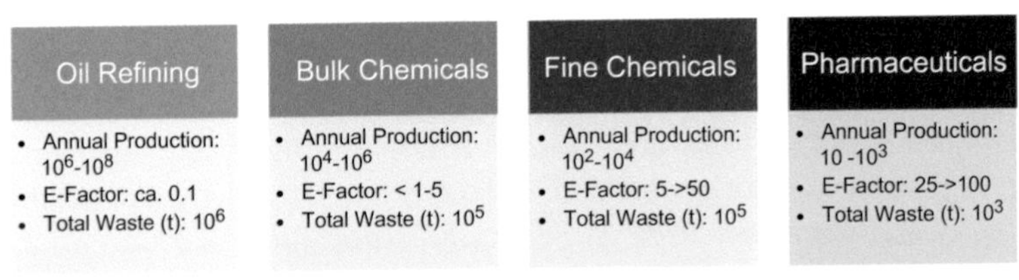

Figure 2.2 Waste produced by different industries as a proportion of product—the *E*-factor.

He found that the waste generated from pharmaceutical industries is more as compared to that from oil refining, fine chemicals, and bulk chemicals. This large volume of pharmaceutical waste is attributed to the requirement of highest levels of purity in the product of drugs and other related materials.

2.2.1.1 Pharmaceutical wastes

The pharmaceutical industry as we know is the leading commercial industry responsible for discovering, developing, producing, and marketing drugs for use as different types of medicines. During the twentieth century, the quality of life improved tremendously owing to better life-saving medicines. However, with the increase in population, the demand for drugs also increased rapidly. To cope up with the growing demands, the industries strived hard to increase the production, which led to more and more waste generation. It has been observed that the synthesis of 1 kg of an active pharmaceutical ingredient (API), which is the active part of the drug responsible for its effects, generates about 50–100 kg of waste. By this waste, we refer to the by-products, unreacted starting materials, spent catalysts, air pollutants, water pollutants, solvents, etc. The synthesis of phloroglucinol, which is an important intermediate and produced chiefly from 2,4-trinitrotoluene (a well-known explosive), exemplifies the production of enormous amounts of solid waste in the form of $Cr_2(SO_4)_3$, NH_4Cl, $FeCl_2$, and $KHSO_4$ (the by-products), elucidated in Scheme 2.1 [7]. An atom economy of about 5% and an E-factor of 40 indicates the production of large amounts of waste. As a result of the enormous quantities of waste generated in this process, further manufacture of this drug was discontinued.

Apart from this, pharmaceutical and personal care products (PPCPs) end up forming waste. Some of these include expired drugs, which are usually thrown away and eventually end up reaching the landfills from our waste bins. Also, syringes, tubing, vials, chemotherapy drug residues, open containers of drugs containing acute hazardous waste drugs, contaminated garments, etc., result in waste generation. In fact, quite shockingly, the extent of water pollution arising primarily from the pharmaceutical waste materials in Andhra Pradesh (India) was estimated to be so high that it surpassed even the highest level of pollution in the United States [8]. Reports have documented that different classes of pharmaceutical compounds, including steroids, analgesics, contraceptives, hormones, antidepressant drugs, are found in water samples in the mg/L to μg/L range [9]. Further, according to the estimates by the World Health Organization, 85% of the hospital waste is nonhazardous while around 10% falls under the category of infectious, and the rest, which is neither hazardous nor infectious, may be considered noninfectious though it may contain some noxious chemicals, such as CH_3Cl and HCHO [10].

Depending on the nature and composition, the pharmaceutical wastes have been classified as hazardous and nonhazardous wastes (Fig. 2.3) [11].

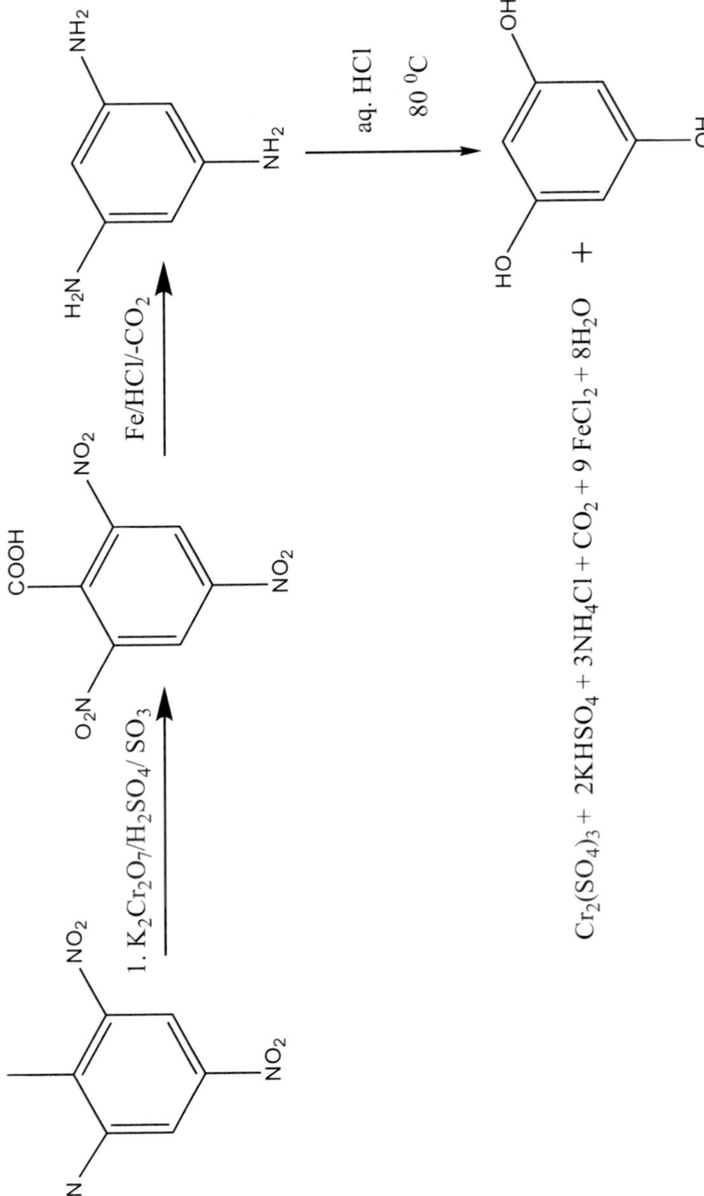

Scheme 2.1 Traditional approach to the synthesis of phloroglucinol.

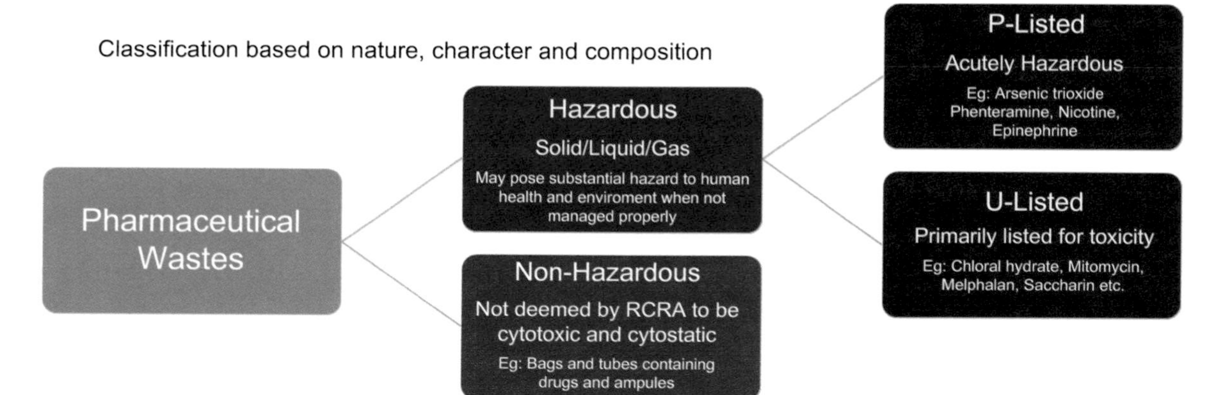

Figure 2.3 Classification of pharmaceutical waste. Here RCRA denotes the Resource Conservation and Recovery Act, which is the principal federal law establishing and regulating a proper system for managing both hazardous as well as nonhazardous solid waste.

One may consider waste as hazardous if it possesses any of the following material characteristics [12]:

(*Note*: They pertain to chemical wastes in general and are not restricted to pharmaceutical wastes.)

- **Ignitability**: If a flammable solvent is used in any of the manufacturing steps, then a formulation may be considered as ignitable. During routine storage, disposal, and transportation, it is important to identify whether wastes present a fire hazard. An example of ignitable waste is a liquid with a low flash point (<60°C).
- **Corrosivity**: A waste may be called corrosive if it can readily corrode or dissolve flesh, metal, or other materials. Corrosivity may be ascertained by measuring the pH of a waste matter (pH < 2 = highly acidic and pH > 12.5 = highly basic). The Environmental Protection Agency (EPA) has given another criterion for classifying whether a waste is corrosive or not, which is based on a steel corrosion test: if a waste is able to corrode steel according to a specific EPA-approved protocol, then we may call it corrosive. Glacial acetic acid has been identified as one of the corrosive acid wastes while potassium and sodium hydroxides have been demarcated as corrosive bases.
- **Reactivity**: A waste may be called reactive if it shows the liability to explode, react violently, or release toxic wastes when heated, compressed, or mixed with water. Discarded munitions and explosives are examples of reactive wastes. Any reactive waste is assigned the waste code D003. Nitroglycerin is the sole example falling in this category.
- **Toxicity**: If a waste contains hazardous constituents such as heavy metals, pesticides, volatile organic compounds, and semivolatile organic compounds that may leach into wastewater streams, then it may be classified as toxic. Examples include arsenic, carbon tetrachloride, 2,4-dichlorophenoxyacetic acid (2,4-D), hexachloroethane, pyridine, and vinyl chloride.

Apart from the hazardous and nonhazardous wastes, there may be mixed waste that contains both radioactive as well as hazardous waste components, generated from diverse operations, including medical diagnostic testing, pesticide research, nuclear operations, and pharmaceutical and biotechnology development.

An important point to understand here is how pharmaceuticals enter the environmental matrices (Fig. 2.4). The two major pathways via which this happens are excretion and direct disposal. The excretion of different drugs and metabolites by humans and animals follows sewage, septic tank, or surface run-off pathways to wastewater. On the other hand, the direct disposal of unused pharmaceutical products may also lead to contamination. But what happens after that? Once a pharmaceutical waste is in groundwater, it ends up entering the food chain and even a trace amount can cause serious harm to all forms of life on earth. As per studies, incessant exposure to subtoxic concentrations of PPCPs may lead to unforeseen and undesirable effects on different living organisms. For instance, they may alter the sex functions of fish and affect the human reproductive system. Surface water has been found to contain nonsteroidal and anti-inflammatory drugs like indomethacin, ibuprofen, naproxen, ketoprofen, diclofenac, and phenazone, among which diclofenac, which is sold under the brand name Voltaren, has particularly proven to be highly toxic for vultures and cattle [13].

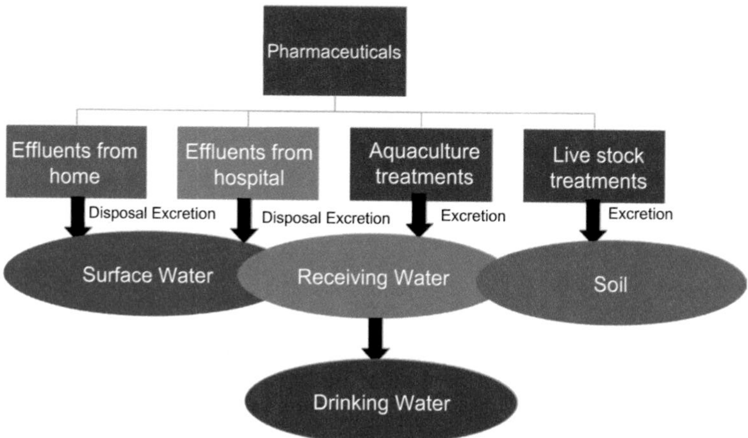

Figure 2.4 How pharmaceuticals enter our environment?

Fortunately, with the adoption of green chemistry practices, there is a remarkable change in the way pharmaceutical industries are operating today. Chemical processes are redesigned to reduce their hazardous impact on the environment. In Chapter 6, we shall take up specific case studies on how the industries could reduce waste by implementing green chemistry in their practices.

2.2.1.2 Wastes from the academic research sector

Not only industries, but also academia has been encountering thorny problems in dealing with hazardous waste materials. Although in comparison to the industries, colleges and universities generate lesser amounts of waste, yet this cannot be ignored as every single operation performed in a practical laboratory contributes a certain amount of waste. We know that an undergraduate chemistry laboratory runs for about 40 hours per week and the average amount of liquid waste (a complex mixture of substances that are difficult to separate and reuse) generated daily is about 1000 L. Most of this is thrown down the sink, which eventually contaminates water. Commercial disposal of this waste also presents an expensive proposition. Besides, the chemical reactions performed in the laboratories generate exorbitant amounts of obnoxious fumes, causing deterioration of indoor air quality. In view of rising environmental cognizance, ensuring a safe and stress-free chemistry laboratory for students, teachers, and supporting staff is highly significant.

The University of Illinois, Urbana-Champaign, US, disposed of an enormous 27,500 kg of waste chemicals in 1984. It would be interesting to know what academic wastes precisely include. Almost all the toxic chemicals highlighted by the EPA have been found in academic wastes [14], a few of which are primarily used as solvents or additives: hydrochloric acid (HCl), methanol (CH_3OH), and polychlorinated biphenyls (PCBs).

The Resource Conservation and Recovery Act, Subpart K

On December 1, 2008, the EPA added a new subpart, subpart K, to the Resource Conservation and Recovery Act, according to which incentives would be provided to eligible academic entities, including colleges, universities, teaching hospitals, and other nonprofit research institutes owned by or formally affiliated with a college or university for cleaning out old and expired chemicals that may pose unnecessary risk.

As a classic example of attempts toward waste minimization, many academic institutes have started following microanalysis methods/spot tests for cation-anion detection in inorganic qualitative analysis [15]. The spot tests greatly help in reducing enormous amounts of reagents used that would have eventually ended up as waste, contaminating the environment and ultimately leading to health problems.

2.2.2 Plastic Wastes

Plastics—this miracle material has made modern life possible. But more than 40 percent of it is used just once, and it's choking our waterways.

—Laura Parker

The story of plastic wastes raises awareness about the potential damaging effects it could have on living beings and the environment. Each year, millions of marine beings are killed by the plastics found in ocean and as per estimates, nearly 700 species, including endangered ones, are affected by (i) getting entangled in plastic objects, (ii) ingesting the wastes, and (iii) getting exposed to the chemicals present in the plastic wastes that eventually interfere with their physiology. Reports have revealed that sea turtles and fish are often not able to distinguish between plastic waste and food and end up ingesting the plastic waste, which causes obstruction in their digestive systems, eventually leading to their deaths. Apart from this, the seabirds diving for fish often end up with plastic. The irony of the situation is that every second, the oceans witness the entry of a quarter of a ton of plastic—microplastics and microfibers forming the eventual by-products. Astonishingly, as per the worldwide data available, almost 83% of tap water is found to contain plastics, microfibers, and microplastics that have been detected in species residing in seas, including fish and shellfish, and sea salt. The problem of plastic is so big that the pollution caused by it and the deaths occurring among sea creatures that had ingested plastic garbage continue to be in the headlines (Fig. 2.5). The Ellen McArthur Foundation has warned that if we continue to throw away plastic and not pay attention to this issue, then our oceans are going to have more plastic than fish by the year 2050 [16].

Figure 2.5 The picture of a 50 ft beached whale containing an excessive amount of plastic waste. Adopted from Greenpeace.

It is important to understand what plastic waste comprises and why it is so problematic. Plastics are polymeric materials that have proven to be ubiquitous as well as integral to our society, by showing their utility in diverse sectors, such as packaging, disposable equipment, electronic, and cookware. Although there are innumerable consequences of using plastics and disposing of them in the environment directly, all these problems arise because plastics fail to degrade readily and remain where they were thrown away unless removed/cleared up.

The environmental and health impacts of plastic wastes are elucidated below [17, 18]:

- Plastics contain hazardous compounds, such as bisphenol A (BPA) and polystyrene (PS), which show a propensity to trickle into water. BPA is a known endocrine disruptor that can imitate the body's hormones, interfering with the natural functioning of hormones. Besides, reports have indicated the link of BPA exposure to fertility issues, male impotence, heart diseases, and other conditions.
- PS pieces and nurdles (petroleum-based plastic) are the prime plastic ocean pollutants. In fact, styrene has been declared to be a human carcinogen by the EPA and the International Agency for Research on Cancer.

- Although not visible, plastic is also used in clothing and cosmetics (Fig. 2.6).

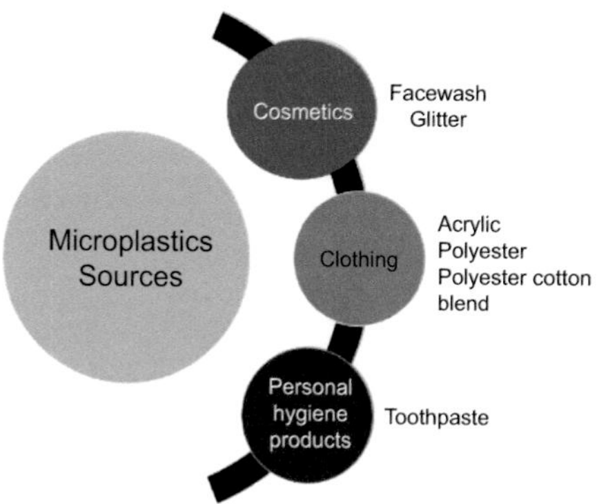

Figure 2.6 Sources of microplastics.

Clothing made of materials such as polyesters, nylon, and acrylic release not thousands but millions of minute fibers consisting of plastic when washed. These are so small and ultrafine that they trickle through the sewage treatment plants, entering water channels and finally disrupting the food chain. Not just clothes, but even cosmetics and modern toiletries consist of microplastics.

Nevertheless, there is a ray of hope toward resolving the plastic issue as people all over the world are taking appropriate actions and there has been an overwhelming reaction against the current global plastic litter. Considering the fact that 94% of microplastics emerge from only four sources, (i) synthetic clothing, (ii) car tyres, (iii) city dust, and (iv) road paint, and 95% of the marine litter arises from only 10 rivers, this should not be perplexing. As a step toward preventing problems arising due to microplastics, many large cosmetic companies have made voluntary commitments to phase out microbeads by 2020 [19]. Despite all these attempts, we are still a long way from effectively handling the most seemingly inexorable pollution of our time.

Also, now extensive research is being carried out by academia and industry to synthesize biodegradable plastics derived from renewable resources. Biodegradable plastics have emerged as a viable solution to disposing of the plastic waste in landfills, as they can be decomposed by microorganisms, which break down the hydrolysable bonds [20].

Plastic Man of India

Rajagopalan Vasudevan, an Indian scientist, also known as the "plastic man of India," has laid the foundation for reusing plastic waste in the construction of improved, durable, and cost-effective roads. For making roads faster and saving the environment from dangerous consequences, he innovated the idea of shredding plastic waste, by mixing it with bitumen (a black viscous mixture of hydrocarbons, often a residue of petroleum distillation), and thereafter using the polymerized mix in road construction.

2.2.3 Electronic Wastes

Electronic wastes (e-wastes) have been categorized as the most rapidly growing hazardous solid wastes across the globe. More than 20 million tons of e-waste are produced every year [21, 22]. In 2016, the estimated quantity of hazardous e-waste circulating in the world was found to be 44.7 million metric tons, a massive figure. As per the latest report provided by the World Economic Forum, dated January 19, 2020, e-waste had contributed to an estimated waste stream of 48.5 million tons, a huge number, in 2018.

So, before we get into further details about problems related to e-waste, let us understand what e-waste is. Electronic products, including computers, printing devices, DVD and CD players, washing machines, and mobile phones, that have become undesirable and nonfunctional constitute e-waste. Figure 2.7 reflects the percentage composition of e-waste.

Such huge amounts of e-waste are attributed to the rapid technological development that have taken place over the course of time, and today, they are polluting the soil and the drinking water and harming ecosystems around the world. The obsolete electronic goods are piling up in the landfills across the globe. The big question is, what do we do with piles and piles of discarded electronics?

While 60% of the generated wastes are trashed, the remaining 40% is recycled. If we consider the data from the United States alone, over 100 million computers are dumped without being subjected to proper recycling. The majority of the electronic materials that are indecorously discarded in landfills may comprise of toxic heavy metals such as beryllium, cadmium, mercury, and lead, which even if present in trace amounts, add up to make large volumes, may further seep into groundwater, and cause a threat to the environment and human health. Apart from this, there might be other heavy metals and potentially hazardous flame retardants present. For instance, liquid-crystal displays (LCDs) contain lamps made up of mercury and circuit boards made of Pb-, Sn-, and bromine-based flame retardants. Apart from this, the plastic casings of many of the electronic goods are made up of polyvinyl chloride (PVC). Table 2.1 enlists the various toxic materials present in e-waste, with their damaging effects to human health. Initiatives are being directed in this regard by various international agencies as they address e-waste management and trade concerns globally, while simultaneously giving due emphasis to critical problems caused by this category of waste.

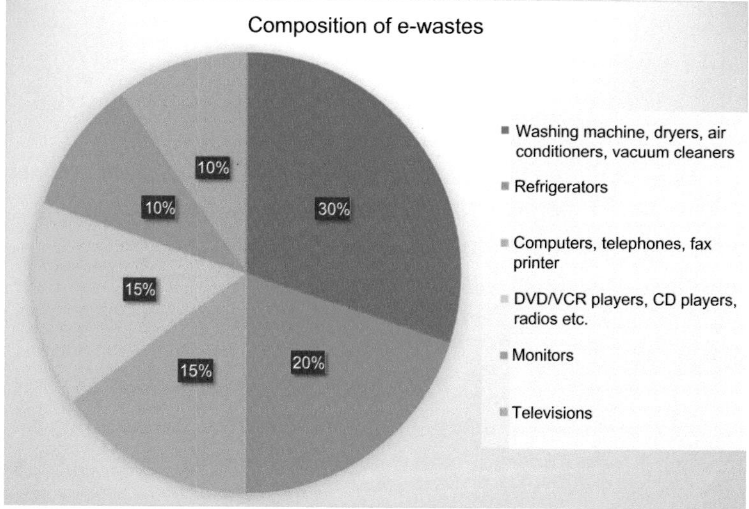

Figure 2.7 Percentage composition of e-waste.

Table 2.1 Toxic components of e-waste and the damage they inflict on human health

Hazardous component	Birth defect	Brain	Heart, liver, lung, and spleen damage	Kidney damage	Nervous/ Reproductive system damage	Skeletal system damage
Ba		Yes	Yes			
Cd	Yes		Yes	Yes	Yes	Yes
Pb	Yes	Yes		Yes	Yes	
Li	Yes	Yes	Yes	Yes	Yes	
Hg	Yes	Yes	Yes	Yes		
Ni	Yes		Yes	Yes	Yes	
Pd	Yes	Yes	Yes	Yes		
Rh			Yes			
Ag	Yes	Yes	Yes	Yes	Yes	

Without recycling, most of these toxic metals enter the soil and the water system, poisoning these sources.

E-waste Management Campaign by Nokia India

India occupies an integral position in the list of those few countries that have implemented a law for adequate handling and proper disposal of e-waste. Mobile manufacturer Nokia India set up an example for many through its e-waste management campaign that it undertook in the year 2008. During the first phase of the campaign, the company set up drop boxes across the country for taking back/retrieving the used phones as well as other accessories regardless of the brand. This Take Back campaign had very positive results right from the beginning, as evident from the enormous number of waste mobile phones and accessories retrieved from this campaign. In the second phase, the discarded phones were sent to authorized recyclers, who first broke up the phones, then segregated and retrieved the metal along with the plastic components, and then crushed them separately with the aid of innovative recycling strategies. The as-produced crushed components were thereafter utilized for the manufacture of valuables. Appreciably, 100% of the materials in a phone could be put to reuse.

2.2.4 Paper Wastes

According to the EPA, MSWs have been defined as items we use on a daily basis and eventually end up discarding. Surprisingly, out of the total MSW, 29% is paper and paper products. In fact, we are all heavily dependent on paper, right from our newspapers to our wrappings, and as such it would not be wrong to say that paper is everywhere, but today, it has become a major source of waste instead of being just a precious commodity. Most of this paper waste ends up in our landfills across our planet, causing environmental pollution (affecting air, water, and soil simultaneously) [23]. For instance:

- When the paper starts rotting, it leads to the emission of methane gas, which is 25 times more toxic than CO_2.
- The use of chlorine-based bleaches during their preparation causes the generation of hazardous materials in various environmental compartments, including air, water, and soil.
- Dioxin is another toxic by-product that results from the manufacture of paper and it is a known carcinogen.
- In addition to this, the mindless use of paper has led to problems related to deforestation. Thus, the critical need of the hour is to "recycle paper wastes." Indeed, it is believed that we can all contribute toward saving 41,000 trees from being cut down and eventually destroyed by simply recycling one morning newspaper every day. In this, way we will be able to reduce the carbon footprint greatly and improve our environment.

Typically, all wastes discussed can be classified into three categories:

- Solid waste: This includes wastes that are visible and result from various industrial operations as well as other construction-related work, mining, domestic work, etc.
- Liquid waste: The liquid industrial chemical effluents and the wastewater from our homes are liquid wastes.
- Gaseous waste: This category of waste includes poisonous gases from various industries.

On the basis of the amount of moisture present, waste may be classified as:

- Wet waste: This refers to the organic waste generated usually by eating establishments.
- Dry waste: This includes wood and related products, metals, and glass.

2.3 Problems Associated with the Generation and Mismanagement of Waste

In the previous section, we saw that waste can be divided into chemical waste and MSW, including paper, plastics, food, and electronics, the distribution of which would differ from country to country. Table 2.2 reflects the variation in composition of Indian MSW progressively, beginning from 1996, which shows that there has been a drastic increase in the amount of solid waste generated.

Table 2.2 How the composition of Indian MSW has been varying over the years (data provided in percentage)?

Year	Biodegradables	Paper	Plastic waste	Glass	Rags	Others	Inert
1996	42.21	3.63	0.6	0.6	-	-	45.13
2005	47.43	8.13	9.22	1.01	4.49	4.02	26.16
2011	42.51	9.63	10.11	0.96	-	-	17

Source: Ref. [24]

In pharmaceutical industries, copious quantities of waste are generated and solvents alone account for more than 80% of the nearly 200 million pounds of waste emitted. Apart from this, pharmaceutical packaging materials, especially blister packaging, consist of PVC, polypropylene (PP), polyester (polyethylene terephthalate [PET]), or aluminum (Al), which when disposed of improperly may lead to the generation of harmful products. Literature reports have revealed that burning of blister pack, due to the combustion of PET, results in global warming due to the greenhouse effect while the combustion of PVC releases HCl, which harms human beings.

Traditional waste management techniques include incineration, wherein combustion is carried out so that the solid waste can be transformed into gaseous products and other residues. Although this technique shows good prospects for the disposal of solid residues, it may give rise to new problems, including the emission of gaseous pollutants. Besides, the ash from these incinerators requires disposal to a secure landfill and thus such incinerators are often associated with a high operating cost and require highly skilled operating personnel. Some of the key challenges associated with waste management have been highlighted in Fig. 2.8.

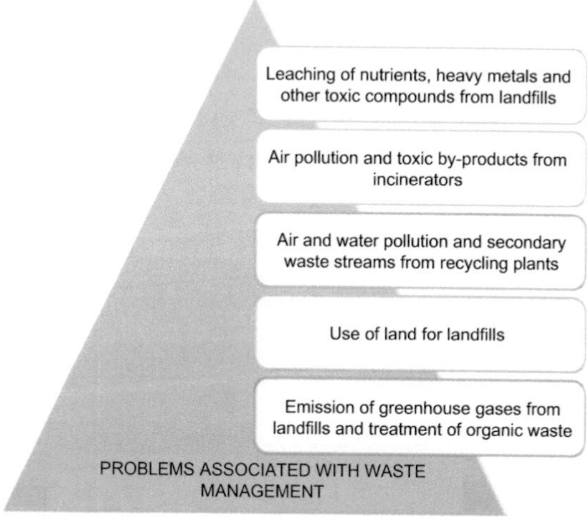

Figure 2.8 Challenges of waste management (solid incineration methods).

Currently, recycling of materials is seen as a possible solution to combating problems associated with waste generation, but recycling can also have some environmental impact. For instance, scrapping cars leads to the generation of huge amounts of shredder waste that is poisoned with oil and heavy metals. Also, the smelting of these metals may lead to the emission of toxic metals and other chemical wastes from secondary steel works and Al smelters. To add to the complexity of waste management, the variability in the type of waste materials, along with the complicated waste treatment methodologies, including some illegal ones, really causes difficulty in getting a comprehensive overview of the waste produced. The fact that the strategies devised for waste management are not for a single substance/product, analysis of every substance may seem incoherent when considering the overall design for a waste management system.

2.3.1 Global Case Studies Reflecting Mismanagement of Waste

2.3.1.1 Minamata mercury poisoning incident

If we talk about chemical wastes particularly, then the problem is not new. Recalling the Minamata incident that took place in the northwest region of Japan around the year 1956, more than 100 cases of mercury poisoning were reported, out of which around 45 people died while at least 30 developed cerebral palsy [25]. The root cause of the Minamata disease was ingestion of fish poisoned by methyl mercury (CH_3Hg^+), released into wastewater by Chisso Corporation, a chemical company located in the Minamata City of Japan. The Chisso Corporation had initially started a chemical company that was known for producing fertilizers. However, within a few years, the factory expanded and began synthesizing pivotal chemicals such as acetylene, acetaldehyde, acetic acid, and vinyl chloride. The synthesis techniques for obtaining some of these chemicals involved catalysts. What happened was that year after year, this company kept disposing of the catalyst residues containing Hg, which was being utilized to produce PVC chemical, into the Minamata River Bay. The water soon became contaminated with Hg, which was consumed by the fish in the river and then eventually entered the human food chain, resulting in the severe outbreak of this disease. What had happened clearly reflected either ignorance or lack of knowledge pertaining to wastes. Some of the other incidents related to the mismanagement of waste have already been highlighted in Chapter 1.

2.4 Waste as a Resource

"Waste mountains . . . are actually misplaced treasures," uttered Shao Zheru, who was from Everbright International (China), while delivering talk at the Twenty-Sixth Session of United Nations Economic Commission for Europe's Committee on Sustainable Energy in Geneva on September 27, 2019. Reading between these lines, a very important question creeps up: Who knew once that the wastes we have been discarding over the years could turn out to be a resource?

Over time, with the accelerated pace of industrialization and urbanization, humans have been utilizing the precious and limited natural resources and generating waste. Also, to treat the wastes, they are developing methods that in turn are damaging the environment and simultaneously affecting the health. But we know that if this is continued, probably life on earth will cease completely. To solve the inherent challenges associated with waste generation, the birth of a "circular economic model" has taken place, which is a viable alternative to the linear economic model [26]. This model relies on the concept of reconverting waste into a precious resource and proves that "green" and "growth" need not be binary alternatives. It ensures that most of the green chemistry principles are followed in a process design. A circular economy provides an immense opportunity for growth and innovation and clearly shows that waste is the "biggest economic opportunity of our time." As apparent from Fig. 2.9, we can avoid waste as it is converted into a value-added product.

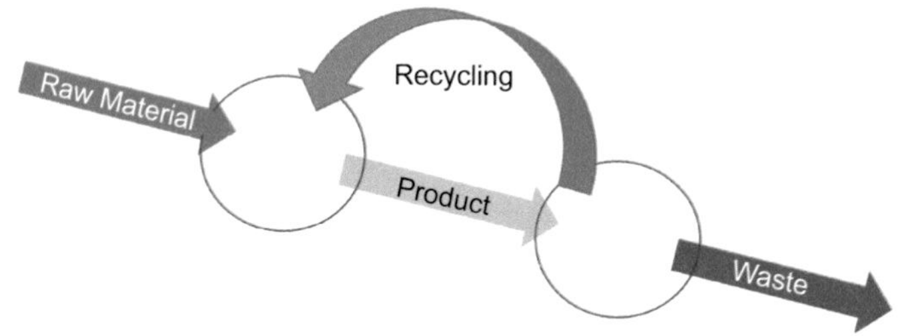

Figure 2.9 Illustration of the concept of circular economy.

Today, circularity lies at the core of ecodesign, wherein waste is repurposed for a valuable cause and motive and environmental impacts involving the use of raw materials are significantly diminished via a reuse and recycling strategy. Thus, we can perceive waste as a challenge waiting to be renovated into a prospect with the aid of the various goals of the 2030 Agenda for Sustainable Development (Fig. 2.10).

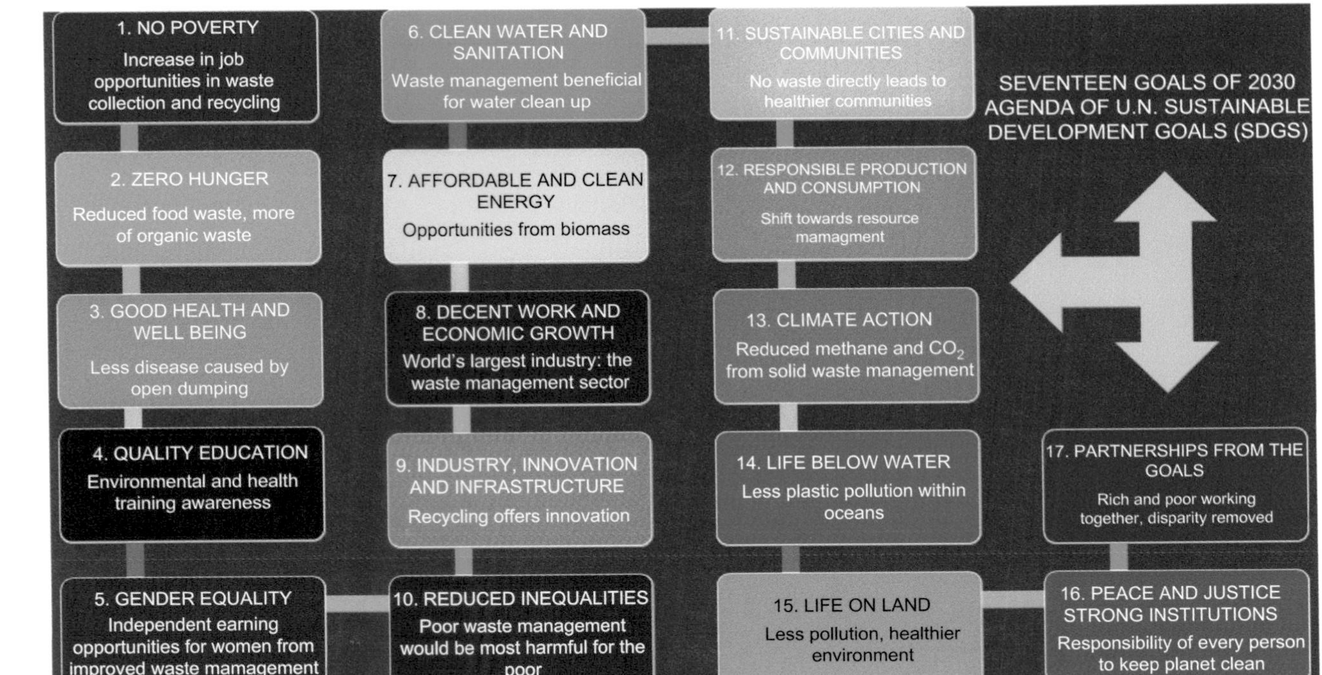

Figure 2.10 Waste management: Key to delivering the global goals for sustainable development.

Consistent with the goals of sustainable development, efforts are being directed toward the conversion of different waste materials into wealth (value-added products) using green chemistry approaches (Table 2.3). The benefits have already been realized. The stimulating and sustainable "waste-to-resource" opportunities include food and plastic wastes, which are utilized as chemical feedstocks and fit well within the premises of a circular economic model.

Table 2.3 Trash to treasure: Converting wastes into useful materials

No.	Type of waste	Waste material	Chemical process (waste valorization process)	Value-added product (derived)
1	Plastics	Plastic wastes	• Pyrolysis by direct heating • Pyrolysis-catalytic cracking • Solid-state chemical vapor deposition method	Paraffin and crude oil Gasoline and diesel oil Graphene foil
2	Paper	Newspaper	Deinking and fermentation	Bioadsorbent (for removal of heavy metals)
3	Food	Lignocellulosic waste such as corncobs, rice straw, sugarcane bagasse Fruit extracts prepared from waste, including banana peels, orange peels, watermelon Tea extract Citrus peels Waste mineral oils Corn	Solid-state fermentation Green-/Plant-mediated method Green synthesis Gelation Pyrolysis Fermentation	Cellulose, hemicellulose, and lignin Silver nanoparticles Gold nanoparticles Sheer thinning gels C_1-C_{12} hydrocarbons Bioethanol

No.	Type of waste	Waste material	Chemical process (waste valorization process)	Value-added product (derived)
4	E-waste	LCDs	Microbial fuel cells	Treatment of wastewater/ Electricity generation Recovery of heavy metals
6	Chemicals	Glycerol	Saponification	Liquid handwash

The following section will throw light on the different types of waste materials that have turned out to be valuable resources.

2.4.1 Biomass: A Renewable Feedstock

Biomass is a renewable and sustainable source of energy that includes organic plant and animal materials, such as wood from forests; agricultural wastes (such as sugarcane bagasse; wheat, barley, and rice straw; corncobs; maize cane trash; and rice husk); animal wastes, such as cow manure, and poultry litter; and other food processing and industrial waste.

It is used to create electricity and other forms of power. Originally, the energy stored in biomass emanated from the natural storehouse the sun through photosynthesis, wherein CO_2 in the air was trapped and converted into molecules such as starch, sugar, and cellulose. The chemical energy trapped in the plant and animal waste is referred to as biomass energy, or bioenergy.

In the present scenario, biomass has been acquiring significance in the production of clean fuels that can be utilized for cooking, heating, and other domestic work in addition to its use for running vehicles. To a great extent, it has helped us reduce our dependence on fossil fuels, leading to an overall decrease in greenhouse gas emissions, pollution, and other waste issues. A biorefinery has different units capable of converting biomass into value-added substances, including chemicals, diverse sources of energy, and fuels (Fig. 2.11).

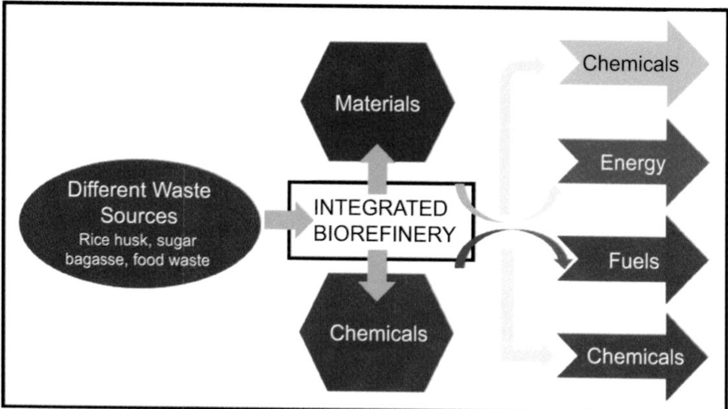

Figure 2.11 Biorefinery processes involving conversion from waste to wealth in the form of useful chemicals, fuels, and energy.

2.4.2 Biodiesel

Biodiesels have emerged as one of the cleanest alternatives to fossil fuel–based energy resources, such as petrol, diesel, and gasoline, and are obtained by chemically changing the fats present in different oil sources—like soya, canola, mustard, and waste cooking oil—through a process called transesterification (Scheme 2.2).

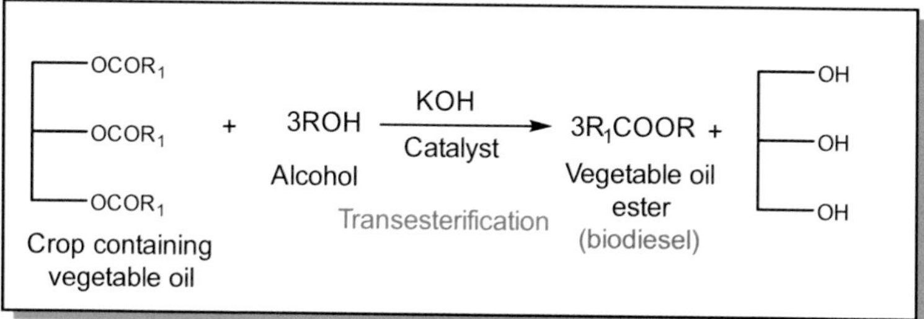

Scheme 2.2 Biodiesel synthesis via the transesterification process.

During the transesterification process, the long-chain fatty acids forming the essential component of waste cooking oil are cleaved to form monoalkyl esters, which form biodiesel in the presence of a suitable base and methanol. The waste (by-product) obtained in this synthesis is glycerol, which may be again utilized for making liquid handwash or soap via saponification. This is one of the best examples of waste reduction at the source or utilization of waste by generating energy.

In comparison to the conventional energy sources, biodiesel possesses numerous benefits:

- It is obtained from renewable domestic resources of energy, such as vegetable oil containing crops like soya, canola, and mustard.
- Waste cooking oil, which is usually discarded after cooking, can also be utilized to produce biodiesel (we can derive wealth out of waste).
- It is more energy efficient.
- It is nontoxic and biodegradable.
- It possesses the potential to reduce global warming/CO_2 emissions.
- There is lesser sulfur emission when it is burned, eventually helping in the reduction of acid rains.
- It degrades rapidly in the environment, thereby reducing the possibility of an accidental spill, as often witnessed in the case of crude oil.

2.4.3 Polymers from Renewable Raw Materials: Thinking Green

If we go back in history, we will find that polymeric substances have been utilized from renewable resources since decades. In fact, naturally occurring polymers were among the initial few materials exploited by humans, and if the records are to be believed, some of the natural materials that were transformed to generate polymeric materials were natural rubber, casein, and cellulose. Today, biobased polymers are synthesized using a two-step process from biomass, which includes lignin, cellulose, starch, and plant oils. Some of the renewable resources, along with their sources and principal industrial sources, have been highlighted in Table 2.4.

Table 2.4 Examples of some important renewable biopolymers: their sources, types, and fundamental industrial applications

Type of biopolymer	Renewable source	Industrial applications
Starch	Corn, potato, cassava, wheat, etc.	Adhesives, foams, food, plastics, gums, and pharmaceutics
Cellulose	Trees, plant biomass, and by-products from bioprocessing	Textiles, wood manufacturing, and composites
Protein	Soybean, vegetables, fruits, and animals	Plastics, adhesives, and composites
Lignin	Trees, plants, pulping processes	Adhesives, coatings, paints, and plastics
Oils and waxes	Vegetable crops and specialty crops	Adhesives, resins, coating, and paints
Chitin	Fish waste	Gums, food, pharmaceutics, and cosmetics
Pectin	Citrus fruit waste	Food, gum, emulsifiers, pharmaceuticals, and cosmetics

2.4.3.1 Bioplastics

Visualization of a future where the entire packaging is made from waste material that gets biodegraded readily and completely to form innocuous, nontoxic by-products itself is highly tempting. Slowly but steadily, driven by the economic and environmental pressure to substitute conventional plastics manufactured from synthetic petrochemicals, bioplastics are entering the mainstream.

Bioplastics, often referred to as the next generation of biodegradable and compostable plastics, are polymers made up of renewable raw materials such as vegetable fats, oils, cornstarch, straw, woodchips, and food waste, that score better on the sustainability scale in comparison to plastics of plastic origin. Some of these are even biodegradable. Examples include polyethylene (PE), polypropylene (PP), polyethylene terephthalate (PET), polytrimethylene terephthalate (PTT), and thermoplastic polyester elastomers. Among all the potential biobased plastics that are obtained from annually renewable resources, polylactic acid (PLA) has particularly garnered immense attention as it is biobased as well as biodegradable. PLA has been in the market for quite some time (almost a decade), primarily as disposable packaging. Although a wide spectrum of raw materials, such as sugarcane, corn, and bitter cassava, have been utilized for PLA-based bioplastic production, fermentable sugar has emerged as the cheapest and most abundant raw material in the world. Cargill Dow LLC, an international company popular for manufacturing bioplastics—polymers derived entirely from plant resources—is building a world-scale manufacturing facility for PLA production, with a capacity of 300 million pounds a year at a cost of several hundred million dollars [27]. The technique for PLA production has been illustrated and explained elaborately in Chapter 6. It is due to the innovation of Cargill Dow that a growing percentage of plastics is now synthesized from corn, sweet potato, and even tomatoes. For this innovation, they utilized corn (maize in UK and India) and received the Presidential Green Chemistry Challenge Award 2002.

2.4.3.2 Bioadhesives

Until now, most of the adhesives have been manufactured from petroleum-based products. However, today, adhesives made from renewable raw materials are drawing intensive research interest. Examples of pioneering products featuring these new class of bioadhesives are wallpaper pastes and glue sticks.

2.5 Waste Minimization Techniques

A few countries have devised waste management policies whereby a significant proportion of the waste is valorized, while other greener approaches are also being utilized for achieving either zero waste or reducing the total waste generated in a process.

2.5.1 Minimizing the Use of Derivatives in Chemical Processes: A Way toward Improving the Environmental Credentials of Chemical Synthesis

Principle 8 says, "Unnecessary derivatization (use of blocking groups, protection/deprotection, temporary modification of physical/chemical processes) should be minimized or avoided if possible, because such steps require additional reagents and can generate waste."

In Scheme 2.3, we have shown how a reaction can be selectively accomplished with the aid of a temporary blocking agent, also called the protecting group.

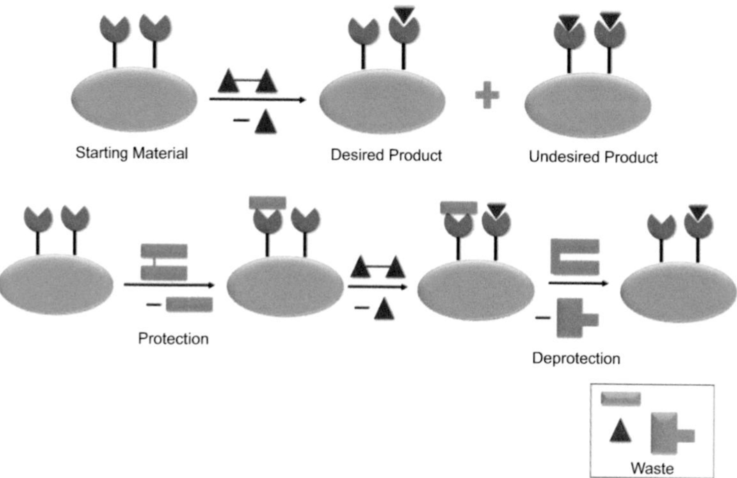

Scheme 2.3 Schematic illustration of how derivatization leads to waste generation.

Although the product formed here is the desired product, to only make one change (avoid the undesired product), we had to carry out three different reactions, which made the process inefficient as each step led to the generation of undesirable waste products. Undeniably, protecting agents are useful tools that help chemists in synthesizing target (desirable) molecules. However, the reaction sequences involving such additional reagents make them inefficient and wasteful. Thus, the design of selective pathways that would eradicate the need to use protecting groups has become an active area of research. The prime concept of selectivity today revolves around the area of exploiting a molecule's chemistry in such a way so as to make the reaction occur at a particular active site by virtue of which a synthesis can be made greener, sustainable, and effective as the number of steps required and the amount of waste produced could be minimized drastically.

Recently, with the escalating interest in green chemistry, pharmaceutical chemists are directing their creative energies toward minimizing the environmental impact of their craft. In a report by GlaxoSmithKline, a life-cycle analysis study of waste generated from their API manufacturing facilities divulged that 80% of their waste is solvent based. Thus, an attempt toward solvent waste reduction has been made by several companies. The success stories clearly reveal the fact that green chemistry principles help in redesigning the existing wasteful processes in the pharmaceutical industry. For instance, Pfizer won the Crystal Faraday Award for its contribution to minimizing the environmental footprint of the synthesis process of the active ingredient of Viagra by reducing the amount of solvent waste generated from 1300/kg to 22/kg of the product.

2.5.1.1 Sitagliptin

The story of sitagliptin—a drug used for treating diabetes mellitus type 2 by inhibiting the DPP-IV enzyme—provides an interesting example of a case where green chemistry was manifested by minimizing the derivatives [28]. Merck, which is one of the largest pharmaceutical companies in the world, established in the year 1891, utilized a first-generation synthetic pathway to develop about 100 kg of the drug for clinical trials, after which an asymmetric hydrogenation of β-ketoester was carried out in the presence of (*S*)-binapRuCl_2–triethylamine complex, which worked as a catalyst (Scheme 2.4).

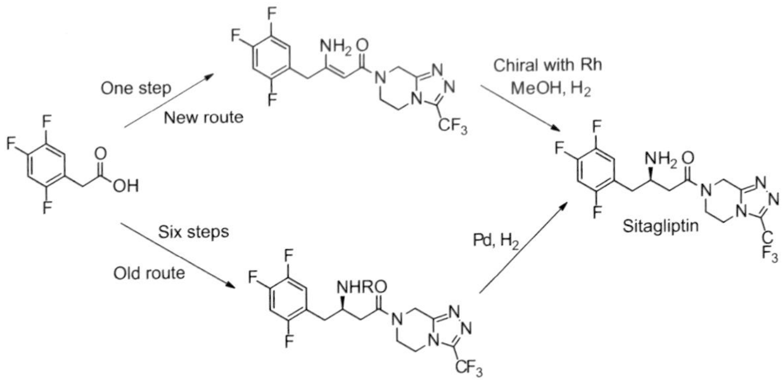

Scheme 2.4 Comparison of old and new routes for the preparation of sitagliptin.

Thereafter, hydrolysis of the ester was done to give β-hydroxy carboxylic acid, which was eventually transformed to protected β-lactam, followed by coupling to triazole. Finally, after carrying out the deprotection, sitagliptin could be obtained in the form of phosphoric acid salt. Although the conventional route showed promising prospects, yet its large-scale industrial applicability was hindered by the fact that it involved multiple steps, which led to an increase in the waste generated as enormous amounts of reagents as well as volumes of solvents were being used. Hence, an improved/reformed process was sought for upscaling. To overcome the problem of waste generation, an alternative strategy was devised that would effectively eliminate the additional steps required for protecting the initial substrate, followed by deprotecting, to render the targeted compound. This notion was implemented during the second-generation manufacture of this drug.

Process chemists succeeded in discovering a completely unprecedented transformation: they utilized a rhodium catalyst with a ferrocenyl phosphine ligand, which boosted the asymmetric hydrogenation of unprotected enamines, rendering the final product with a very improved reaction yield. The new way of synthesis consisted of only three steps, which also enhanced the overall yield by almost 50%. What was worth noting was that the crafting of a new reaction step allowed the reduction to occur right at the end, leading to a reduction in the use of copious quantities of expensive catalyst, allowing the recovery of more than 95% of rhodium.

Also, this new route on the manufacturing scale offered other benefits, such as:

- A significant amount of waste reduction, by greater than 80%
- Complete elimination of aqueous waste streams

Through the development of this greener pathway, Merck anticipated a drastic reduction in the amount of waste generated over the entire lifetime of the drug.

2.5.2 Recycling Reagents

Recycling has emerged as one of the most viable and sustainable techniques of waste management as it provides an opportunity to convert various forms of waste into potentially useful materials while simultaneously enabling a reduction in the consumption of fresh raw materials. It also helps in meeting the growing economic demand of a nation along with reducing the overall energy involved and simultaneously reducing the pollution that might otherwise arise if a waste is subjected to a conventional process, like incineration or landfilling. It is an important part of the "reduce, reuse, and recycle" waste hierarchy, which primarily aims at environmental sustainability.

The practice of recycling goes as far back as the fourth century BC, as evident from the archaeological studies of ancient waste dumps. These dumps showed lesser household wastes, such as broken tools and pottery, which in turn implied that more waste was being recycled. History reveals that recycling of paper was first accomplished in AD 1031, when Japanese shops started selling repulped paper. Today, the list of recyclable materials spans a wide variety of daily use items, such as glasses, metals, plastics, textiles, batteries, and electronics.

2.5.2.1 Recycling reagents in chemical industries and laboratories

Currently, the pharma sector is directing its efforts toward searching for strategies to make sure that the entire product cycle, from manufacture to distribution to disposal, is conducted in an eco-friendly way. There is a paradigm shift in the thoughts of researchers, who have started to realize that the solution lies not just in safely disposing of waste but also in generating minimal waste. This can be achieved through the effective integration of green chemistry and engineering models. If efforts are directed toward recycling reagents such as the solvents and catalysts employed commonly during the drug manufacturing process, then there may be a drastic enhancement in the overall efficacy. To add to the environmental credentials, recyclability can also help in minimizing waste and reducing energy usage.

2.5.2.1.1 *Solvent recycling*

Solvents play an essential role in the formulation of life-saving drugs, accounting for roughly 56% of the total mass utilized in a typical pharmaceutical batch chemical operation. However, the problems associated with the use of solvents in industries, such as high cost and the resulting waste, can be easily minimized with an effective solvent recycling system. There are several methods of recovering solvents, like fractionation, azeotropic distillation, and extractive distillation. The recovered solvents can be recycled back in a process that eliminates the need to produce new solvents while drastically reducing the waste generated. In addition, it helps remove potentially hazardous substances from what's left.

In light of minimizing solvent wastes, a number of companies have emerged. For instance, Maratek Environmental has gained prominence as the world leader in solvent recycling and solvent recovery equipment manufacturing [29]. A few examples of the solvents this company recycles are acetone, dimethylformamide, butanol, octane, and pentanol—and the list goes on.

ChemGenes Corporation, an ISO-certified company located in Wilmington, Massachusetts, US, is involved in supplying products such as building blocks for DNA/RNA synthesis for the biotechnology industry [30]. The core product line of this company is nucleoside phosphoramidites, which should be synthesized in a highly pure form. However, unfortunately, it is not possible to crystallize the final product and the only way to purify it is column chromatography, which utilizes copious quantities of solvents, leading to waste generation. Therefore, the company directed its efforts toward solvent minimization and recycling. In its first attempt toward this goal, it designed a 100 L plant for the separation of the by-product, to permit the reuse of the solvents. But this was quite a labor-intensive operation and not even cost effective. On the recommendation of the Office of Technical Assistance and Technology, ChemGenes considered an automated solvent recycling system manufactured by CBG Biotech. The solvent recycler TechnoCleanF-2500 consists of an auto tank with a 25-gallon capacity, fitted with a complete automated system. Using this system, ChemGenes could recycle 1340 L of solvent, which led to savings of approximately US$3500.

The academic sector is not behind in contributing to solvent waste minimization through recycling programs. The Columbia University, situated in the City of New York, US, began a solvent recycling program in 2001 at its medical center, when Environment, Health and Safety (EH&S) Department collaborated with the Department of Pathology to recycle xylene and ethanol. EH&S also began a pilot program at Morningside in order to examine the efficacy of the recycled spent acetone from laboratory glass washing activities in reuse. This program has now expanded to include methanol and ethanol in chemistry, applied physics and applied mathematics, biology, and engineering laboratories. This recycling program is available free of charge to the campus community.

2.5.2.1.2 *Catalyst recycling*

The use of catalysts has seen an exponential increase since a "catalyst" was first utilized by J. Roebuck in the synthesis of sulfuric acid. The pharma sector has particularly been benefitted tremendously by the use of catalysts as they have enabled the rapid production of active ingredients and intermediates for drug synthesis. Mostly, precious metal salts such as palladium, ruthenium, rhodium, and platinum are used as catalysts and offer significant advantages over traditional reaction chemistry, including better activity, higher selectivity, and higher turnover numbers. Despite these benefits, they suffer from severe drawbacks, such as high cost and difficulty in separation from the product stream. Thus, a few catalyst recovery techniques, like solvent extraction, nanofiltration, chemical precipitation, and adsorption, have been utilized so far.

The heterogenization approach, wherein homogeneous metal complexes are tethered onto a solid support material with good recoverability, has recently been recognized as a potent tool to overcome the limitations of conventional homogeneous and heterogeneous catalysts. The Novartis scientists designed a new heterogeneous Pd catalytic system by immobilizing palladium onto a Smopex-111 support matrix (a fiber-based benzenethiol) for a Suzuki coupling reaction. This Smopex-based catalyst not only exhibited excellent activity in the concerned coupling reaction but also allowed effective recovery of the precious palladium metal from the process stream. As evident from Scheme 2.5, a nearly quantitative yield of the desired coupling product has been obtained using the palladium Smopex catalyst. The most striking feature of this novel route was that the catalyst could be reused for four consecutive runs with consistent activity (Scheme 2.5) [31].

+ $B(OH)_2$

2.5 Pd-Smopex-111, 2 equiv K_3PO_4

Toluene/H_2O(4:1)

95°C, 2.5 h, 97%

Scheme 2.5 Suzuki cross-coupling using the palladium Smopex catalyst.

2.5.3 Miniaturization

"Small is beautiful." Indeed small is beautiful because small is quick, inexpensive, and highly profitable. Miniaturization has proven this true. The word "miniaturization" stands for a process of making something very small using modern technology. This term was coined long back by a Nobel laureate Richard Feynman as early as 1959, when he was delivering his inspiring talk entitled "There's Plenty of Room at the Bottom." In his talk, he introduced a field focused on the problem of manipulating and controlling things on a small scale. Apparently, till now, this field has been relatively less explored and very little has been accomplished, but it holds enormous potential to combat the various challenges, as principally a lot can be done through this. The field he was referring to was "miniaturization!" Also, notably, his talk laid the cornerstone for the buzzword "nano." This talk was delivered by him way back, almost 40 years ago! Although this great man had advocated the concept of miniaturization, the real technological advancements in this area were instigated only after the innovative creation of transistors by the famous Nobel Prize winners W. Schockley, J. Bardeen, and W. H. Brattain in the year 1947.

This significant innovation gave birth to the concept of integrated circuits (ICs) in 1955, and the first IC was manufactured by Jack St. Clair Kilby, an American electrical engineer, within a span of three years since the term was first introduced. Today, advances in miniaturization technologies have dramatically affected our lives and the results can be seen in every field—mobile phones, computers, and vehicle engines—that has taken the downsizing route. As a result of the miniaturization of electronic equipment, enormous amounts of waste could be reduced. The prime benefits of the miniaturization technology are:

- Reduction in sample amount and decreased consumption of chemicals and solvents: The sample volume required for carrying out an appropriate analysis can be drastically reduced by scaling down the various sample preparation and detection techniques. This is especially useful while dealing with scarce or precious samples.

  One classic example of miniaturization that is being followed in the academic sector is the adoption of micro and semimicro qualitative inorganic analysis technique for the detection of basic and acidic radicals in different samples. As we know, qualitative analysis of inorganic mixtures is a vital part of the laboratory curriculum that generally involves the use of huge amounts of corrosive acids and toxic chemicals. Besides, the traditional methods rely on heating with acids, which in turn generate obnoxious fumes, degrading the indoor air quality. Also, in this process, a considerable amount of liquid waste is generated, which results in the contamination of water. Therefore, a paradigm shift toward microanalysis and semimicroanalysis is a major step toward waste reduction at source (in accordance with Principle 1 of green chemistry). Microanalysis, especially in the form of selective eco-friendly spot tests, has helped in cutting down the consumption of chemicals as well as minimizing the waste produced.
- Reduction of associated wastes: Owing to a notable reduction in the quantities of starting materials, the waste generated during the entire analytical process can also be reduced enormously. Hence, miniaturization is a step toward sustainability.
- Improved sensitivity: Sensitivity of various analytical techniques can also be greatly enhanced by making effective use of the miniaturized-sample preparation technique.
- Rapidity: The development of miniaturized-sample preparation, separation, and detection can help in decreasing the time required to perform any analysis. This can, in turn, have direct benefits: lower energy requirements and smaller amounts of waste.

Also, the effects are apparent in the technological sector. For instance, the miniaturization of transistors helps them operate faster.

- Portability: The development of field portable instrumentation has remained a long-standing challenge for the analytical chemists. Miniaturization of a part or whole of the process steps could contribute significantly toward the portability of analytical systems.

2.5.4 Reduce, Reuse, and Recycle

As discussed in the previous section, recycling is an environmentally benign way to manage waste. However, a more comprehensive way to do so has evolved in the form of the three Rs (3Rs): reduce, reuse, and recycle [32]. For overall reduction in waste sent to the landfills, it is imperative to devote attention to first minimizing the total amount of waste generated, then utilizing the prevailing materials for further use, and finally recycling the materials/products. The 3R initiative is being propagated to build the foundation of a sound society via the best use of resources, and today it has been recognized worldwide that the reduce, reuse, recycle paradigm can contribute to a decrease in the greenhouse gas emissions.

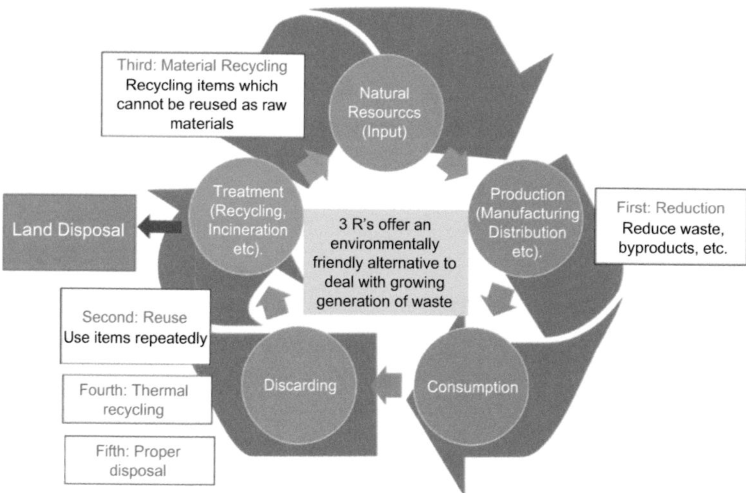

Figure 2.12 The 3Rs offer an environmentally friendly alternative to deal with the growing generation of waste.

The waste management hierarchy explained earlier has gained international acceptance with regard to judicious waste management practices. It gives due emphasis to waste reduction at source wherever possible; one must prevent waste to the maximum extent possible, and focus on utilizing the reuse and recycle options if waste has been generated. For proper waste management, the EPA and the Solid Waste Act favored an integrated model of the solid waste management strategy (Fig. 2.12) [33]:

The 3R approach is explained next.

2.5.4.1 Reduce

The first step in the waste management model involves preventing waste generation by implementing a source reduction strategy in every manufacturing activity undertaken. It is principally related to the first principle of green chemistry, which aims at complete prevention/avoidance of waste by redesigning existing strategies. Just to outline a few steps that can be taken in this regard:

- Improvise the conventional waste generating production processes by transforming the protocols; for example, packaging could be reduced to a considerable extent to minimize the related waste.
- Convince and influence the consumers in such a way that they buy high-quality products with less packaging.

To promote awareness and educate general public, various awareness raising campaigns are being run wherein the increasing emphasis is on helping people understand the significance of generating less waste, which would eventually drive the creation of a more resource-efficient market and ultimately a safer planet. But the big question that arises here is, how can we reduce?

This will not be very difficult. All of us can contribute at the basic domestic level by adopting the following steps to minimize waste:

- Choosing items that you need, not want
- Shopping for high-quality items
- Using minimal packaging

2.5.4.2 Reuse

Reuse stands out as the next-best option for solid waste management, and it embraces the effective reuse of any article/product/material for either the same purpose or for an entirely new cause/objective. So the question is, how can we reuse to the best extent possible? This can be accomplished readily by effective repair, sale, or donation of these goods; if we follow this, then we can all contribute toward waste minimization. It is worth highlighting here that as compared to recycling, reusing is the preferred option as it eliminates the requirement of any reprocessing. In addition to this, reuse also helps in saving our economy, conserving resources, and most importantly gratifying the human urge to remain inventive.

As simple but wonderfully effective examples to what can be done, we could refill a bottle (whether it was once containing water or any other drink) instead of opting for a brand-new one; we could also reuse papers and towels utilized for cleaning goods and soaking liquids; instead of paper napkins, we can go for reusable ones; also we can substitute the use of paper cups with reusable cups and utilize reusable grocery bags for collecting garbage.

2.5.4.3 Recycle

Developed countries are increasingly incorporating the recycling strategy, which involves collecting, segregating, and reprocessing waste with the aid of modern technologies in order to recover valuables [34].

- Recycling generates industry
- Recycling creates jobs
- Recycling cuts down cost

2.6 Design for Degradation

Principle 10 says, "Chemical products should be designed so that at the end of their function they break down into innocuous degradation products and do not persist in the environment." Daily, we use several chemicals for a variety of purposes, and these chemicals need to be stable during use. Once their use is over, we send them either to landfills or they end up entering the wastewater treatment plants, reaching our diverse environmental strata, including water and soil. So, the challenge before the chemists and researchers is to design chemicals that remain stable while being used but do not persist in the environment, that is, designing chemicals that are stable on one hand but will degrade on the other. And the very important point to remember here is that the degraded products must be nontoxic and not persistent.

A common example of biodegradable products is biodegradable soaps. The goal must be to design soaps and detergents that would readily break down in the environment. One of the most frequently employed detergents is sodium dodecylbenzenesulfonate, which is referred to as "linear alkylbenzene sulfonate" (LAS) [35]. LAS has found use in several things and is especially known as a laboratory detergent. The most fascinating property of this detergent is that it breaks down rapidly under aerobic oxidative conditions. What happens during this process is that microbes utilize the linear alkyl chain through a process called β-oxidation and enable the breakdown of the carbon chain. As soon as the long carbon chain degrades, the remaining components of the compound also get degraded.

If LAS is compared to a branched chain version, one would find that (Fig. 2.13) β-oxidation cannot be performed due to the lack of sites for initiation of this reaction. Hence, these days, branched-chain detergents are no longer used as they are too persistent and nonbiodegradable. In fact, the branched version has been found to be four times more toxic than its linear counterpart because of its tendency to persist.

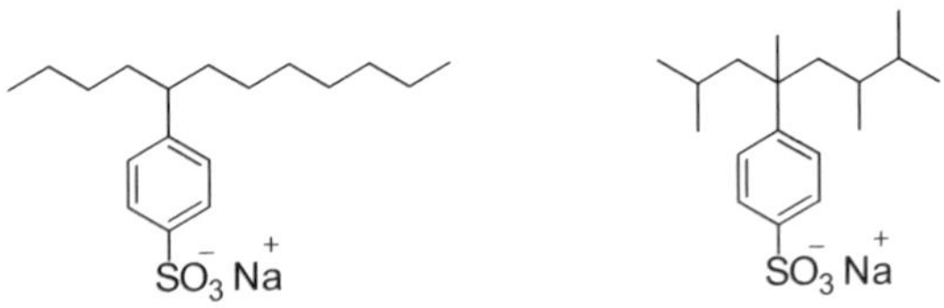

$SO_3^- \ Na^+$ $SO_3^- \ Na^+$

Figure 2.13 (a) Sodium dodecylbenzenesulfonate, an example of a linear alkylbenzene sulfonate, which is biodegradable and (b) a branched alkylbenezene sulfonate (does not biodegrade).

As we just saw in the example of detergents, they undergo microbial degradation when oxygen is present. However, if these soaps reach water sources directly, even the linear version would not break down rapidly, the primary reason being the presence of lesser amount of microbes in water. Thus, designing chemicals that would break down can be quite challenging. However, if Principle 10 could be followed, then the major problems related to waste generation could be avoided.

2.7 Conclusion

To summarize, waste is one of most critical challenges we are facing today. The 2030 agenda of SDGs underscores the significance of waste management. This chapter reflects the problems associated with the traditional techniques to deal with waste and how there is a shift in efforts toward eliminating/eradicating waste at the outset rather than disposing of toxic by-products. Elimination of waste can be achieved at the molecular level by redesigning existing processes using the various principles of green chemistry. Pharmaceutical industries have benefitted tremendously by doing so, and many of the redesigned pathways for drug synthesis have won recognition in the form of Presidential Green Chemistry Challenge Awards. Also very importantly, today, there is an increasing emphasis on utilizing the concept of circular economy, which has slowly started replacing the linear economic model. Reduce, reuse, and recycle have emerged as the 3Rs that advocate the significance of zero waste generation. This chapter sheds light on all the latest green technologies being employed for waste reduction/minimization. At the end, what has been projected is how important it is to design products that can degrade easily. Ultimately, a waste-free world would be cleaner, greener, and healthier!

2.8 Learning Outcomes

At the end of this chapter, students will be able to:

- Understand what is waste and why waste generation has been recognized as a critical issue, especially in recent times
- Identify and analyze the different sources of waste—chemical, pharmaceutical, plastic, food, electronics, etc.
- Comprehend the problems related to the generation and mismanagement of waste and realize how green chemistry provides a viable solution for this
- Creatively think about utilizing waste as a resource by incorporating the concept of circular economy
- Learn the various innovative waste management techniques, such as minimizing derivatives, recycling reagents,

miniaturization, the 3R concept of waste management hierarchy "reduce, recycle, and reuse," that could provide potential solutions for the problems pertaining to waste

- Realize the challenges associated with Principle 10—design for degradation—which would play an instrumental role in resolving major problems associated with waste
- Students will be fostered to think about designing and following synthetic approaches that would lead to no waste or less waste

2.9 Problems

1. Why do you think waste has been referred to as a misplaced resource?
2. What do you understand by returning safe substances to the environment? Which principle does the statement rely on?
3. How can one think about greening wastes? Elucidate any waste management technique that could potentially resolve critical issues associated with waste.
4. List some of the hazardous chemicals released into the environment and their toxic effects.
5. What are biodegradable plastics? How are they manufactured?
6. Discuss how pharmaceutical industries have been able to reduce the amount of waste generated significantly by practicing green chemistry.
7. Explain the zero-waste concept. Is it the same as the 3R concept?
8. The story of sitagliptin drug provides an interesting example of a case where green chemistry was manifested. Compare the conventional and modified synthesis schemes for this drug and highlight the advantages of the modified green synthetic pathway.
9. How can you correlate atom economy, *E*-factor, and waste generated? Enlighten using example.
10. Give examples of polymers obtained from renewable raw materials.
11. Do you think miniaturization could ultimately resolve waste-related issues? If so, how?

12. Name a few valuable resources generated from the following waste materials and the technique involved in waste valorization:
    (i) Citrus peels
    (ii) Cornstarch
    (iii) Discarded electronic parts
    (iv) Discarded paper
    (v) Glycerol

References

1. Jambeck, J. R., Geyer, R., Wilcox, C., Siegler, T. R., Perryman, M., Andrady, A. and Law, K. L. (2015). Plastic waste inputs from land into the ocean. *Science*, **347**, pp. 768–771.
2. Environmental Protection Agency (2017). Sustainable materials management: Non-hazardous materials and waste management hierarchy.
3. Clark, J. H., Pfaltzgraff, L. A., Budarin, V. L., Hunt, A. J., Gronnow, M., Matharu, A. S. and Sherwood, J. R. (2013). From waste to wealth using green chemistry. *Pure Appl. Chem.*, **85**, pp. 1625–1631.
4. Yadav, M., Dutta, S. and Sharma, R. K. (2017). *Hazardous Reagent Substitution*, Chapter 8, New directions from academia (Royal Society of Chemistry, UK) pp. 130–167.
5. Singhal, S. (2003). India 2025 - Environment. Government of India. http://planningcommission.gov.in/reports/sereport/ser/vision2025/env2025.pdf
6. Sheldon, R. A. (2017). The E factor 25 years on: The rise of green chemistry and sustainability. *Green Chem.*, **19**, pp. 18–43.
7. Sheldon, R. A. (2012). Fundamentals of green chemistry: Efficiency in reaction design. *Chem. Soc. Rev.*, **41**, pp. 1437–1451.
8. Reddy, A. G. S., Saibaba, B. and Sudarshan, G. (2012). Hydrogeochemical characterization of contaminated groundwater in Patancheru industrial area, southern India. *Environ. Monit. Assess.*, **184**, pp. 3557–3576.
9. Patneedi, C. B. and Prasadu, K. D. (2015). Impact of pharmaceutical wastes on human life and environment. *Rasayan J. Chem.*, **8**, pp. 67–70.
10. Yadav, M. (2001). Hospital waste: A major problem. *JK Practitioner*, **8**, pp. 276–282.

11. Sharma, S., Dutta, S. and Sharma, R. K. (2017). *Hazardous Reagent Substitution*, Chapter 2, Recyclability of reagents (Royal Society of Chemistry, UK) pp. 18–52.

12. Liu, D. H. and Lipták, B. G. (1999). *Hazardous Waste and Solid* (CRC Press, US).

13. Patneedi, C. B. and Prasadu, K. D. (2015). Impact of pharmaceutical wastes on human life and environment. *Rasayan J. Chem.*, **8**, pp. 67–70.

14. Environmental Protection Agency (2017). Managing hazardous waste at academic laboratories rulemaking.

15. Sidhwani, I. T. and Chowdhury, S. (2008). Greener alternative to qualitative analysis for cations without H_2S and other sulfur-containing compounds. *J. Chem. Educ.*, **85**, p. 1099.

16. Wearden, G. (2016). More plastic than fish in the sea by 2050, says Ellen MacArthur. *The Guardian*, 19 Jan 2016.

17. Thompson, R. C., Moore, C. J., Vom Saal, F. S. and Swan, S. H. (2009). Plastics, the environment and human health: Current consensus and future trends. *Philos. Trans. R. Soc. B*, **364**, pp. 2153–2166.

18. Huang, Y. Q., Wong, C. K. C., Zheng, J. S., Bouwman, H., Barra, R., Wahlström, B. and Wong, M. H. (2012). Bisphenol A (BPA) in China: A review of sources, environmental levels, and potential human health impacts. *Environ. Int.*, **42**, pp. 91–99.

19. McCormack, N. (2018). The problem with microbeads; or how the cosmetics you use one day end up in the sushi you eat next. *Austl. L. Libr.*, **26**, p. 171.

20. Romeo, J. (2019). Plastics that do not last forever: Engineered bioplastics. *Plast. Eng.*, **75**, pp. 36–41.

21. Hussain, M. and Mumtaz, S. (2014). E-waste: Impacts, issues and management strategies. *Rev. Env. Health*, **29**, pp. 53–58.

22. Widmer, R., Oswald-Krapf, H., Sinha-Khetriwal, D., Schnellmann, M. and Böni, H. (2005). Global perspectives on e-waste. *Environ. Impact Assess. Rev.*, **25**, pp. 436–458.

23. Villanueva, A. and Wenzel, H. (2007). Paper waste–recycling, incineration or landfilling? A review of existing life cycle assessments. *J. Waste Manage.*, **27**, pp. S29–S46.

24. Joshi, R. and Ahmed, S. (2016). Status and challenges of municipal solid waste management in India: A review. *Cogent Environ. Sci.*, **2**, p. 1139434.

25. Sharma, R. K. (2005). Bioinorganic Chemistry E-Book Module. National Science Digital Library, http://niscair.res.in/ispui/handle/123456789/230
26. Hester, R. E. and Harrison, R. M. (2013). *Waste as a Resource* (Royal Society of Chemistry, UK).
27. Gruber, P. R. (2003). Cargill dow LLC. *J. Ind. Ecol.*, **7**, pp. 209–213.
28. Cue, B. W. and Zhang, J. (2009). Green process chemistry in the pharmaceutical industry. *Green Chem. Lett. Rev.*, **2**, pp. 193–211.
29. https://www.maratek.com/en/solvent-recycling-equipment-sp-bu
30. https://www.cbgbiotech.com/about/customer-successes/chemgenes-corporation
31. Jiang, X., Sclafani, J., Prasad, K., Repič, O. and Blacklock, T. J. (2007). Pd–Smopex-111: A new catalyst for Heck and Suzuki cross-coupling reactions. *Org. Process Res. Dev.*, **11**, pp. 769–772.
32. Dijkers, M. (2019). Reduce, reuse, recycle: Good stewardship of research data. *Spinal Cord*, **57**, pp. 165–166.
33. Liu, A., Ren, F., Lin, W. Y. and Wang, J. Y. (2015). A review of municipal solid waste environmental standards with a focus on incinerator residues. *Int. J. Sustainable Built Environ.*, **4**, pp. 165–188.
34. Mulholland, K. L., Sylvester, R. W. and Dyer, J. A. (2000). Sustainability: Waste minimization, green chemistry and inherently safer processing. *Environ. Prog.*, **19**, pp. 260–268.
35. Anastas, P. and Eghbali, N. (2010). Green chemistry: Principles and practice. *Chem. Soc. Rev.*, **39**, pp. 301–312.

Chapter 3

Catalysis: A Promising Green Technology

Manavi Yadav,[a,b] Radhika Gupta,[a] Gunjan Arora,[a] and Rakesh K. Sharma[a]

[a]*Green Chemistry Network Centre, Department of Chemistry, University of Delhi, Delhi 110007, India*

[b]*Department of Chemistry, Hindu College, University of Delhi, Delhi 110007, India*

manavirgo@gmail.com

> *Chemistry without catalysis, would be a sword without a handle, a light without brilliance, a bell without sound.*
>
> —Alwin Mittasch

3.1 Introduction

Owing to the growing needs of mankind, manufacturing activity is taking place at a faster rate in all sectors and catalysis science and technology has been a great contributor in this. Long before the emergence of green chemistry principles, several industrial processes successfully employed catalysis, including oil refining, polymer production, and bulk and fine chemicals. Undoubtedly,

Green Chemistry for Beginners
Edited by Rakesh K. Sharma and Anju Srivastava
Copyright © 2021 Jenny Stanford Publishing Pte. Ltd.
ISBN 978-981-4316-96-5 (Hardcover), 978-1-003-18042-5 (eBook)
www.jennystanford.com

the chemical industry of the twentieth century could not have advanced to its current status solely on the basis of noncatalytic and stoichiometric reactions. In fact, 90% of the products are synthesized by catalytic processes in the chemical industry. Thus, catalysts are considered as workhorses of chemical reactions [1].

3.1.1 What Is a Catalyst?

A catalyst provides an alternative, energetically favorable mechanism that involves a transition state of lower energy and enables the overall process to be carried out under mild conditions of temperature and pressure. As depicted in Fig. 3.1 the activation energy of the catalytic reaction is much lower in comparison to that of the uncatalyzed reaction, thereby facilitating an enhanced rate of reaction. It should be noted that a catalyst can change the kinetics of a reaction but cannot alter its thermodynamics. Thus, for a thermodynamically favorable reaction, if a small amount of catalyst is added, there is an increased rate of attainment of chemical equilibrium, whereas if a reaction is thermodynamically unfavorable, a catalyst cannot change the situation.

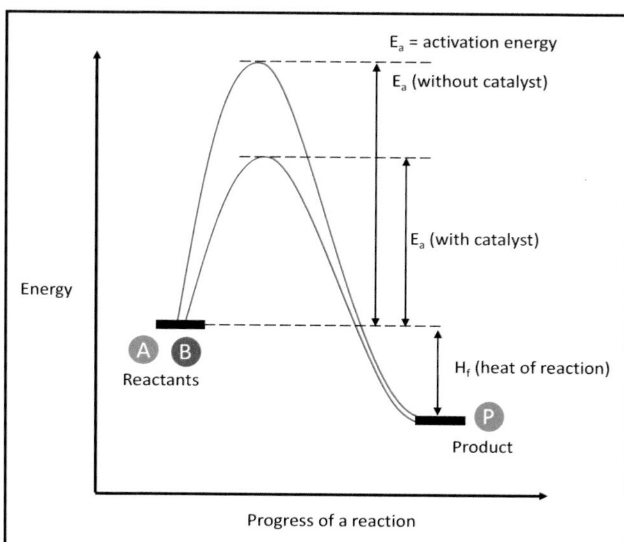

Figure 3.1 Comparison of the energy profile diagrams of a catalyzed and an uncatalyzed reaction.

A catalytic reaction can be described as a cyclic event in which a catalyst participates and can be recovered in its original state at the end of the cycle. A catalytic mechanism has been illustrated in Fig. 3.2, where two reactant molecules A and B bond to the catalyst and form a complex in which A and B react with each other to form the product P. This product is also bound to the catalyst and finally separates from the catalyst, thereby regenerating the catalyst, which can undergo further reaction.

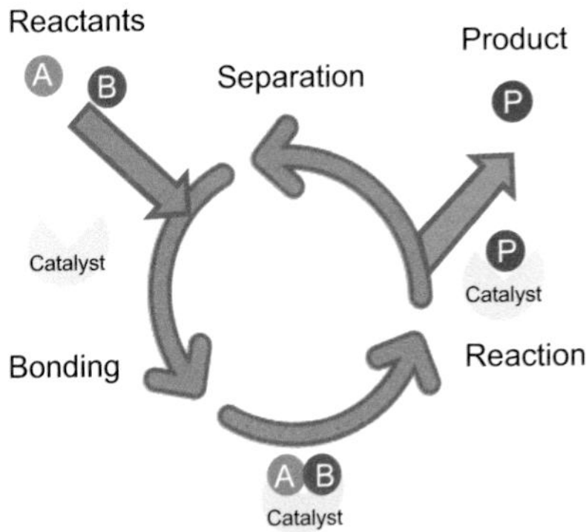

Figure 3.2 A catalytic reaction consisting of a sequence of elementary steps where reactant molecules bind to the catalyst and react to form a product that finally detaches from the catalyst.

3.1.2 History of Catalysis

Use of catalysts goes back several centuries, when catalysts ranging from inorganic compounds to enzymes, under different names, were used for cheese and wine production as also for other beverages and food products. In 1875, catalytic technology was employed in industries for the large-scale production of sulfuric acid using platinum. Later, Ostwald developed a method for producing nitric acid using a platinum gauge. A major breakthrough was achieved by Mittasch, Bosch, and Haber in 1908–1914, when ammonia was synthesized using iron as a catalyst. In subsequent years, major development was seen in the energy sector when catalytic cracking was introduced. Soon after this, the Ziegler–Natta catalyst was developed, which changed the scenario of the polymer industry. Afterward, the homogeneous catalytic process gained significant attention and soon catalysts were developed for controlling environmental pollution. Thus, catalysis is a steadily evolving field, with new and efficient systems continuously replacing the older less efficient ones.

3.1.3 Catalytic Route vs. Stoichiometric Route: The Greener Aspect

A technology is termed "green" only if it uses resources and raw materials efficiently, avoids toxic solvents and hazardous reagents, and minimizes the formation of undesirable products. Catalysis has proven to be a key tool for green chemistry, where catalytic materials with improved properties and performance have been facilitating energy-efficient processes. In fact, a heterogeneous catalyst can be easily recovered at the end of the catalytic cycle and can be reused multiple times. This saves the use of additional reagents, thereby reducing the amount of waste produced at each step [2]. Thus, catalysis has the potential to contribute to the guiding principle of "benign by design" and in many ways to achieving the global goals of sustainable development (Fig. 3.3).

While comparing traditional stoichiometric syntheses with catalytic methods, catalysis offers several benefits over stoichiometric reagents. Some of them are listed below.

- A catalyst can carry out multiple runs per mole of catalyst, whereas stoichiometric reactions require at least 1 mole of reagent per mole of substrate.
- Stoichiometric reactions can generally be very slow and may require substantial amounts of energy to carry out the transformation. Raising the temperature will enable the reaction to proceed smoothly, but for this, special conditions are required along with safety precautions, which are both difficult and expensive to maintain. On the other hand, catalytic reactions are fast and therefore do not require special reactors to carry out the syntheses.

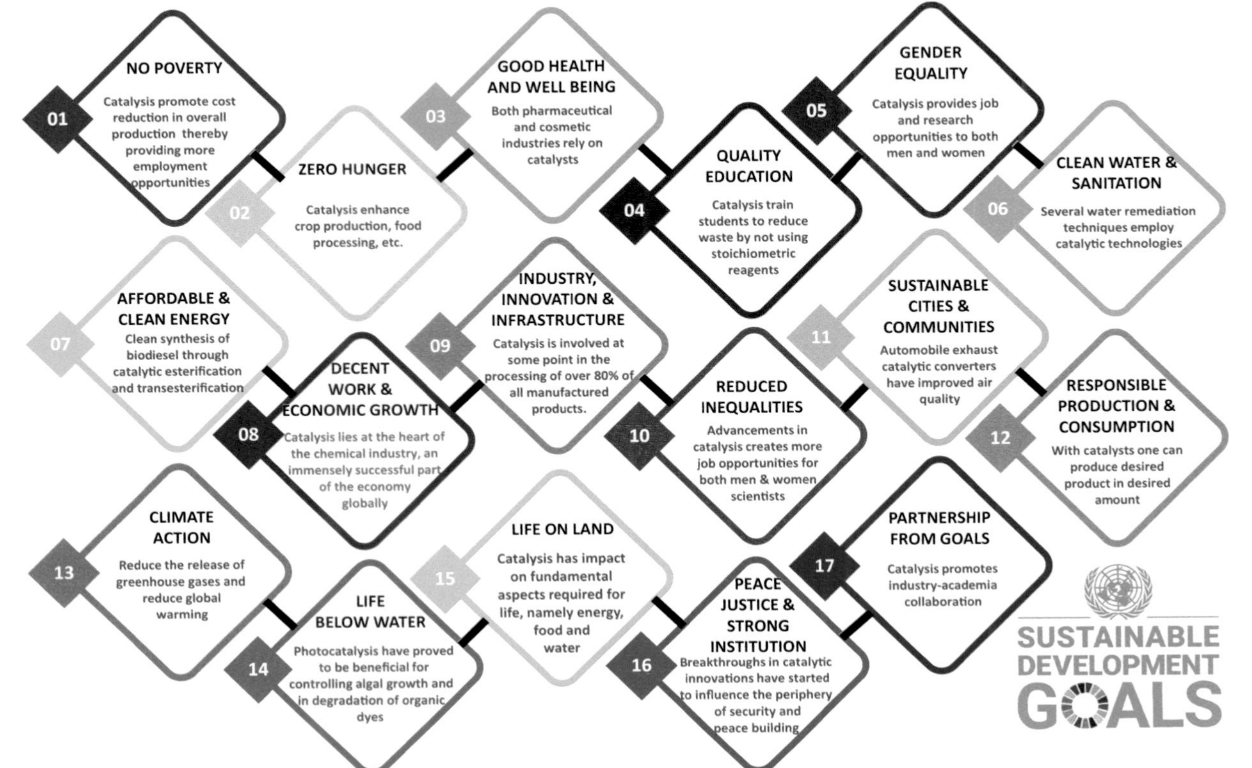

Figure 3.3 How catalysis aids in achieving the various UN Sustainable Development Goals.

- Stoichiometric reactions are usually accompanied by the production of unwanted side-products that could be detrimental to the environment and their disposal can be very costly, while reactions employing catalysts surpass these shortcomings. For example, in stoichiometric reduction with metals such as Na, Mg, Zn, and Fe and metal hydrides $LiAlH_4$ and $NaBH_4$ or oxidation with $KMnO_4$ and MnO_2; reactions like nitration, sulfonation, Friedel-crafts reactions, and diazotizations; and halogenation employing various mineral acids and Lewis acids, a considerable amount of waste is generated consisting primarily of inorganic salts. In this respect, catalytic oxidation, hydrogenation, and carbonylation have appeared as promising low-salt solutions with high efficiency.

Therefore, employing catalytic reactions instead of stoichiometric methods can dramatically improve atom economy, save time and energy, minimize the amount of reagents, and reduce the waste generated. Thus, both the economic and environmental statuses can be significantly improved with the development and application of catalytic methods. A few examples have been discussed below.

Example 1: In the hydrogenation of ketones, the stoichiometric reaction involves the addition of sodium borohydride, followed by water addition. However, borane and sodium hydroxide are formed as waste in the overall reaction. But when the same reaction is carried out in the presence of palladium on carbon as a catalyst, the ketone directly reacts with hydrogen and gives the desired product without generating any waste (Scheme 3.1).

Stoichiometric method

1. $NaBH_4$; 2. H_2O → + BH_3 + NaOH

Catalytic method

H_2, Pd/C →

Scheme 3.1 Stoichiometric route versus catalytic route toward the hydrogenation of ketones.

Example 2: The selective oxidation of ethylene to ethylene epoxide is an important intermediate in the chemical industry. Previously, it was synthesized by a noncatalytic route that followed three steps (Scheme 3.2), but for each molecule of ethylene oxide, one molecule of salt was formed, creating a lot of waste that was finally dumped into the rivers—something totally inadmissible. On the other hand, the catalytic route is clean, simple, and effective, although it produces a small amount of carbon dioxide. In this method, 90% of ethylene oxide can be selectively formed from ethene and dioxygen by employing silver promoted by chlorine as a catalyst.

$$Cl_2 + NaOH \longrightarrow HOCl + NaCl$$

$$C_2H_4 + HOCl \longrightarrow ClH_2C - CH_2OH$$

$$ClH_2C - CH_2OH + 1/2\ Ca(OH)_2 \longrightarrow 1/2\ CaCl_2 + C_2H_4O + H_2O$$

Scheme 3.2 Three-step selective oxidation of ethylene to ethylene epoxide.

Example 3: Conventionally, hydroquinone was synthesized from benzene by nitrating it first to form nitrobenzene, followed by reduction with iron and hydrochloric acid, also known as Béchamp reduction, to produce aniline. Then, aniline was oxidized with stoichiometric amounts of MnO_2 to form benzoquinone, which was again reduced by Béchamp reduction. However, the overall reaction produced more than 10 kg of inorganic salts as waste (consisting of $MnSO_4$, Na_2SO_4, NaCl, and $FeCl_2$) per kilogram of hydroquinone synthesized. This process has now been replaced by a modern method that involves auto-oxidation of *p*-diisopropylbenzene formed by Friedel–Crafts alkylation of benzene, followed by acid-catalyzed rearrangement of the bis-hydroperoxide, which generates 1 kg of inorganic salts as waste per kilogram of hydroquinone. Another method involves the hydroxylation of phenol with aqueous H_2O_2 using titanium silicate as the catalyst (Scheme 3.3).

Example 4: Benzoin condensation is an addition reaction between two aromatic aldehydes to produce α-hydroxy ketones. It is a highly atom economic reaction as all the atoms of the reactants are retained in the final product. The original benzoin condensation was catalyzed by cyanide anion. However, due to the extremely hazardous and toxic nature of the cyanide-containing reagents, safer alternatives were required. One such greener reagent is thiamine. Thiamine, or vitamin B_1, is a coenzyme that is used by living systems for the catabolism of sugars and amino acids. It has also been explored by synthetic chemists as a nontoxic, biodegradable, and inexpensive catalyst for various organic transformations. In the present reaction, thiamine hydrochloride has been utilized as a green catalyst for the synthesis of benzoin (Scheme 3.4). The reactive part of thiamine is the thiazole heterocyclic ring.

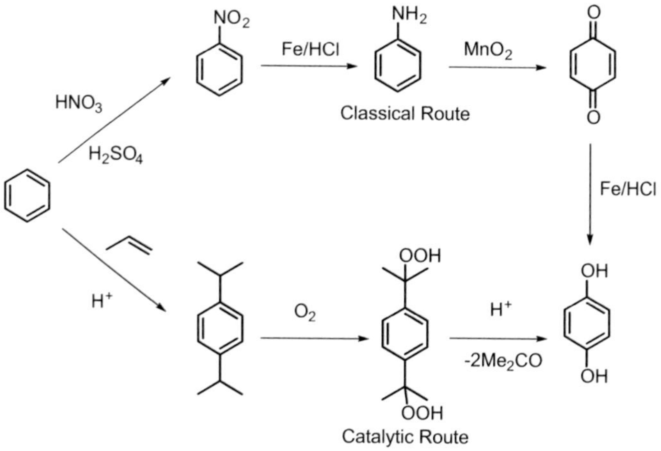

Scheme 3.3 Classical and catalytic route for manufacturing hydroquinone.

Ibuprofen, a widely used anti-inflammatory drug, was conventionally synthesized in six stoichiometric steps and accompanied by lots of waste and by-products and thus was considered an inefficient process. Nowadays, the production of this drug involves three catalytic steps with improved efficiency and reduced waste.

Sitagliptin, an active ingredient in Januvia™, used for the treatment of type 2 diabetes, was developed by Merck and Codexis by an enzymatic process that enhances yield, reduces waste, and eliminates the use of a metal catalyst.

Simvastatin, a leading drug for treating high cholesterol levels, was earlier synthesized by a multistep reaction employing large amounts of hazardous reagents and huge amounts of toxic waste was generated during the process. Later, Codexis, a biocatalysis company, optimized the enzyme and chemical process to produce this drug, significantly reducing the use of reagents and formation of noxious waste.

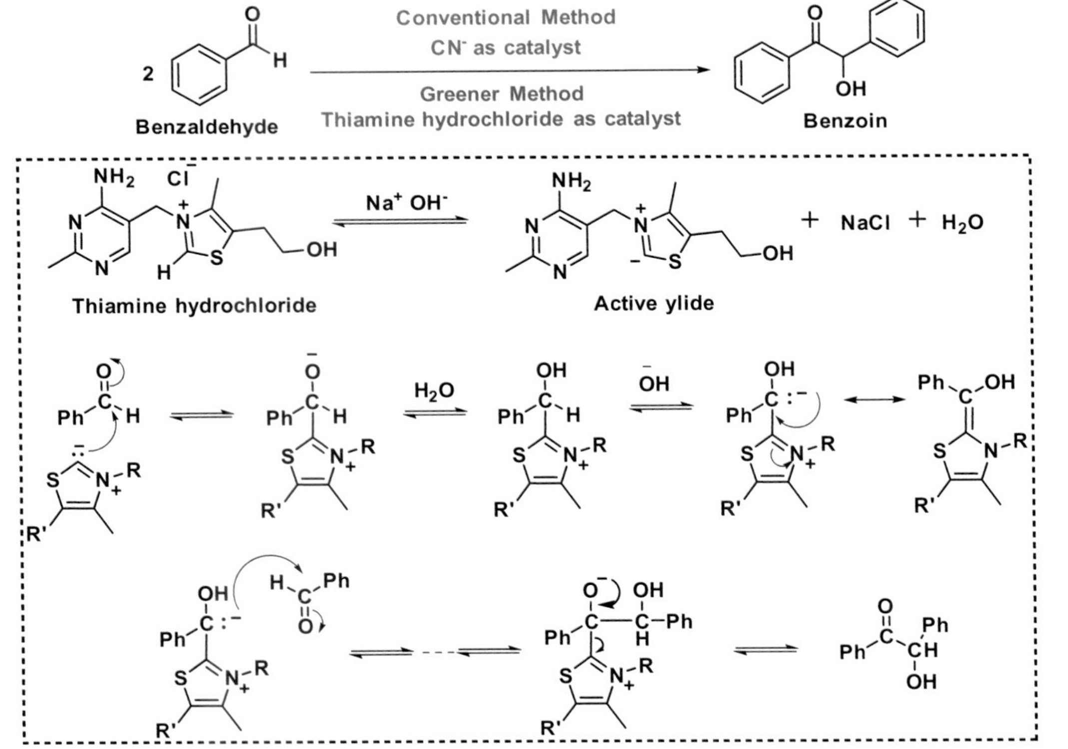

Scheme 3.4 Benzoin condensation using thiamine—an example of greener catalysis.

3.1.4 Nobel Prize Awards in the Development of Catalysis

Since the early 1900s, many achievements in the development of enzymatic and chemical catalysis have been recognized by the Nobel Foundation [3]. Table 3.1 lists the Nobel Prize winners, with their notable contributions in the field of catalysis.

Table 3.1 List of Nobel Prize winners in the catalytic field

Year	The notable contribution in catalysis for which the Nobel Prize was conferred	Winners		
2010	Palladium-catalyzed cross-coupling reactions	Richard F. Heck	Ei-ichi Negishi	Akira Suzuki
2005	Catalytic chemical process known as metathesis	Yves Chauvin	Robert H. Grubbs	Richard R. Schrock
2001	Chirally catalyzed hydrogenation reactions	William S. Knowles		Ryoji Noyori
2001	Chirally catalyzed oxidation reaction	K. Barry Sharpless		
1989	Discovery on catalytic properties of RNA	Sidney Altman		Thomas R. Cech
1975	Stereochemistry of enzyme-catalyzed reactions	John Warcup Cornforth		
1972	Understanding of the connection between the chemical structure and catalytic activity of the active center of the ribonuclease molecule	Stanford Moore		William H. Stein
1909	Catalysis and investigations on the fundamental principles governing chemical equilibria and rates of reaction	Wilhelm Ostwald		

3.2 Role of Catalysis

Catalysis plays a crucial role in producing environmentally benign chemicals, both new and existing, while saving energy, time, raw materials, and cost related to the overall process [4, 5]. The role of a catalyst has been summarized below.

- Catalysis decreases the activation energy required for a chemical reaction and thus enhances the rate of reaction.
- It allows reactions to proceed under milder reaction conditions; therefore, catalytic reactions are considered efficient.
- The greater activity of catalysts can sometimes lead to the conversion of million times their own weight.
- Catalysts allow high product selectivity in multifunctional compounds by enabling site-specific transformations and diastereomeric control. This allows the effective utilization of resources and minimization of waste.
- Catalytic methods also circumvent the need for activation and deactivation of starting materials in the prefunctionalization steps, thereby lessening the number of steps in a reaction.
- Catalysis prevents pollution by avoiding the formation of unwanted side-products in a reaction.
- Catalysts are also applied for improving air quality by removing and controlling NO_x emissions, decreasing the use of volatile organic solvents, and replacing chlorine-based chemicals by developing alternate catalytic methodology.
- Catalysis is employed for the production of transportation fuels and bulk and fine chemicals all over the world.
- Catalysts that can be reused provide economic feasibility and facilitate their industrial application.

Parameters affecting both commercial usefulness and greenness of a particular catalyst are depicted in Fig. 3.4.

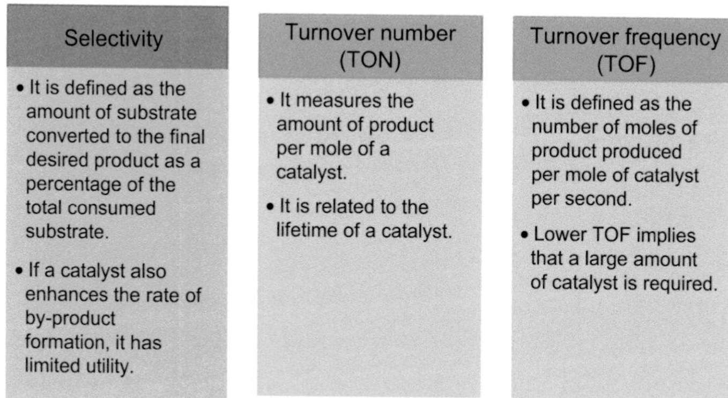

Figure 3.4 Important parameters for a catalyst.

3.3 Next-Generation Catalysts

A catalyst is termed "green" only if it exhibits certain characteristics. A recent example includes the use of Nobel Prize–winning metathesis catalytic technology by Elevance for the production of low-cost greener chemicals with high performance [6]. Till date, it is the first and only company that has commercialized the natural oil metathesis technology. Another green catalyst was developed that reduces the ecological footprint associated with the manufacture of propylene oxide. This work also won the prestigious Dow Chemical Award. Several green catalysts have been developed so far, and many of them have won the Presidential Green Chemistry Challenge Awards [7]. These award-winning works will be discussed later, in Chapter 6. Some of the features of an ideal catalyst have been depicted in Fig. 3.5.

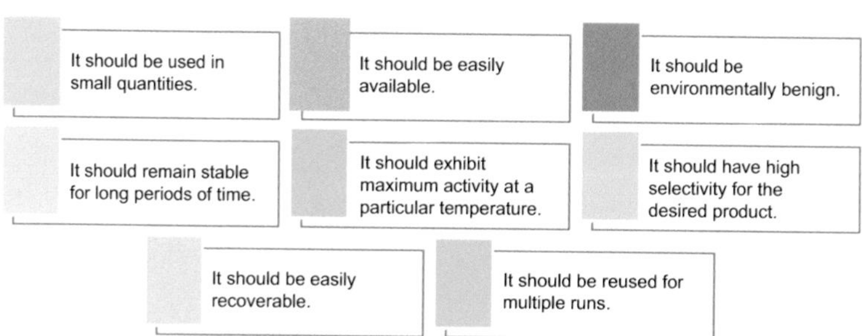

Figure 3.5 Properties of an ideal catalyst.

3.4 Classification of Catalysts

All catalysts can be broadly classified under two categories, homogeneous catalysts and heterogeneous catalysts (Fig. 3.6). Within these categories further classification is done. From the perspective of green chemistry, there are some important catalysts that will be discussed later in this chapter.

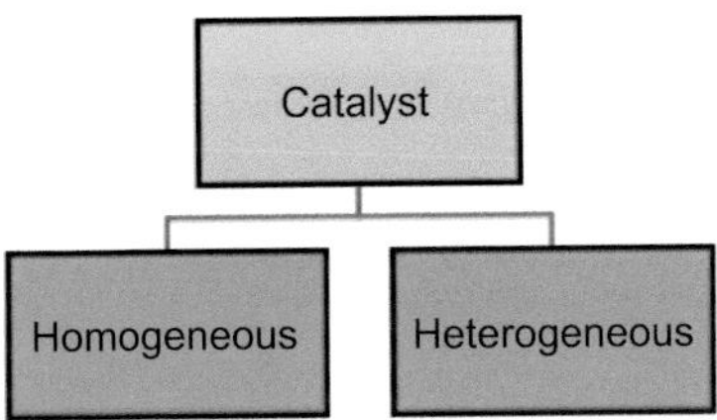

Figure 3.6 Classification of catalysts.

3.5 Homogeneous Catalysis

Homogeneous catalysis involves the participation of the reactants and catalyst in the same physical state, commonly liquid. This implies that both reagents and the catalyst are dissolved in a solvent for carrying out the reaction. Examples of homogeneous catalysts are liquid-phase acid/base catalysts and transition metal complexes in solution. Some of the most significant industrially established homogeneous reactions are discussed next [8].

3.5.1 Hydroformylation Reaction

One of the major industrial applications that use homogeneous catalysts is hydroformylation/oxo reaction. This reaction was discovered by German chemist Otto Roelen in 1938 (hence also known as Roelen reaction). It is an addition reaction in which a formyl group (CHO) and a hydrogen atom is added to a carbon–carbon double bond in the presence of transition metal catalysts; bulk processes usually apply cobalt or rhodium complexes (Scheme 3.5). Cobalt catalysts completely dominated the industrial hydroformylation until the early 1970s, when rhodium catalysts were commercialized. In 1992, approximately 70% of all processes were carried out using rhodium triarylphosphine catalysts as they led to the regioselective synthesis of straight-chain aldehydes. Ever since its invention, this reaction has seen continuous growth in its demand, primarily because the aldehydes so formed can be easily converted into many other industrially important secondary products, such as alcohols, acids, esters, amines, and diols. Currently, the largest hydroformylation plants are being operated by companies such as Shell, Mitsubishi, BASF, Imperial Chemical Industries, Kuhlmann, Ruhrchemie, UCC, Eastman, BP, DOW, and Evonik.

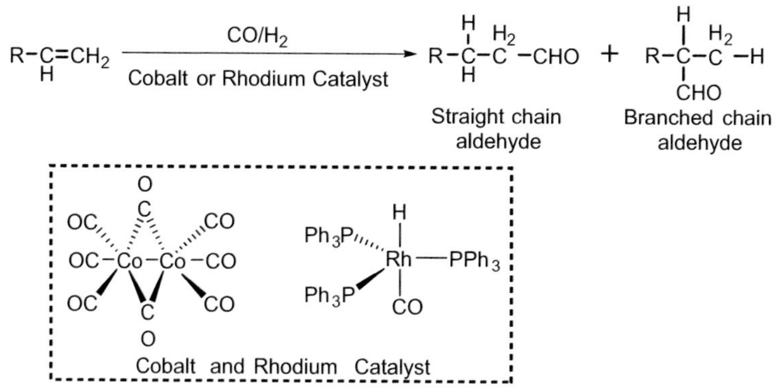

$R{-}CH{=}CH_2 \xrightarrow[\text{Cobalt or Rhodium Catalyst}]{CO/H_2} R{-}CH_2{-}CH_2{-}CHO + R{-}CH(CHO){-}CH_3$

Straight chain aldehyde   Branched chain aldehyde

Cobalt and Rhodium Catalyst: $Co_2(CO)_8$, $RhH(CO)(PPh_3)_3$

Scheme 3.5 Hydroformylation reaction.

3.5.2 Olefin Hydrogenation Using Wilkinson's Catalyst

$RhCl(PPh_3)_3$, or Wilkinson's catalyst, is one of the active homogeneous hydrogenation catalysts discovered by Geoffrey Wilkinson in 1964. It is a square planar 16-electron complex. It is widely used for the selective hydrogenation of alkenes and alkynes, such as the selective hydrogenation of exocyclic double bonds over endocyclic, isolated double bonds over dienes, and terminal alkynes over terminal alkenes (Scheme 3.6).

3.5.3 Monsanto and Cativa Process

Both of these are industrial methods being used for the manufacture of acetic acid by catalytic carbonylation of methanol. This methodology was first commercialized in 1960 by BASF and involved iodide-promoted cobalt catalyst, $[HCo(CO)_4]$. However, because this process involves the use of very high temperature (230°C) and high pressure (600 atm), with low selectivity (90%), the Monsanto process was developed by Monsanto Company in 1966. It uses the *cis*-$[Rh(CO)_2I_2]^-$ anion as the catalytically active species to catalyze the reaction under mild conditions) 150°C–200°C temperature and 30–60 bar pressure) to give a selectivity over 99%. The Cativa process was developed by BP Chemicals and uses iridium-based catalysts such as the $[Ir(CO)_2I_2]^-$ complex. It is operated at a reduced water content of <8 wt.% in comparison to the 14–15 wt.% required by the Monsanto process.

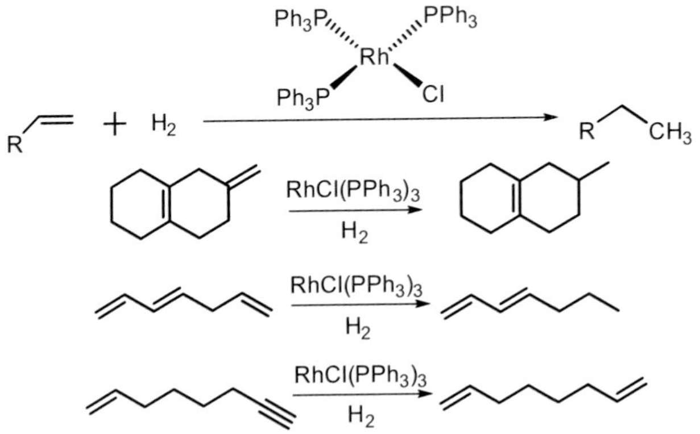

Scheme 3.6 Olefin hydrogenation using Wilkinson's catalyst.

3.5.4 Reppe Carbonylation Process

In this process, carbon monoxide is inserted into unsaturated hydrocarbons (such as alkenes or alkynes) to form a variety of saturated and unsaturated acids, esters, anhydrides, thioesters, and amides by using a CO complex of transition metals (most commonly iron, cobalt, or nickel) as a catalyst with nucleophilic compounds having a reactive hydrogen atom, such as H_2O, HOR, HSR, or HNR_2 (Scheme 3.7). BASF is one of the major companies that utilize the Reppe process for the production of propionic acid (from ethylene and water), acrylate (from acetylene and alcohol), and acetic acid (from methanol).

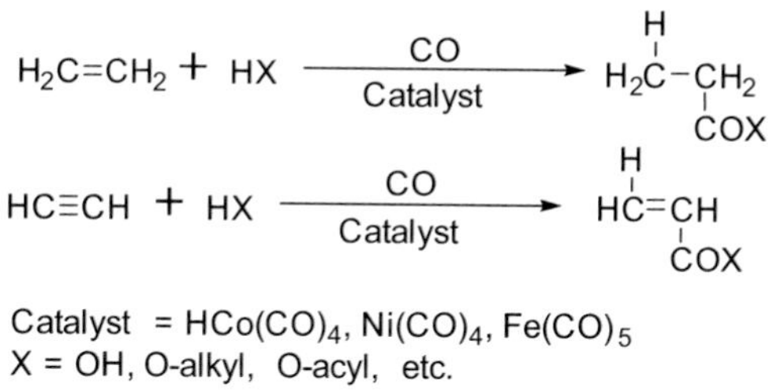

$$H_2C{=}CH_2 + HX \xrightarrow[\text{Catalyst}]{CO} H_2\overset{H}{\overset{|}{C}}-\underset{\underset{COX}{|}}{CH_2}$$

$$HC{\equiv}CH + HX \xrightarrow[\text{Catalyst}]{CO} H\overset{H}{\overset{|}{C}}{=}\underset{\underset{COX}{|}}{CH}$$

Catalyst = $HCo(CO)_4$, $Ni(CO)_4$, $Fe(CO)_5$
X = OH, O-alkyl, O-acyl, etc.

Scheme 3.7 Reppe carbonylation.

3.5.5 Koch Reaction

The Koch reaction is another industrially significant process. It involves the synthesis of tertiary carboxylic acids from alcohols or alkenes with simultaneous carbonylation using carbon monoxide; when formic acid is used as a carbonylating agent, the reaction is known as the Koch–Haaf reaction. These reactions are catalyzed by simple homogeneous Lewis acids, such as H_2SO_4, H_3PO_4, BF_3, and HF. Approximately 150,000 tons of Koch acids and their derivatives, including pivalic acid, 2,2-dimethylbutyric acid, and 2,2-dimethylpentanoic acid, are produced annually by leading companies such as Shell, ExxonMobil, Enjay Chemical Company, and DuPont. A schematic illustration of the formation of pivalic acid is shown in Scheme 3.8.

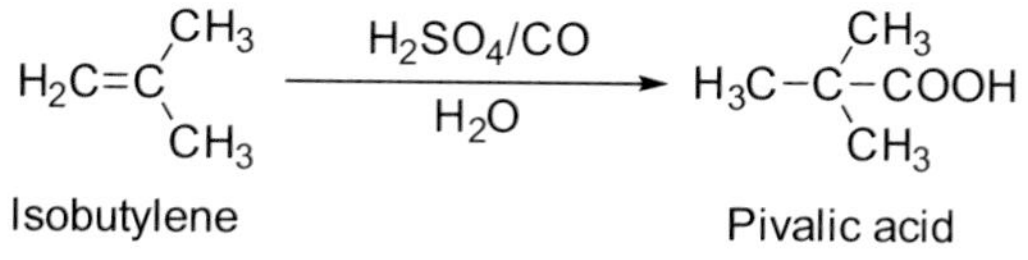

$$\underset{\text{Isobutylene}}{H_2C{=}C(CH_3)_2} \xrightarrow[H_2O]{H_2SO_4/CO} \underset{\text{Pivalic acid}}{H_3C-C(CH_3)_2-COOH}$$

Scheme 3.8 Koch reaction for the formation of pivalic acid.

3.6 Heterogeneous Catalysis

In contrast to homogeneous catalysis, heterogeneous catalysis involves phase differences between reactants/products and catalysts [9]. Most commonly, the reactants/products are in a liquid or gaseous phase, whereas the catalyst is in the solid phase. Examples of heterogeneous catalysts are metal oxides, supported metal complexes/acids/bases, and zeolites. A heterogeneous catalyst usually consists of the following parts:

- Catalytically active species: Metal salts/complexes or acids/bases.
- Solid support: This is the component that provides retrievability to catalytically active species. The most commonly used catalytic supports are carbon, alumina, and silica. Some important features that an ideal support should exhibit are:
    - A high surface area so that a high loading of active phase per unit weight or per unit volume can be achieved. Typical surface area ranges from 100 to 1000 m^2g^{-1} or even more.
    - Resistance to abrasion and chemical attack.
    - A high melting point to provide usability at higher temperatures.
- Linker: It helps in the effective immobilization of active species. This may be optional as the catalytic species can also be directly attached onto the support.

Some of the most significant industrially established heterogeneous reactions are discussed next.

3.6.1 Haber–Bosch Process

It is one of the most significant examples in the field of industrially practiced heterogeneous catalysis. This process is used for the production of ammonia using elemental nitrogen and hydrogen. In this case, the two gaseous reactants are passed over catalytic beds that consist of iron oxide supported on a mixture of other metal oxides, such as calcium, aluminum, potassium, and silicon, under enormously high temperature (400°C–500°C) and pressure (150–250 bar) (Scheme 3.9). Currently, 450 million tons of ammonia is produced using the Haber process, which is mainly used as a nitrogen fertilizer in the form of ammonia itself, ammonium nitrate, and urea.

$$N_2 + 3H_2 \xrightleftharpoons{\text{Catalyst}} 2NH_3$$

Scheme 3.9 Haber–Bosch process.

3.6.2 Ziegler–Natta Polymerization

Since 1956, Zeigler–Natta catalysts have been used in the commercial manufacture of a variety of polyolefins (Scheme 3.10). The catalyst consists of two parts: halides of titanium/chromium/vanadium/zirconium (catalyst) and an alkyl aluminum compound (cocatalyst). However, it was in the 1970s when heterogeneous Zeigler–Natta catalysts were introduced. Since then, they have dominated the industrial world for polymerizing propylene and other terminal alkenes. They involve silica-supported titanium and magnesium chloride activated by alkyl aluminum compounds. Both titanium compound and $MgCl_2$ remain packed into the pores of silica.

$$n\ H_2C{=}CHR \xrightarrow{\text{Zeigler-Natta Catalyst}} \left[H_2C{-}CHR \right]_n$$

Scheme 3.10 Zeigler–Natta polymerization.

3.6.3 Ostwald Process

It is the commercial process for the synthesis of nitric acid. It is carried out in two stages:

Stage 1: Ammonia is oxidized by heating with oxygen over a Pt-Rh gauze to form nitric oxide and water. The use of gauze maximizes the surface area of the solid catalyst (Scheme 3.11).

$$4\ NH_3\ (g) + 5\ O_2\ (g) \xrightarrow[\text{600-800°C}]{\text{Pt-Rh gauze}} 4\ NO\ (g) + 6\ H_2O\ (g)$$

Scheme 3.11 Catalytic oxidation of ammonia during the Ostwald process.

Stage 2: This stage is performed in an absorption apparatus containing water. Nitric oxide is oxidized to nitrogen dioxide, which is further absorbed by the water, yielding nitric acid, which can be concentrated to the desired level using distillation (Scheme 3.12).

$$2\ NO\ (g) + O_2\ (g) \longrightarrow 2\ NO_2\ (g)$$

$$3\ NO_2\ (g) + H_2O\ (l) \longrightarrow 2\ HNO_3\ (aq) + NO\ (g)$$

Scheme 3.12 Oxidation of nitric oxide during the Ostwald process.

3.6.4 Contact Process

Currently, the high industrial demand for sulfuric acid is fulfilled using the contact process. The manufacturing process involves five major steps (Scheme 3.13):

1. Oxidation of sulfur to form sulfur dioxide (Scheme 3.13a)
2. Purification of SO_2 to remove dust particles and oxides of iron and arsenic impurities present in the sulfur feedstock
3. Oxidation of sulfur dioxide in the presence of solid catalyst vanadium pentoxide at 450°C and 1–2 atm (Scheme 3.13b)
4. Conversion of sulfur trioxide into oleum (Scheme 3.13c)
5. Dilution of oleum to sulfuric acid (Scheme 3.13d)

$$\text{(a)}\quad S + O_2 \longrightarrow SO_2$$

$$\text{(b)}\quad 2\,SO_2 + O_2 \xrightarrow[450\ ^{\circ}C]{V_2O_5} 2\,SO_3$$

$$\text{(c)}\quad SO_3 + H_2SO_4 \longrightarrow H_2S_2O_7$$

$$\text{(d)}\quad H_2S_2O_7 + H_2O \longrightarrow 2\,H_2SO_4$$

Scheme 3.13 Steps involved in the contact process.

Some of the leading companies in the market that utilize this process for sulfuric acid production are BASF, Dupont, PVS Chemicals, Chemtrade Logistics, and Jacobs.

3.6.5 Catalytic Converters

The most remarkable achievement of heterogeneous catalysis can be seen in controlling automotive emissions. Application of a catalytic converter in the exhaust system of an automobile helps in reducing the emission of toxic gaseous pollutants into the environment by converting them into less toxic substances with the help of catalyzed chemical reactions [10]. A catalytic converter has three basic components:

- Catalytic support: A metal or ceramic catalytic support with a honeycomb-like structure is designed in a manner so as to maximize the surface area for contact with the gases.

- Wash coat: The honeycomb structure is covered with a layer of wash coat, which is a mixture of silica and alumina. It is used as a carrier to disperse the catalytic material over a large surface area. The catalytic material is suspended into this wash coat prior to applying it to the catalyst support.
- Catalytic material: A combination of platinum or palladium or both with rhodium is used as a catalytic material for the deep oxidation of unburned hydrocarbons and carbon monoxide and reduction of oxides of nitrogen. Because they perform three simultaneous chemical reactions, such catalytic converters are known as "three-way catalytic converters."

A simplified explanation of the working of catalytic converters is illustrated in Fig. 3.7.

Table 3.2 Comparison between homogeneous and heterogeneous catalysis

Parameter	Homogeneous catalysis	Heterogeneous catalysis
Catalyst phase	Generally liquid	Generally solid
Solubility of catalyst	Soluble	Insoluble
Catalytic effectiveness	All metal atoms/catalytic entities available for the reaction	Only surface atoms available as active centers
Activity and selectivity	High activity and high selectivity due to lack of phase boundaries	High activity and low selectivity
Diffusion problems	Not present	Slow transport of reactants and products
Product separation	Difficult	Easy
Catalytic recovery	Extraction, distillation, and filtration after chemical decomposition; usually costly and inefficient	Filtration, centrifugation, and magnetic; easy recovery
Catalytic loss	High	Small
Catalyst modification	Easy	Difficult
Reaction mechanism	Well understood	Poorly understood

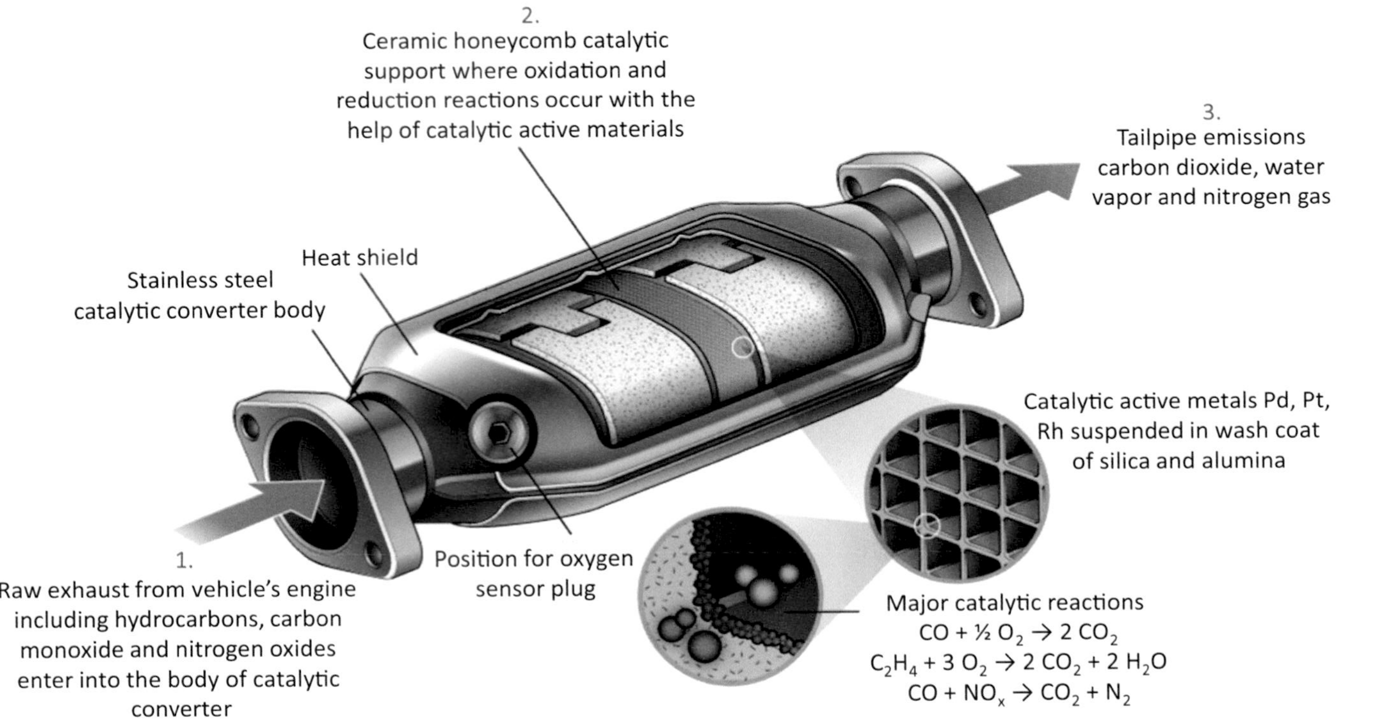

Figure 3.7 Schematic illustration of the working of catalytic converters. Reproduced with permission from ClearMechanic.

After this illustrative discussion of homogeneous and heterogeneous catalysis, you will get a brief overview of the two in Table 3.2.

3.7 Phase Transfer Catalysts

Many desirable reactions cannot be brought about because the reactants have opposite solubility preferences and hence are not accessible to each other. In such a case, each reactant is dissolved in the respective compatible solvent. Because the two solvents are immiscible with each other, addition of an agent is required that promotes the transport of one reagent from one phase to another where the other reagent exists and makes the reaction feasible. Such agents are known as phase transfer catalysts (PTCs) [11, 12]. Any compound for being a PTC should fulfill the following criteria:

- It must be cationic, having sufficiently enough organic structure to be able to partition the nucleophilic anion into the organic phase.
- It must have loose enough cationic-anionic bonding to ensure high anionic reactivity.
- It should be stable under reaction conditions, easily available, and easy to separate or recover.

Some of the most commonly used PTCs are quaternary ammonium and phosphonium salts (Fig. 3.8). Other compounds, such as macrocyclic ethers (crown ethers), aza-macrobicyclic ethers (cryptands), and open-chain polyethers (polyethylene glycols), can form stable complexes with alkali and alkaline earth metal cations and, hence, act as active PTCs.

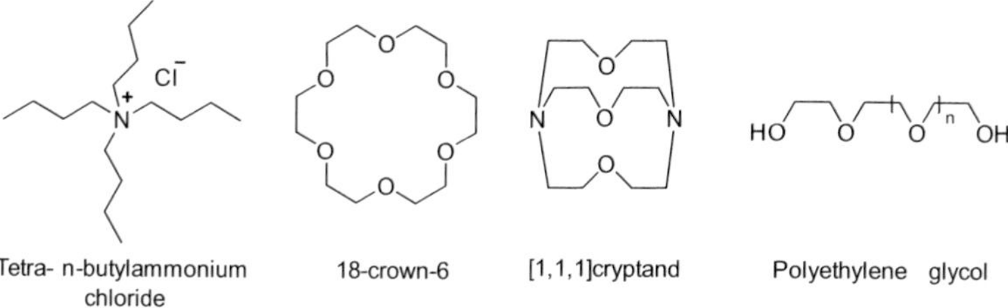

Figure 3.8 Structures of some of the most common PTCs.

Here are some of the advantages associated with the use of PTCs:

- There is no need for expensive anhydrous or aprotic solvents, such as dimethyl sulfide or dimethyl sulfoxide.
- They are extremely useful from the viewpoint of green chemistry as they allow the use of water, thereby reducing the need for organic solvents.
- They require mild reaction conditions, such as low energy and temperature.
- They provide better yields with less side-product formation.
- The reaction rates are faster.
- There is greater chemo- and regio-selectivity.
- The work-up procedures are easy in most of the cases.

Phase transfer catalysis is a special form of heterogeneous catalysis. It is estimated that PTCs are being used in more than 500 commercial processes, including pharmaceuticals, agricultural chemicals, petrochemicals, monomers, polymers, dyes and pigments, flavors and fragrances, solvents, and other organic chemicals. Here are some examples:

Example 1: High-yield displacement reaction of 1-chlorooctane with aqueous NaCN using hexadecyl-tributylphosphonium bromide as the PTC (Fig. 3.9)

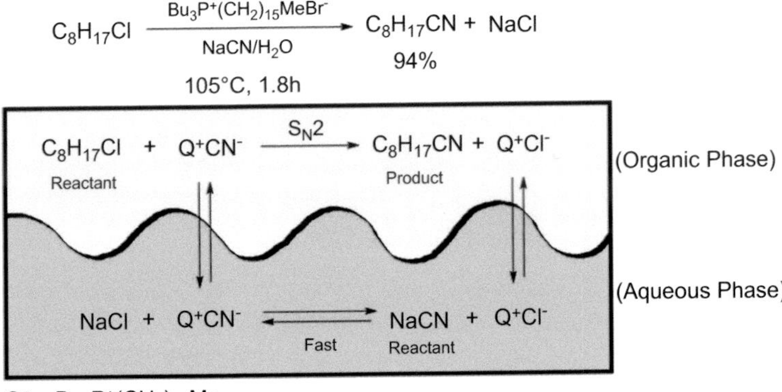

Figure 3.9 Displacement reaction using hexadecyl-tributylphosphonium bromide as the PTC.

Example 2: Continuous dehydrochlorination of dichlorobutene to produce chloroprene, a monomer for synthetic rubber, on a large scale using cocoalkyl benzyl bis[beta-hydroxypropyl] ammonium chloride as the PTC (Scheme 3.14)

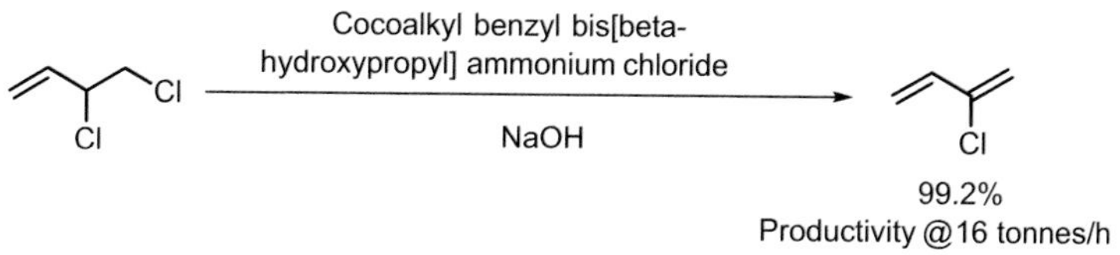

Cocoalkyl benzyl bis[beta-hydroxypropyl] ammonium chloride

NaOH

99.2%
Productivity @16 tonnes/h

Scheme 3.14 Dehydrochlorination using cocoalkyl benzyl bis[beta-hydroxypropyl] ammonium chloride as a PTC.

Example 3: Synthesis of bisphenol A-polycarbonate using tetrabutylammonium hydroxide as the PTC, which greatly improves the process from the standpoint of safety and environmental concerns by reducing the high-volume utilization of hazardous raw material, phosgene, by 94% (Scheme 3.15)

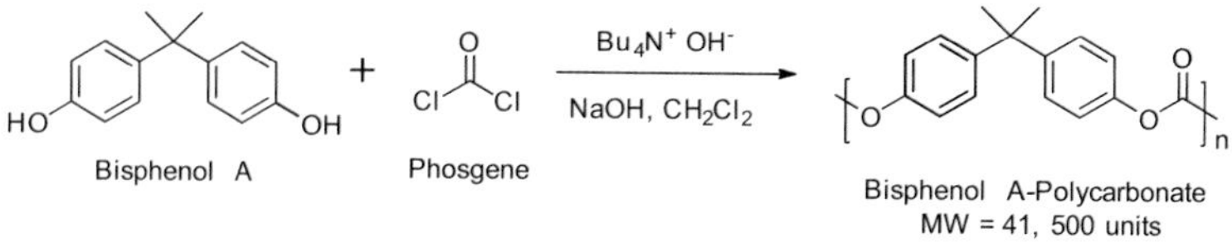

$Bu_4N^+ OH^-$

NaOH, CH_2Cl_2

Bisphenol A

Phosgene

Bisphenol A-Polycarbonate
MW = 41, 500 units

Scheme 3.15 Bisphenol A synthesis using tetrabutylammonium hydroxide as a PTC.

In spite of these many high-performance real-world applications, use of PTCs is accompanied by certain limitations, such as energy-intensive separation processes for product purification and their expensive nature (crown ethers and cryptands). Hence, there is urgent requirement of further advancements in the field of development of more efficient and affordable PTCs.

In 2003, a new concept of self-separation of homogeneous catalyst was introduced by Dioumaev and Bullock [13]. They designed a tungsten catalyst for the solvent-free hydrosilylation of ketones. The enclosed figure illustrates that the catalyst remains soluble in the liquid reactants, but as soon as the reaction progresses, the catalyst self-separates by converting itself into a liquid clathrate, which can be readily recovered by simple decantation of liquid products. Hence, the method is highly suitable from the viewpoint of green chemistry and catalysis.

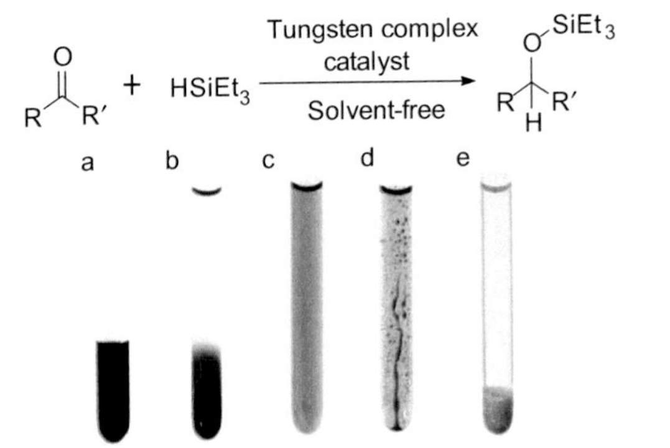

Self-separation of homogeneous catalyst. (a) $Et_2C = O$ + tungsten complex before adding $HSiEt_3$, (b) $HSiEt_3$ added but liquid not mixed, (c) mixed homogenous mixture, (d) formation of liquid clathrate, and (e) end of reaction and precipitation of catalyst. Reprinted by permission from Springer Nature Customer Service Centre GmbH: Springer Nature, Nature, Ref. [13], Copyright (2003).

3.8 Asymmetric Catalysis

Since the existence of living organisms, molecular chirality has played a key role in controlling biological functions. Different enantiomers can cause completely different effects in vivo. Long back, the drug thalidomide was marketed in racemic form as a tranquillizer and as a sleeping pill. It was also advertised as a completely safe sedative, even for pregnant women. Indeed, one of the two enantiomers, (*R*)-thalidomide, was a sedative, but the other one was a teratogenic compound, which caused severe congenital defects in children (Fig. 3.10). This boosted the need for the synthesis of enantiomerically pure products.

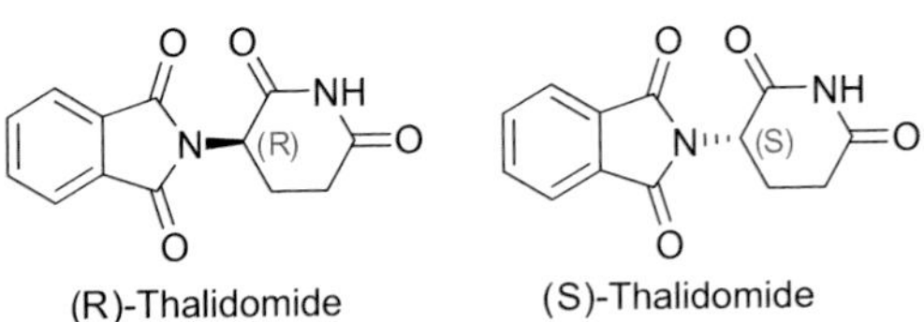

Figure 3.10 Enantiomers of thalidomide.

In this regard, synthesis using asymmetric catalysis has been accepted as the most elegant method [14]. It involves transformation of prochiral starting materials into enantioenriched products with the help of chiral catalysts. As far as enantioselective chemocatalysts are concerned, homogeneous and heterogeneous metal complexes modified with chiral auxiliaries have proven to be synthetically useful. There are numerous industrially practiced asymmetric processes.

One of them is the synthesis of the (–) menthol intermediate. (–) Menthol is an organic compound widely known for its ability to chemically trigger the cold-sensitive receptors present in the skin. The key step in its synthesis is the enantioselective isomerization of an allyl amine to the chiral enamine (Scheme 3.16). Using this asymmetric catalyst, 28,700 tons of (–) menthol was produced with only 250 kg of catalyst between 1983 and 1996.

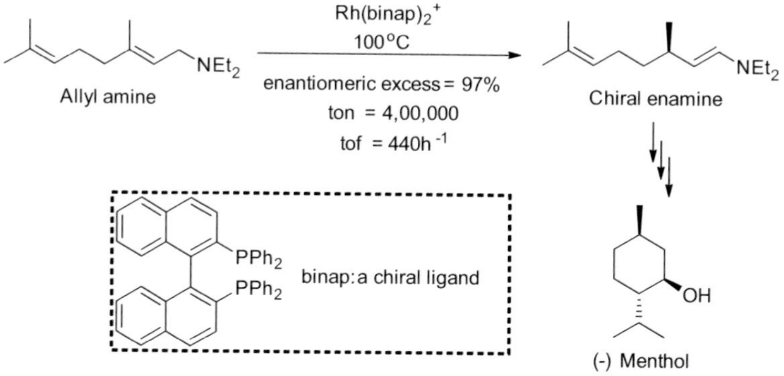

NEt_2
Allyl amine
$Rh(binap)_2^+$
100°C
enantiomeric excess = 97%
ton = 4,00,000
tof = 440h^{-1}
NEt_2
Chiral enamine
PPh_2
PPh_2
binap:a chiral ligand
OH
(-) Menthol

Scheme 3.16 Enantioselective key step in the synthesis of (–) menthol.

3.9 Nanocatalysis: Emerging Hybrid Catalysis

3.9.1 What Is Nanocatalysis?

As discussed in the previous sections, homogeneous catalysts have the benefit of, among other things, easily accessible catalytic sites, high selectivity, and ease of modification. However, low thermal stability and difficulty in product purification and recovery of catalyst impose certain economic and environmental constraints.

Heterogeneous catalysts, on the other hand, are easy to separate and can withstand high temperatures but offer low activity and selectivity. Therefore, the current green chemistry era demands an environmentally benign catalyst that can be easily recovered from the reaction medium and that can be recycled many times without being compromised in terms of efficiency. These demanding conditions led to the development of "nanocatalysis." In nanocatalysis, the presence of nanosized particles (particles ranging from 1 to 100 nm in size) increases the exposed surface area of the catalytically active species and, therefore, the contact between the reactants and catalyst is enhanced dramatically. Furthermore, these nanocatalysts can be easily isolated and regenerated as they are insoluble in reaction solvents. Hence, nanocatalysts provide an interface between homogeneous and heterogeneous catalysts, offering advantages of both [15–17]. Nanoparticles (NPs) can occur in a great variety of sizes, shapes, morphologies, and compositions (Fig. 3.11) [18]. These physical and chemical properties can be easily altered to tune the activity and selectivity of the nanocatalyst.

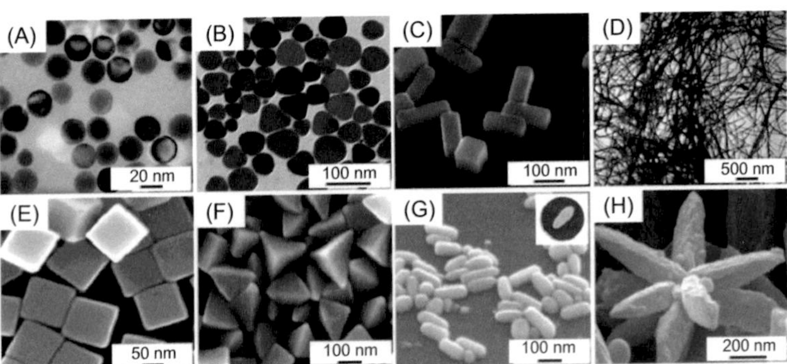

Figure 3.11 TEM images of silver nanoparticles with different shapes: (a) nanospheres, (b) nanoprisms, (c) nanobars, and (d) nanowires. SEM images of (e) nanocubes, (f) pyramids, (g) nanorice, and (h) nanoflowers. Adapted from Ref. [18].

3.9.2 Synthetic Approaches

NPs are generally synthesized by a top-down or a bottom-up approach (Fig. 3.12). In the top-down approach, larger materials are deconstructed physically or chemically to form NPs. For this, ball milling, grinding, crushing, laser ablation synthesis in solution, and heat-up methods are usually employed. Contrary to this, the bottom-up approach involves build-up of a material from the bottom up, that is, smaller components are arranged to form targeted NPs. High-temperature flame process, chemical vapor deposition, and template synthesis are common bottom-up approaches [19]. This route is preferred over the top-down approach because nanomaterials with uniform size, shape, and distribution are achieved.

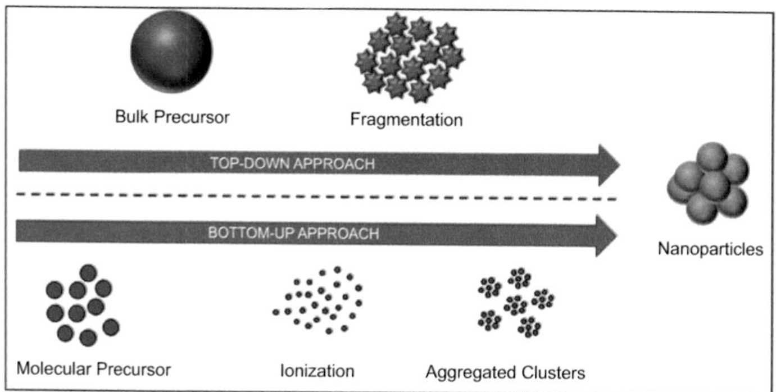

Figure 3.12 Schematic representation of a top-down (above) and a bottom-up (below) approach to synthesize nanoparticles.

3.9.3 Catalytic Applications

Catalysis using NPs has been in the limelight of chemical research and has been recognized as a vital tool for a wide range of organic transformations.

Till date, numerous nanocatalysts having well-defined morphologies have been designed. In this section, we will discuss various inorganic nanomaterials and their application in catalysis.

3.9.3.1 Metal nanoparticles

Metallic NPs are defined as elemental nanoclusters of a metal. Generally, the activity of these NPs increases with decreasing particle size. As size decreases, the exposed surface atoms and the number of defective atoms (at the edges and in the corner) increase. Apparently, small NPs with defects will demonstrate high catalytic activity. Therefore, transition metal NPs, such as gold, platinum, silver, and palladium, have been extensively used to catalyze reactions. However, metallic NPs have high surface free energy and thus low stability. Further, the separation and recovery of these NPs is very challenging. Thus, to overcome these limitations, immobilization of NPs over some support is the most reliable dispersion and stabilization technique [15]. Silica, zeolites, carbon nanotubes, polymers, alumina, etc., have been extensively used as solid supports for the stabilization of NPs (Fig. 3.13).

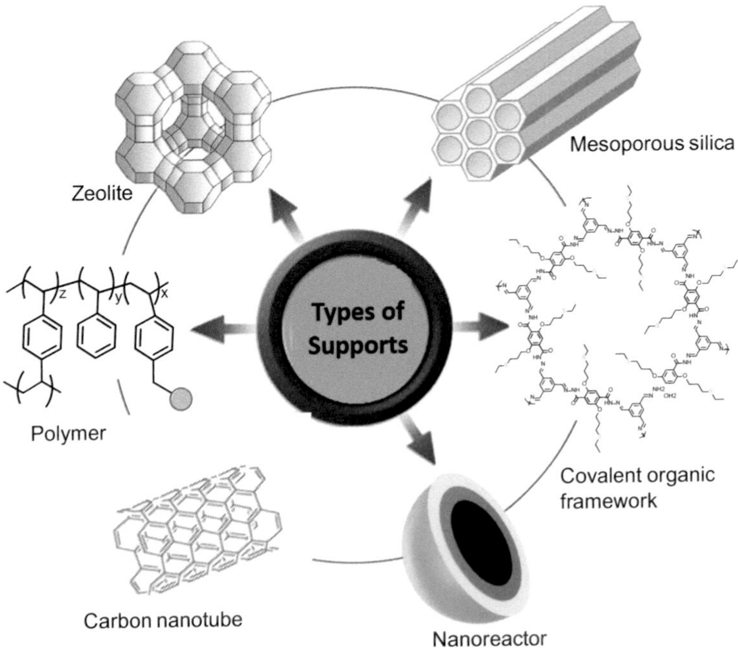

Figure 3.13 Various types of supports used for the stabilization of NPs.

The transformation of CO_2 into valuable chemicals is an urgent priority of the present era. In this regard, the reaction of CO_2 with epoxide to form cyclic carbonates is a well-known example. Polymer-supported Au NPs have shown excellent catalytic efficiency for this reaction [20]. The functional group in the polymer was quaternary ammonium, which stabilizes the gold NPs against deactivation (Scheme 3.17).

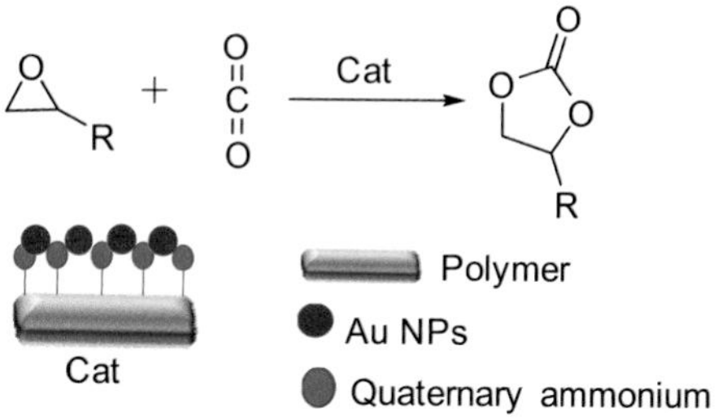

O
R
O
C
O
Cat
O
O
O
R
Cat
Polymer
Au NPs
Quaternary ammonium

Scheme 3.17 Polymer-supported Au NPs as a catalyst for fixation of CO_2.

Au NPs heterogenized over a silica support have been used for the alkylation of aromatics by alcohols, with excellent activity and selectivity (Scheme 3.18) [21].

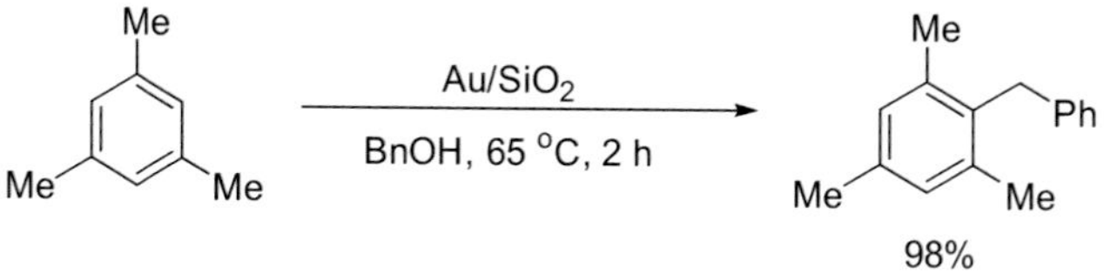

Me
Me
Me
Au/SiO2
BnOH, 65 °C, 2 h
Me
Ph
Me
Me
98%

Scheme 3.18 Benzylation of mesitylene by benzyl alcohol (BnOH) catalyzed by Au/SiO_2.

Thiolate-stabilized Pd NPs decorated on multiwalled carbon nanotubes have been used as a catalyst for carbon-carbon bond formation reaction (Scheme 3.19) [22]. The catalyst shows high activity, stability, and recyclability.

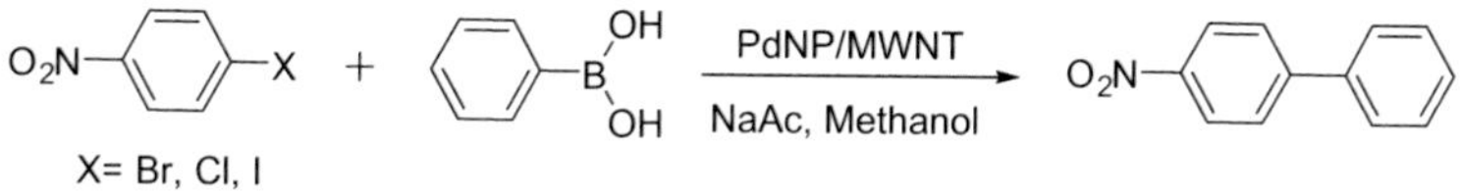

O2N
X
OH
B
OH
PdNP/MWNT
NaAc, Methanol
O2N
X= Br, Cl, I

Scheme 3.19 Suzuki–Miyaura reaction.

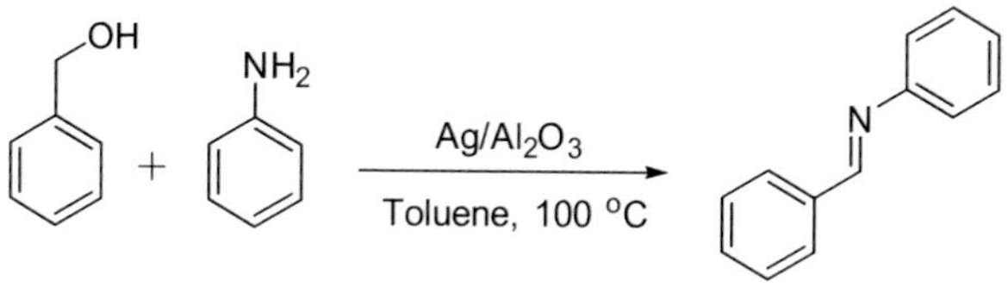

OH
NH2
Ag/Al2O3
Toluene, 100 °C
N

Scheme 3.20 Formation of imines using silver NPs supported on Al_2O_3.

Silver supported on Al_2O_3 is a highly selective catalyst for the formation of imines under oxidant-free conditions (Scheme 3.20) [23].

The reactivity, selectivity, and stability of NPs are highly influenced by the choice of support as metal-support interactions, anchoring properties, surface area, and thermal stability vary with the type of support used. For instance, oxidation of benzyl alcohol to benzaldehyde using Au NPs supported over a variety of metal oxides showed variation in end-product selectivity (Fig. 3.14) [24].

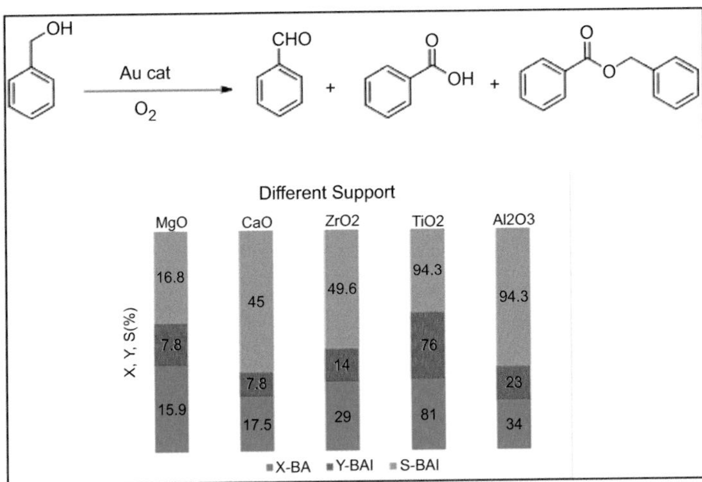

Figure 3.14 Influence of support on oxidation of benzyl alcohol to benzaldehyde over 1% Au/M catalysts (M = MgO, CaO, ZrO_2, TiO_2, Al_2O_3; BA = benzyl alcohol, BAl = benzaldehyde, X = conversion; Y = yield; S = selectivity). Adapted from Ref. [24].

3.9.3.2 Metal oxide nanoparticles

Metal oxide NPs are potential candidates for catalyzing a variety of organic reactions. Some salient features of these catalysts are high catalytic activity, good recyclability, and improved selectivity. Metal oxide NPs demonstrate analogous or superior activity when compared to metal NPs. Owing to a lower surface energy, metal oxide NPs are more mechanically and structurally stable than their bulk counterparts, for example, TiO_2.

Manganese oxide NPs are effective catalysts for the oxidation of olefins to form pharmaceutically important epoxides [25]. Nanosized MnO_2 demonstrates high activity in the epoxidation of aromatic olefins under milder reaction conditions (Scheme 3.21).

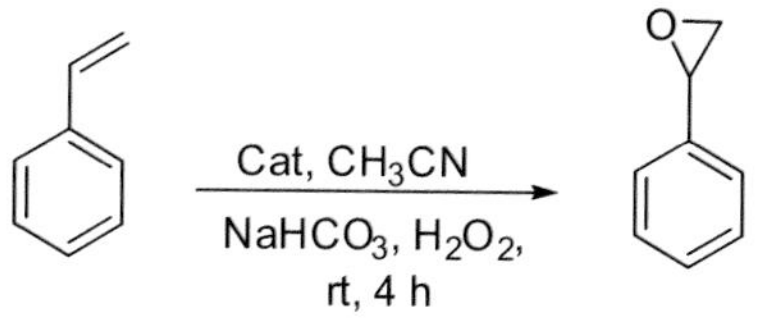

Cat, CH_3CN

$NaHCO_3$, H_2O_2, rt, 4 h

Scheme 3.21 Nano-sized MnO_2-catalyzed epoxidation of aromatic olefins.

Copper oxide NPs are known to catalyze an Ullmann cross-coupling reaction of phenols with aryl halides (Scheme 3.22) [26]. The catalyst is recoverable and highly reusable and eliminates the presence of harsh and toxic ligands. CuO NPs can also effectively catalyze many other cross-coupling reactions involving the formation of carbon-heteroatom bond formation, that is, C–O, C–S, and C–N bond forming reactions.

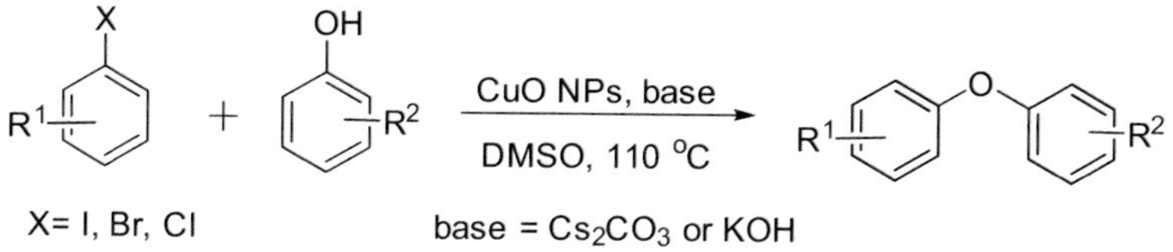

X OH R^1 + R^2 CuO NPs, base DMSO, 110 °C R^1 O R^2

X= I, Br, Cl base = Cs_2CO_3 or KOH

Scheme 3.22 Copper oxide NPs catalyze the Ullmann cross-coupling reaction.

3.9.3.3 Magnetic nanoparticles

The separation and recovery of NPs using conventional methods is not effective due to the small size of the particles. To overcome this limitation, magnetic nanoparticles (MNPs) have emerged as magnificent solid support materials. MNPs can be manipulated easily using external magnetic fields. They generally consist of a magnetic material such as iron, nickel, or/and cobalt. MNPs are insoluble and are paramagnetic in nature, enabling facile and effective isolation from the reaction media using an external magnet. It is possible to control various properties of MNPs, such as size, shape, morphology, and dispersity, thus providing a large room for modifications in order to design nanomaterials for different applications (Fig. 3.15) [27].

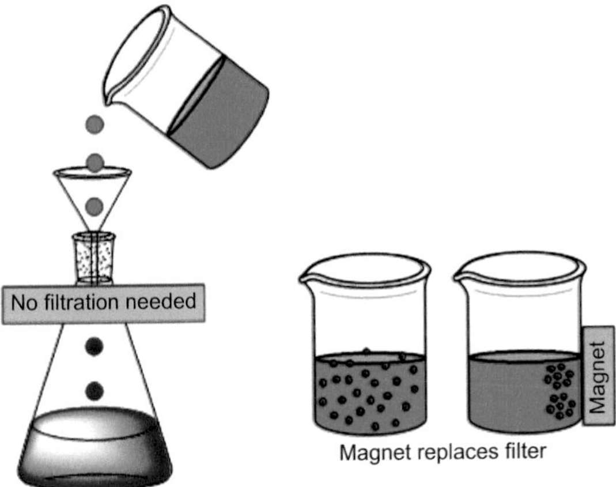

Figure 3.15 Magnetic separation and recycling of the catalyst.

Fe_3O_4 NPs has the ability to act as a peroxidase mimetic catalyst. It has been shown that starch-coated MNPs, synthesized using potato extract, in combination with H_2O_2 as oxidant and ultrasonic radiation can degrade rhodamine B dye in just 30 min. (Fig. 3.16) [28]. Thus, MNPs are effective candidates for fast and easy degradation of the organic dyes present in wastewater.

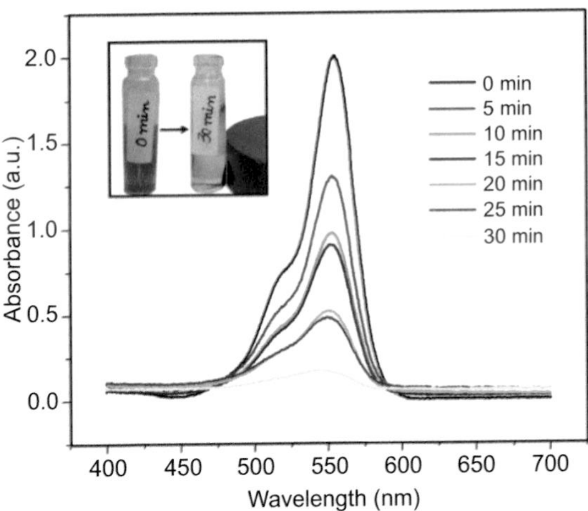

Figure 3.16 Absorption spectra of RhB solution at different intervals of time (3 mL of RhB solution, 10 mg of Fe_3O_4 NPs, 0.2 mL of H_2O_2 solution (30%), ultrasonication, pH = 5, 55°C). Inset: RhB dye solution before and after 30 min. Adapted with permission from Ref. [28]. Copyright (2019) American Chemical Society.

The surface of MNPs can also be easily functionalized. Therefore, they have the potential to be a support for catalyst preparation. A Mn-based magnetic nanocatalyst was designed through covalent grafting of the manganese acetylacetonate complex onto amine-functionalized silica-coated MNPs. The catalyst was successfully used for the oxidation of organic halides and alcohols to corresponding carbonyl compounds (Scheme 3.23) [29]. Excellent yields and high turnover number, along with effortless magnetic recovery and reusability of catalyst, make the protocol environmentally benign.

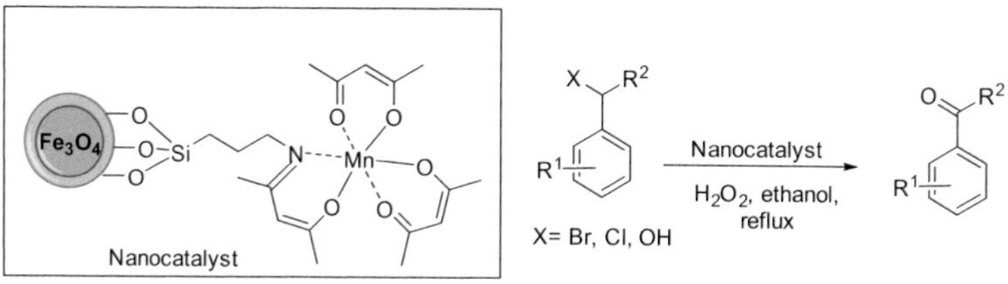

Scheme 3.23 Mn-based magnetic nanocatalyst for the oxidation of organic halides and alcohols.

3.10 Biocatalysis

Use of natural substances, including natural enzymes and enzymes produced in situ from whole cells, to drive chemical reactions is termed as "biocatalysis." Enzymes have the potential to catalyze a wide variety of reactions under mild conditions. The most common examples are the formation of alcohols from fermentation and the formation of cheese by breakdown of milk proteins. Biocatalysts have several advantages over conventional ones, for example, they are highly selective toward substrates, binding sites, and functional groups; reduce the possibility of side reactions; and have shorter reaction times. Moreover, enzymes do not require any external proactive groups, so separation of the product becomes much more convenient. Thus, the introduction of a biocatalyst makes the process green and sustainable. Currently, many pharmaceutical, chemical, and agrochemical industries are adopting biocatalysts. One of the most popular enzyme-based reactions is isomerization of glucose to high-fructose corn syrup (HFCS) using isomerase (Scheme 3.24).

HFCS is widely used in the food and beverage industry as an artificial sweetener [30].

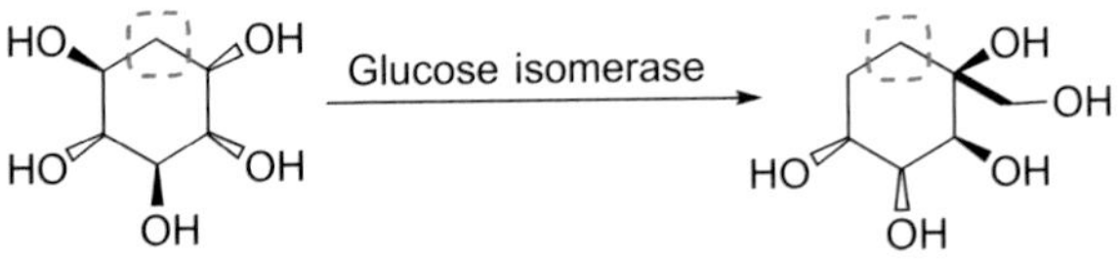

Glucose isomerase

Scheme 3.24 Enzymatic synthesis of HFCS.

3.11 Current Challenges and Future Development in Catalysis

One of the main principles of green chemistry, catalysis has been the driving force behind the production of various chemicals and pharmaceuticals for a long time. Reduction in emissions from cars, development of preservatives to keep our food fresh, and development of drugs and medicines to improve our health, all these are possible because of catalysts. The current era is facing challenges in developing alternative clean fuels, minimizing global warming, ensuring global supply of clean water, protecting living beings from infectious agents, and managing waste. To meet these challenges, a revolution in the protocols for the synthesis of catalysts is highly desirable. This revolution can be achieved by understanding how the structure of a catalyst, especially at the atomic and nanoscale, can control catalytic activity and selectivity. In this regard, advances in computational chemistry, imaging, and new measurement technologies have contributed largely. The selection of appropriate techniques and integration of advantages of nanoscience, homogeneous catalysis, heterogeneous catalysis, and biocatalysis will definitely enable breakthroughs in the development of catalytic materials.

3.12 Learning Outcomes

At the end of this chapter, students will be able to:

- Define a catalyst

- Understand the role of catalysis
- Appreciate the importance of catalysts in comparison to stoichiometric reagents
- Identify the different types of catalysts
- Differentiate between homogeneous and heterogeneous catalysts
- Define nanocatalysts and detail their importance in today's world

3.13 Problems

1. Give one-word answers to the following questions:
   (i) The process of employing a catalyst to speed up a chemical reaction that requires or engages enzymes
   (ii) A particle between 1 and 100 nm in size with a surrounding interfacial layer, with potential applications in catalysis, biomedical, optical, and electronic fields
2. Why are catalysts preferred over stoichiometric reagents?
3. Name one phase transfer catalyst. Give the advantages and principles of a phase transfer catalyst.
4. What are the advantages of thiamine-catalyzed benzoin condensation over the conventional benzoin condensation? Write down the reaction involved.
5. Write short notes on the following:
   (i) Biocatalysts
   (ii) Phase transfer catalysts
   (iii) Asymmetric catalysts
   (iv) Nanocatalysts
   (v) Green catalysts
6. What properties of a biocatalyst make it relevant in green chemistry? Give one industrial application of a biocatalyst.
7. Differentiate between homogeneous and heterogeneous catalysts.
8. Explain the significance of asymmetrical synthesis in the case of the thalidomide drug molecule marketed in 1957 in Germany, correlating the defects produced in babies when this drug was given to pregnant women.

9. How have catalysts played a significant role in the development of a more economical and environment-friendly chemical synthesis?
10. Which is a greener route? Justify your answer.

Ni/H_2
225 °C
route 1
O_2/Co
10 bar
OH
+
O
Ni/H_2
100 °C
route 2
H_2O
High Si H-ZSM--5
OH

References

1. Chorkendorff, I. and Niemantsverdriet, J. W. (2017). *Concepts of Modern Catalysis and Kinetics* (John Wiley & Sons, Weinheim, Germany).
2. Delidovich, I. and Palkovits, R. (2016). Catalytic *versus* stoichiometric reagents as a key concept for green chemistry. *Green Chem.*, **18**, pp. 590–593.
3. https://www.nobelprize.org/prizes/lists/all-nobel-prizes-in-chemistry
4. Onuegbu, T. U., Ogbuagu, A. S. and Ekeoma, M. O. (2011). The role of catalysts in green synthesis of chemicals for sustainable future. *J. Basic Phy. Res.*, **2**, pp. 86–92.
5. Sheldon, R. A., Arends, I. and Hanefeld, U. (2007) *Green Chemistry and Catalysis* (John Wiley & Sons, Weinheim, Germany).
6. https://elevance.com/products/
7. https://www.epa.gov/greenchemistry/green-chemistry-challenge-winners
8. Falbe, J. and Bahrmann, H. (1984). Homogeneous catalysis-industrial applications. *J. Chem. Educ.*, **61**, pp. 961–967.
9. Kamer, P. C. J., Vogt, D. and Thybaut, J. W. (2017). *Contemporary Catalysis: Science, Technology and Applications* (The Royal Society of Chemistry, UK).
10. Heveling, J. (2012). Heterogeneous catalytic chemistry by example of industrial applications. *J. Chem. Educ.*, **89**, pp. 1530–1536.

11. Jiménez-González, C. and Constable, D. J. C. (2011). *Green Chemistry and Engineering: A Practical Design Approach*, Chapter 6, Material selection: solvents, catalysts and reagents (John Wiley & Sons, US) pp. 133–174.

12. Naik, S. D. and Doraiswamy, L. K. (2004). Phase transfer catalysis: chemistry and engineering. *AIChE J.*, **44**, pp. 612–646.

13. Dioumaev, V. K. and Bullock, R. M. (2003). A recyclable catalyst that precipitates at the end of the reaction. *Nature*, **424**, pp. 530–532.

14. Blaser, H.-U. (2007). Industrial asymmetric catalysis: approaches and results. *Rend. Lincei*, **18**, pp. 281–304.

15. Filiciotto, L. and Luque, R. (2018). *Encyclopedia of Sustainability Science and Technology*, ed. Meyers, R. A., Nanocatalysis for green chemistry (Springer New York, New York).

16. Polshettiwar, V. and Varma, R. S. (2010). Green chemistry by nano-catalysis. *Green Chem.*, **12**, pp. 743–754.

17. Sharma, R. K. (2019) *Silica-Based Organic-Inorganic Hybrid Nanomaterials*, (World Scientific Publishing Co. Pte. Ltd., Singapore).

18. Loiseau, A., Asila, V., Boitel-Aullen, G., Lam, M., Salmain, M. and Boujday, S. (2019). Silver-based plasmonic nanoparticles for and their use in biosensing. *Biosensors*, **9**, p. 78.

19. Cao, G. (2004). *Nanostructures & Nanomaterials: Synthesis, Properties and Applications* (World Scientific Publishing, Singapore).

20. Shi, F., Zhang, Q., Ma, Y., He, Y. and Deng, Y. (2005). From CO oxidation to CO_2 activation: an unexpected catalytic activity of polymer-supported nanogold, *J. Am. Chem. Soc.*, **127**, pp. 4182-4183.

21. Mertins, K., Iovel, I., Kischel, J., Zapf, A. and Beller, M. (2006). Gold-catalyzed benzylation of arenes and heteroarenes. *Adv. Synth. Catal.*, **348**, pp. 691–695.

22. Cornelio, B., Rance, G. A., Laronze-Cochard, M., Fontana, A., Sapi, J. and Khlobystov, A. N. (2013). Palladium nanoparticles on carbon nanotubes as catalysts of cross-coupling reactions. *J. Mater. Chem.*, **1**, pp. 8737–8744.

23. Mielby, J., Poreddy, R., Engelbrekt, C. and Kegnæs, S. (2014). Highly selective formation of imines catalyzed by silver nanoparticles supported on alumina. *Chin. J. Catal.*, **35**, pp. 670–676.

24. Alshammari, A. and Kalevaru, V. N. (2016) Supported gold nanoparticles as promising catalysts, Chapter 3, In *Catalytic Application of Nano-Gold Catalysts*, Mishra, N. K., ed. (IntechOpen: Rijeka, Croatia) pp. 57–81.

25. Najafpour, M. M., Rahimi, F., Amini, M., Nayeri, S. and Bagherzadeh, M. (2012). A very simple method to synthesize nano-sized manganese oxide: an efficient catalyst for water oxidation and epoxidation of olefins. *Dalton Trans.*, **41**, pp. 11026–11031.

26. Zhang, J., Zhang, Z., Wang, Y., Zheng, X. and Wang, Z. (2008). Nano-CuO-catalyzed Ullmann coupling of phenols with aryl halides under ligand-free conditions, *Eur. J. Org. Chem.*, **2008**, pp. 5112–5116.

27. Polshettiwar, V., Luque, R., Fihri, A., Zhu, H., Bouhrara, M. and Basset, J.-M. (2011). Magnetically recoverable nanocatalysts. *Chem. Rev.*, **111**, pp. 3036–3075.

28. Sharma, R. K., Yadav, S., Gupta, R. and Arora, G. (2019). Synthesis of magnetic nanoparticles using potato extract for dye degradation: a green chemistry experiment. *J. Chem. Ed.*, **96**, pp. 3038–3044.

29. Sharma, R. K., Yadav, M., Monga, Y., Gaur, R., Adholeya, A., Zbořil, R., Varma, R. S. and Gawande, M. B. (2016). Silica-based magnetic manganese nanocatalyst-applications in the oxidation of organic halides and alcohols. *ACS Sustainable Chem. Eng.*, **4**, pp. 1123–1130.

30. Paul, P. E. V., Sangeetha, V. and Deepika, R. G. (2019). *Recent Developments in Applied Microbiology and Biochemistry*, ed. Buddolla, V., Chapter 9, Emerging trends in the industrial production of chemical products by microorganisms (Elsevier, BV) pp. 107–125.

Chapter 4

Alternative Reaction Media

Radhika Gupta,[a] **Reena Jain,**[b] **and Rakesh K. Sharma**[a]
[a]*Green Chemistry Network Centre, Department of Chemistry, University of Delhi, Delhi 110007, India*
[b]*Department of Chemistry, Hindu College, University of Delhi, Delhi 110007, India*
radgupta123@gmail.com

> *Nothing we use for a few minutes should pollute our environment for decades.*
>
> —Greenpeace, a nongovernmental environmental organization

4.1 Introduction

Solvents are an indispensable part of nearly every sector, including pharmaceuticals, food and flavors, and materials and analytical chemistry. They are used in almost every chemical process to assist in mass and heat transfer as well as to promote the formulation, separation, and purification of chemical products. It would not be surprising to know that they account for more than 80% of the total materials used and consume nearly 60% of the total energy in the

Green Chemistry for Beginners
Edited by Rakesh K. Sharma and Anju Srivastava
Copyright © 2021 Jenny Stanford Publishing Pte. Ltd.
ISBN 978-981-4316-96-5 (Hardcover), 978-1-003-18042-5 (eBook)
www.jennystanford.com

manufacture of an active pharmaceutical ingredient. Therefore, for their all-day use, it is necessary that the solvents meet certain specifications: they should be low in or have no toxicity; be easily retrievable; be stable in reaction conditions; have chemical inertness; avoid product contamination; and be environment friendly during production, use, and disposal. One of the 12 principles of green chemistry also demands the use of safer solvents and auxiliaries. If possible, the use of solvents should be avoided and if that is not possible, safer substances should be used instead. This chapter is an attempt to learn about the replacement of hazardous solvents by more environmentally benign alternatives [1, 2].

4.2 Need for Solvents

Solvents are used in a variety of chemical processes to facilitate the synthesis, separation, and purification of reaction products. In chemical industries, they are used as a medium for carrying out synthetic reactions where they allow the homogenization of the reaction mixture involving solid reactants/reagents and speed up the reaction, which also reduces the energy consumption. They even act as a heat sink during exothermic reactions, providing safety. They can increase the rate of reaction by stabilizing the intermediates or may even act as acids or bases. Not only in industries, solvents are an important constituent of numerous household necessities, such as cleaning agents, coatings, adhesives, fuels, and lubricants. However, as the chapter progresses, it will be realized that reactions can also proceed under solvent-free conditions.

4.3 Problems Related to Traditional Solvent Use

Usually volatile organic compounds (VOCs), such as low-molecular-weight hydrocarbons and ethers, chlorinated hydrocarbons, alcohols, esters, and sulfoxides, are used as solvents due to their high vapor pressures at room temperature. The advantage is that they are easy to remove from the reaction mixture. However, their use is associated with several environmental and health hazards. For example:

- Their vapors contribute to the greenhouse effect, causing global warming. They have the ability to catalyze the destruction of the ozone layer that protects the earth from harmful ultraviolet radiations.
- VOCs are highly flammable and can cause severe accidents on even slight negligence. In 2007, the US Chemical Safety and Hazard Investigation Board examined the reasons behind the massive explosion that occurred at the Barton Solvents Wichita facility in Valley Center, Kansas, USA. The fire broke out due to the interaction of ignitable vapor–air mixtures with the accumulated static electricity produced during storage of a nonconductive and flammable "naphtha liquid," chiefly used as a paint thinner. Some other solvents, such as benzene, toluene, hexane, and heptane, may also cause such accidents. A similar accident was reported in 1992 at Guadalajara, Mexico's second-largest city, where 252 people died and more than 500 people were injured due to 10 consecutive explosions caused by the leakage of gasoline from a private cooking oil factory into the sewage system. The gasoline was used to extract edible oils from seeds.
- Solvents may also enter the human body by inhalation, by swallowing, and through the skin. Exposure to large amounts of solvents can cause a number of adverse health effects, including irritation in the eyes and the respiratory tract. Upon contact with skin, they can cause allergic skin reactions or dermatitis. They can even penetrate the skin and reach the bloodstream, causing damage to vital body organs, including the liver, the heart, kidneys, and the nervous system. Many of the solvents also have a narcotic effect, causing dizziness and fatigue. Besides, some of the solvents are known carcinogens, including benzene.

Due to these risks and hazards, legislations and control measures have been introduced to avoid the use of conventional toxic solvents and to encourage academicians, researchers, and industries to use alternative solvents (Table 4.1). A few of the pharmaceutical companies are also modifying their synthetic pathways in order to either minimize the use of solvents or shift from conventional solvents to their greener alternatives.

Table 4.1 Measures taken to avoid the use of conventional solvents

Purpose	Conventionally used solvent	Risks associated	Alternative solvent/measure taken
Additive in petrol	Benzene	Suspected carcinogen	Amount of benzene added reduced from 5% to <1%
Extraction of caffeine from coffee	Dichloromethane	Suspected carcinogen	Supercritical CO_2
Dry-cleaning purposes	Perchloroethylene	Suspected carcinogen	Supercritical CO_2
Printing inks	Perchloroethylene	Suspected carcinogen	Ethyl acetate

4.4 Criteria for the Selection of Green Solvents

What is a "green solvent"? The answer is absolutely critical because the use of solvents is associated with environmental, health, and safety (EHS) considerations. The idea of a "green" solvent aims to minimize the environmental impact resulting from the use of solvents in chemical processes. Various frameworks have been proposed by different agencies and universities to assess the greenness of solvents. The majority of the frameworks apply two assessment methods [3]. One is the EHS assessment method, which identifies potential hazards of a solvent on the basis of nine parameters: persistency, air and water hazard (environmental hazards), irritation, acute and chronic toxicity (health hazards) and release potential, explosion and reaction/decomposition (safety hazards). The other is the life-cycle assessment method, which performs a detailed assessment of the solvent's environmental emissions. Additionally, as the name suggests, this method takes care of the resource utilization, from its production and use to its recycling and disposal. The results of the combination of the two methods are then used to determine the best solvent. People in the pharmaceutical sector has also formulated their own solvent selection guides by establishing their individual benchmarks according to the company usage policy. For example, Pfizer classified most classical solvents into three categories: preferred (green), usable (yellow), and undesirable (red)

(Table 4.2) [4]. As an extension, Pfizer also provided a substitution table for undesirable solvents (Table 4.3) [4, 5]. Thus, in the context of sustainable and clean chemistry, researchers are working in different directions to eliminate the undesirable solvents and to replace them by developing green alternatives, such as biosolvents, supercritical fluids (SCFs), and ionic liquids (ILs), some of which will be discussed in the later sections. However, there is no perfect solvent that can impart green credentials to all situations. Therefore, choices have to be made accordingly.

Table 4.2 Pfizer solvent selection guide

Water	Cyclohexane	Pentane
Acetone	Heptane	Hexane(s)
Ethanol	Toluene	Di-isopropyl ether
2-propanol	Methylcyclohexane	Diethyl ether
1-propanol	TBME	Dichloromethane
Ethyl acetate	Isooctane	Dichloroethane
Isopropyl acetate	Acetonitrile	Chloroform
Methanol	2-MeTHF	NMP
MEK	THF	DMF
1-butanol	Xylenes	Pyridine
t-butanol	DMSO	DMAc
	Acetic acid	Dioxane
	Ethylene glycol	Dimethoxyethane
		Benzene
		Carbon tetrachloride

Source: Republished with permission of the Royal Society of Chemistry, from Ref. [4]; permission conveyed through Copyright Clearance Center, Inc.

Taking into account the need for developing greener technologies, Pfizer redesigned the process involved in the manufacture of sertraline, an active ingredient of an antidepressant drug called Zoloft, from a three-step sequential process (which earlier involved the use of dichloromethane, toluene, tetrahydrofuran, and hexane) to a single step using ethanol as the sole solvent. This change eliminated the use of and the need to distill and recover these solvents. Due to the development of significant green chemistry innovation in the manufacture of sertraline, Pfizer was given the US EPA Presidential Green Chemistry Challenge: 2002 Greener Synthetic Pathways Award.

Table 4.3 Pfizer solvent replacement table

Solvent	Issues	Alternatives
Pentane	Lower flash point than other similar solvents	Heptane
Diethyl ether	Lower flash point than other similar solvents	2-MeTHF, TBME
Diisopropyl ether	Powerful peroxide formation compared to similar solvents	2-MeTHF, TBME
Hexane(s)	More toxic than other similar solvents	Heptane
Benzene	Carcinogen	Toluene
Chloroform	Carcinogen	DCM
1,2-DCE	Carcinogen	DCM
1,2-DME	Carcinogen	2-MeTHF, TBME
Pyridine	Carcinogenicity (not classifiable)	Triethylamine (base)
1,4-Dioxane	Carcinogenicity (not classifiable)	2-MeTHF, TBME
DCM	Emissions	Application dependent
Carbon tetrachloride	Emissions	DCM
DMF	Reproductive toxicity	Acetonitrile
DMAc	Reproductive toxicity	Acetonitrile
NMP	Reproductive toxicity	Acetonitrile

4.5 Green Solvents for Organic Synthesis

Over the last few years, new solvent alternatives have encouraged the research community into replacing conventional solvents. Described next are some of the widely accepted green solvents that do not compete with each other but, instead, complement each other, with their own advantages and disadvantages.

4.5.1 Water

Living organisms are the most complex forms of organic compounds on our planet, with nearly 70% of the body weight constituted of water. They involve the construction of chemical bonds in a medium that was earlier considered to be highly undesirable for carrying out synthetic chemical reactions. Earlier, solvents were often defined as a medium required for the homogenization of reactants and reagents. However, inspired by Mother Nature, aqueous organic chemistry is now emerging as a green and sustainable alternative to classical synthetic chemistry, where the utility of solvents was limited to a solubilizing medium [6]. Figure 4.1 represents some of the significant advantages of using water as a solvent in organic reactions. One of the major incentives for the development of water-based reactions is to assist in the separation of product from catalyst. Because of the high polarity of water, a majority of the organic compounds are either insoluble or sparingly soluble in it, thereby allowing easy product separation from the solvent by means of simple extraction, provided the catalyst is water soluble. On this note, various metal catalysts having hydrophilic ligands are being developed. Besides, water doesn't always act as an innocent spectator. It may interact with transition states and intermediates through hydrogen bonding, thereby affecting reaction rates. There are a few limitations as well. Reactions that involve the use of water-sensitive reagents must be carried out in nonaqueous conditions. Also, in the case of water-soluble products, it would be difficult to isolate the compounds. Additionally, one cannot neglect the energy-intensive distillation procedure that would be required to concentrate the contaminated water stream. Some of the potential reactions in which water is used as a solvent are discussed next.

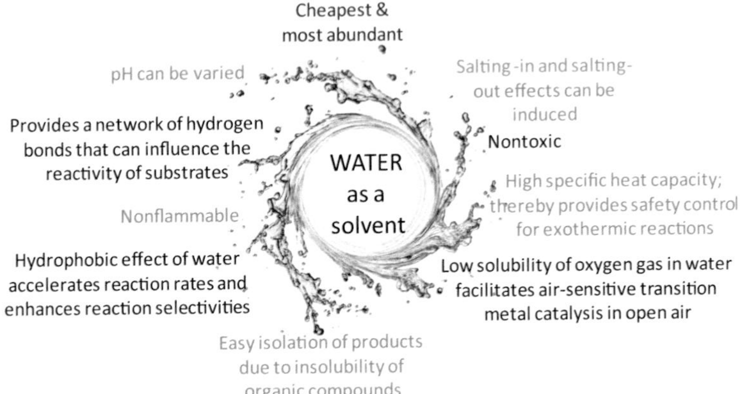

Figure 4.1 Some of the significant advantages of using water as a solvent in organic reactions.

- Diels–Alder reaction

  This is a chemical reaction between a conjugated diene and a substituted alkene to form a cycloaddition product. It was carried out in an aqueous medium in the 1930s. However, it was in the 1980s when Breslow highlighted the rate enhancements when the reaction between cyclopentadiene and butenone was performed in water. As depicted in Scheme 4.1, the reaction rate was enhanced from 75.5 to 4400 when the solvent was changed from methanol to water. This unusual acceleration in the reaction rate was attributed to the hydrophobic effect in water. Due to the difference in polarity between water and reactants, small drops of reactants are formed so as to diminish the interfacial hydrocarbon–water area. This self-association of reactants in water results in accelerated reaction rates. This rate was further altered by the addition of salts to the water. When lithium chloride solution was used, the hydrophobic effect was increased, which "salted out" the nonpolar reactants, further raising the reaction rate to 10,800. On the other hand, when guanidinium chloride was used, the hydrophobic effect was decreased, which "salted in" the nonpolar reactants, decreasing the reaction rate to 4300.

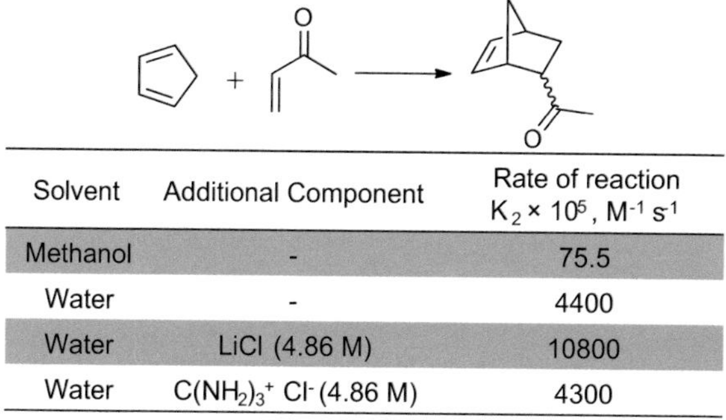

Solvent	Additional Component	Rate of reaction $K_2 \times 10^5$, $M^{-1}\ s^{-1}$
Methanol	-	75.5
Water	-	4400
Water	LiCl (4.86 M)	10800
Water	$C(NH_2)_3^+\ Cl^-$ (4.86 M)	4300

Scheme 4.1 Rate enhancement of the Diels–Alder reaction in water.

- Oxidation reaction

  Many oxidation reactions are frequently carried out in aqueous media using a variety of oxidants. Many a times, special outcomes are revealed when usual organic solvents are replaced by water. For instance, when epoxidation of alcohols, with multiple double bonds, was performed in organic media, a mixture of oxidized products was obtained. However, when the same reaction was carried under controlled pH, in an aqueous medium, only selective products were obtained (Scheme 4.2).

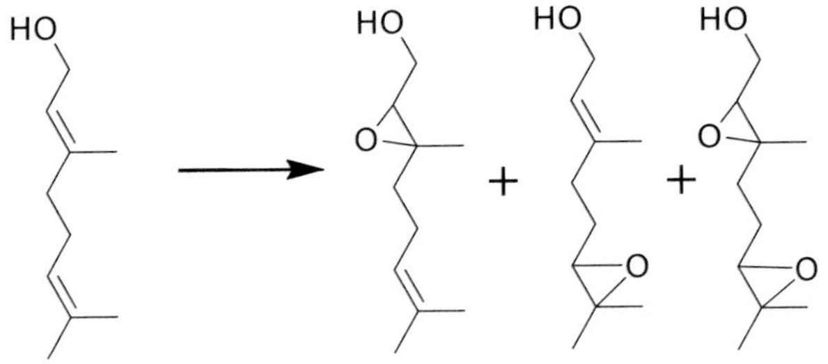

Conditions	Selectivity %		
Oxidizing agent in CH_2Cl_2	30	57	13
Oxidizing agent in H_2O, pH = 12.5	92	-	8

Scheme 4.2 Improvement in the selectivity percentage of oxidation reaction in water.

- Reduction reaction

  Apart from sodium/lithium borohydride, water is used in rare instances as a solvent in reduction reactions due to incompatibility with most reducing agents. Hydrogen too has a lower solubility in water, 0.81 mM at 20°C, than in most of the organic solvents. Being a totally atom-economical reaction, hydrogenation is one of the most widely studied reduction reactions. Therefore, several water-soluble ligands/metal complexes are being developed in order to facilitate reduction in aqueous media to eliminate waste generation and facilitate easy product purification. Scheme 4.3 demonstrates the hydrogenation of olefins in the presence of $RhCl_3$ and TPPTS at room temperature and 1 bar H_2.

  Reduction of carbonyl compounds is another fundamental kind of organic reaction. Mentioned below are some of the examples where reduction has been carried out using ammonia borane as a water-soluble, nontoxic, environmentally benign reducing agent in neat water (Scheme 4.4).

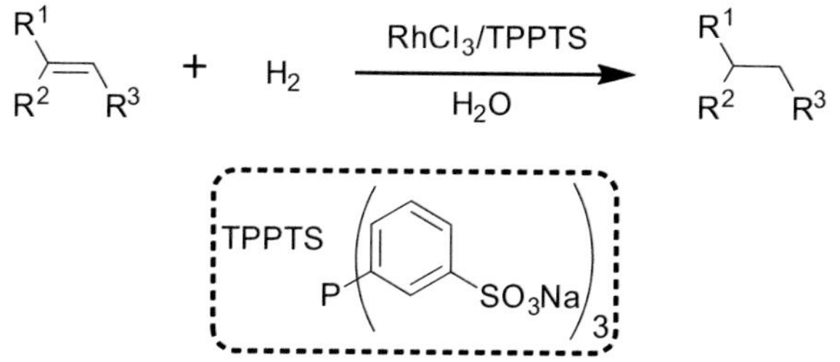

Scheme 4.3 Hydrogenation reaction in an aqueous medium.

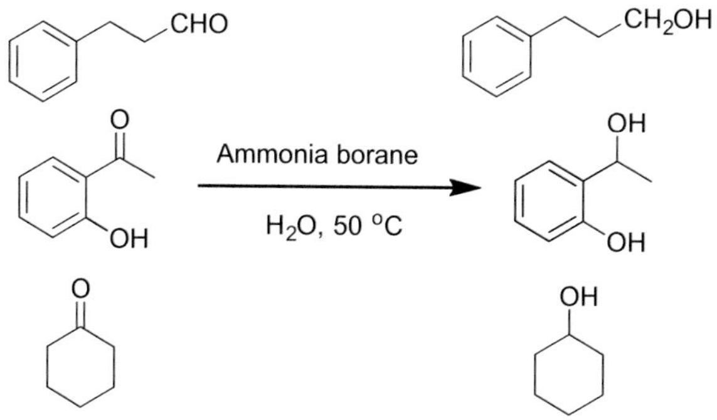

Scheme 4.4 Reduction of carbonyl compounds in water.

- Coupling reaction

  C–C and C–X (where X is a heteroatom) bond formation is one of the most desirable reactions in organic synthesis. Since early times, Grignard-type coupling reactions have gained tremendous attention due to the presence of reactive carbanion species. However, the requirement of strict anhydrous and anaerobic conditions expresses their disagreement with the principles of green chemistry. Hence, there is a necessity to modify the reaction conditions in order to carry out the reactions under aqueous conditions. For example, earlier the Barbier–Grignard-type reactions were carried out in organic solvents using magnesium- or lithium-based Grignard reagents. But, with the use of organometallic reagents with less metallic or softer metals, the same reaction could now take place under aqueous conditions (Scheme 4.5).

Another example of a water-mediated coupling reaction is shown in Scheme 4.6, where an alkene such as propene reacts with a mixture of carbon monoxide and hydrogen in the presence of a water-soluble transition metal catalyst, Rh-TPPTS, to give a high yield of the hydroformylated product. High selectivity, easy product isolation, recycling of the water-soluble catalyst, and use of clean and economical water as a solvent are some of the advantages of this homogeneous two-phase coupling reaction.

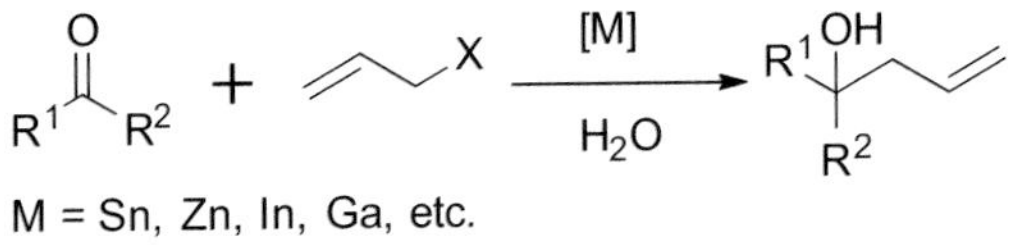

[M], H_2O

M = Sn, Zn, In, Ga, etc.

Scheme 4.5 Carbon–carbon coupling reaction in water.

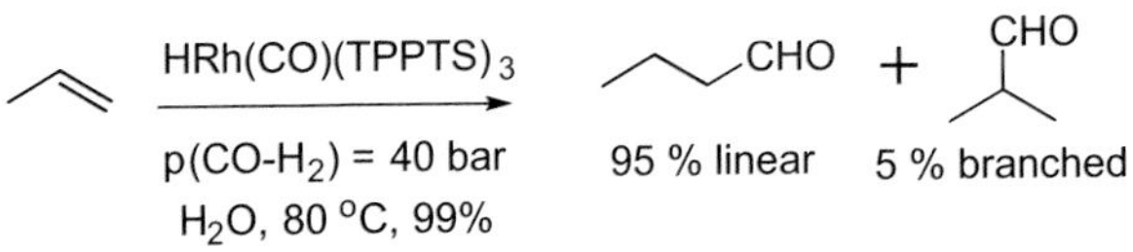

$HRh(CO)(TPPTS)_3$

$p(CO\text{-}H_2) = 40$ bar

H_2O, 80 °C, 99%

CHO + CHO

95 % linear 5 % branched

Scheme 4.6 Hydroformylation reaction in water.

> Those reactions in which water is used as a solvent with water-insoluble compounds are referred as "on water" reactions by Nobel Prize awardee Professor K. Barry Sharpless. Such reactions occur at the interface of the liquid organic oily phase and the bulk water layer that contains no additives. If the organic reactants are solids, at least one of them must liquefy in order to produce the organic oily layer.

4.5.2 Supercritical Fluids

4.5.2.1 Introduction to supercritical fluids

In 1822, Charles Cagniard de la Tour discovered the concept of the critical point of a substance. The critical point is the end point of a pressure–temperature curve above which a liquid and its vapor coexist. On moving upward along the liquid–vapor curve, the density of the liquid decreases due to thermal expansion and that of the vapor increases due to the rise in pressure. At the critical point, the densities of the liquid and vapor phases become identical and the distinction between the two phases disappears. At this point, the substance is regarded as a fluid, neither a liquid nor a gas, having particular values of temperature and pressure known as critical temperature (T_c) and critical pressure (P_c). "Supercritical fluid" is a term for those fluids that have temperatures and pressures higher than their critical values [7]. The phase diagrams in Fig. 4.2 illustrates the supercritical state of H_2O and CO_2. A simple flowchart explaining the appearance of a supercritical stage is also shown in Fig. 4.3.

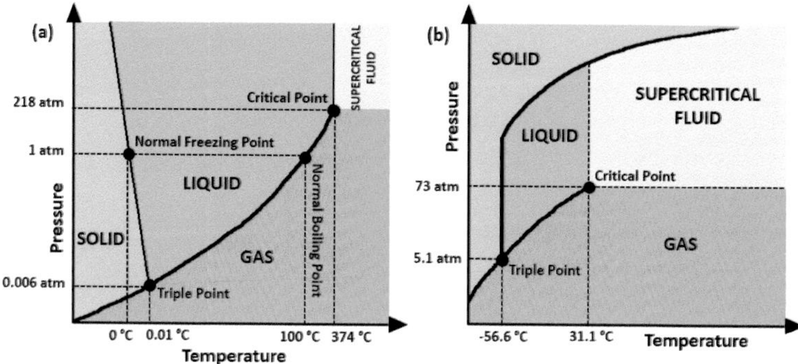

Figure 4.2 Phase diagram depicting the supercritical stages of (a) H_2O and (b) CO_2.

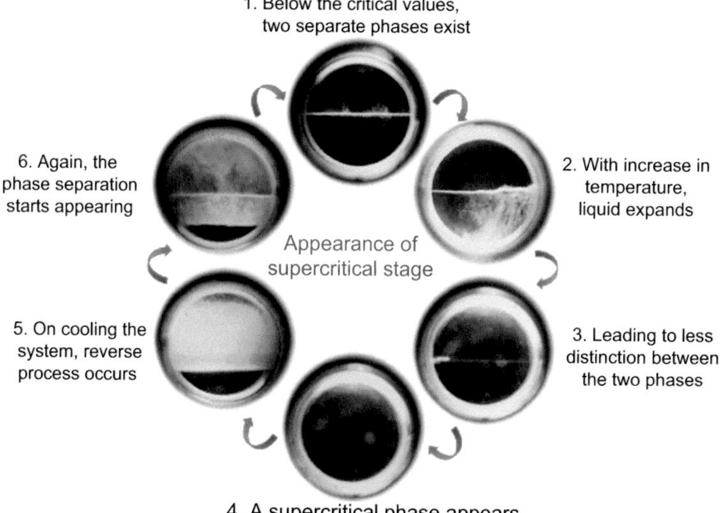

Figure 4.3 Appearance of the supercritical stage of CO_2 [8]. Picture credit: Brian Case. Reproduced with permission from University of Nottingham.

4.5.2.2 Properties of supercritical fluids

The properties of SCFs lie in between those of liquid and vapor phases.

- They have densities closer to liquids but viscosities closer to gases, which leads to high diffusion rates and, thus, improved heat and mass transfers.
- They have the ability to effuse through solids like a gas.
- Another important property is the solubility of materials in SCFs. At constant temperature, the solubility tends to increase with the density of the fluid. Since density increases with pressure, solubility also increases with a rise in the pressure. At a constant density, the solubility increases with an increase in the temperature. However, near the critical point, the density decreases sharply with a slight rise in temperature, thereby decreasing the solubility. Afterward, the solubility again increases with a rise in the temperature.

Their properties can be tuned by varying the temperature and pressure. For instance, on increasing the pressure at the critical temperature, solids can be obtained. However, these pressures are usually very high for most materials. Considering practical limitations, working under conventional conditions is easier than working in supercritical conditions. But if their use substantially enhances the reaction rate or solubility of a substance, it may lead to significant improvements from the viewpoint of green chemistry.

> Supercritical fluids can occur in nature as well. Deep beneath the ocean's surface, there exist high pressures (due to the sheer depth) and temperatures (due to intense heat from erupting volcanoes), leading to the formation of supercritical water. This facilitates the creation of crystals, which are used for making jewelry.

SCFs can be employed as solvents for many useful purposes. They provide numerous valuable advantages, some of which are listed in Fig. 4.4.

Among many of the SCFs, supercritical CO_2 ($scCO_2$) and supercritical H_2O (scH_2O) are the most preferred ones. The next section will provide a brief discussion of the use of these two SCFs as solvents.

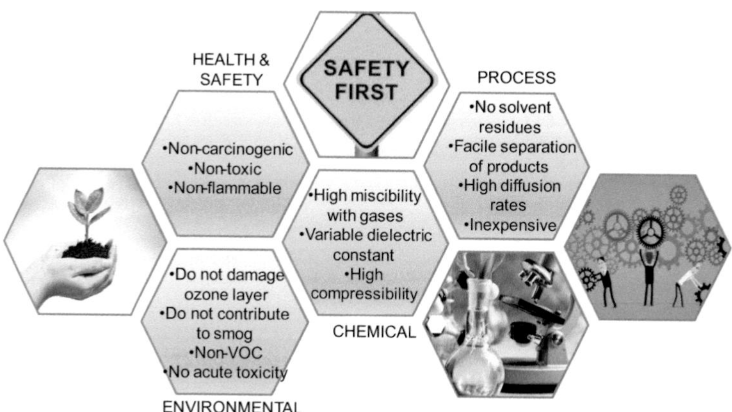

Figure 4.4 Advantages of using supercritical fluids as solvents.

4.5.2.3 Supercritical CO_2 (T_c = 31.1°C, P_c = 73.8 bar)

Apart from the aforementioned advantages of SCFs, the use of scCO_2 as a solvent imparts environmental benefits, for example, the waste CO_2 can be recycled, thereby decreasing dependence on fossil fuels for obtaining carbon-based solvents. Another significant benefit is the elimination of energy-intensive distillation procedures that are required to remove the residual solvents. It can simply be removed by reducing the pressure. scCO_2 is mainly used as an extraction solvent as well as a reaction solvent.

4.5.2.3.1 *scCO_2 as an extraction solvent*

- The first application of scCO_2 as an extraction solvent can be seen in the decaffeination of coffee. Till the 1980s, dichloromethane was the preferred extraction solvent. However, because of its potential detrimental health effects, scCO_2 emerged as the greener alternative. In this process, caffeine is selectively extracted into CO_2 from the green coffee beans.
- Extraction processes using scCO_2 are also used in food industries. For instance, using scCO_2, 50% of the fat content can be reduced from the high-calorific potato crisps without any loss of flavor.
- Another successful application of scCO_2 is in dry-cleaning procedures. Conventionally, chlorinated solvents such as

carbon tetrachloride and perchloroethene were used for dry-cleaning purposes. But due to the suspicion that they are carcinogenic in nature, they have now been replaced by the environmentally benign green solvent scCO_2. It readily penetrates into the fibrous structure of the clothes being cleaned. Besides providing green benefits, it improves the color fastness of certain garments and also increases the domain of items that can be cleaned, including leather, fur, and some synthetics.

4.5.2.3.2 *scCO_2 as a reaction solvent*

scCO_2 has been widely used as a reaction solvent in a variety of industrially significant chemical reactions. The fact that it is a greener alternative to classical solvents, improved reactivity and selectivity, and facile separation of products are some of the important incentives for using scCO_2 as a reaction medium (Schemes 4.7–4.10).

- Esterification reaction

HO–propyl; Biocatalyst, scCO_2, 40 °C, 250 bar → ester, 88.9 %

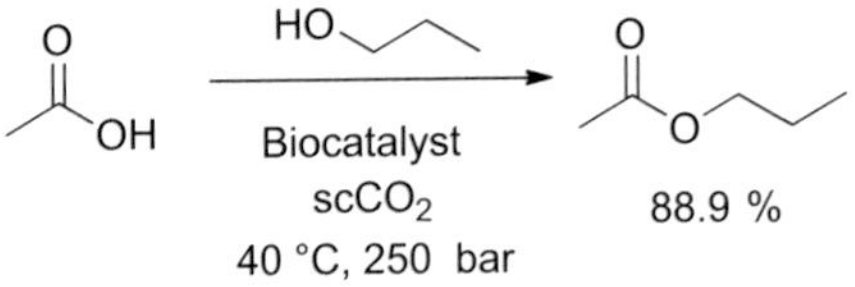

Scheme 4.7 Esterification reaction in scCO_2.

- Selective hydrogenation reaction

CHO; Ruthenium catalyst, 4 MPa H_2, 16 MPa scCO_2, 50 °C, 2 h → CH_2OH, 98 %

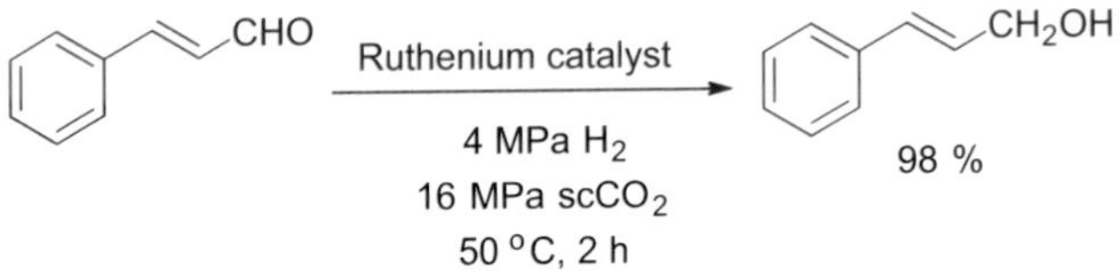

Scheme 4.8 Hydrogenation reaction in scCO_2.

- Polymerization reaction

x O, O $CH_2(CF_2)_nCF_3$; scCO_2, free radical initiator, 60 °C, 200 bar → $*\!-\!\![CH_2-CH]_x$, O O, $CH_2(CF_2)_nCF_3$

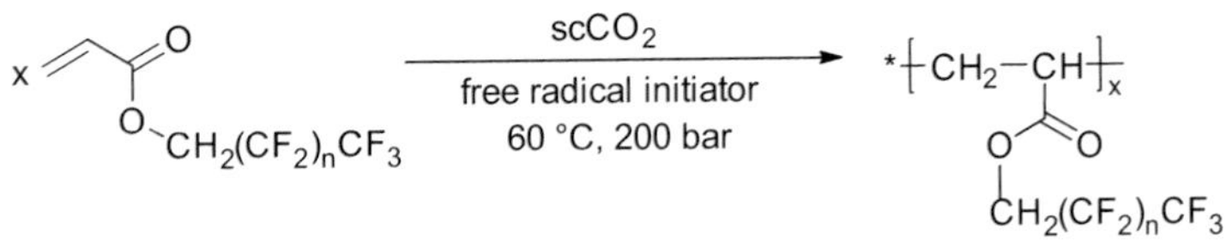

Scheme 4.9 Polymerization reaction in scCO_2.

- Reactions forming carbon–carbon bonds

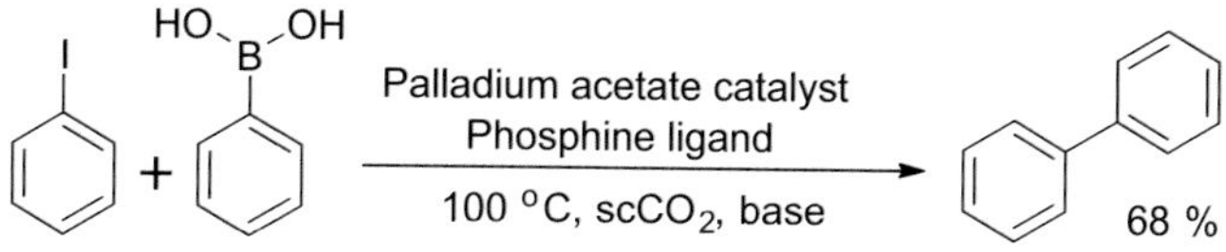

Palladium acetate catalyst
Phosphine ligand
100 °C, $scCO_2$, base
68 %

Scheme 4.10 Formation of a carbon–carbon bond in $scCO_2$.

4.5.2.4 Supercritical H_2O (T_c = 374.2°C, P_c = 220.5 bar)

Although stringent conditions are required to produce scH_2O, the wide utility it provides makes it worth obtaining it. Above the critical point, water has a very reactive nature that permits the occurrence of a reaction even in the absence of strong acids or bases. Described next are some examples where scH_2O is used as a reaction solvent (Schemes 4.11–4.14).

- Oxidation reaction

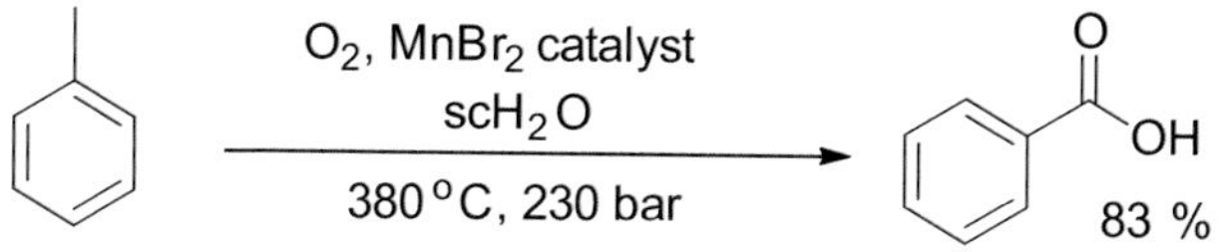

O_2, $MnBr_2$ catalyst
scH_2O
380 °C, 230 bar
OH
83 %

Scheme 4.11 Oxidation reaction in scH_2O.

- Hydrolysis-cyclization reaction

H_2N CN
scH_2O
400 °C, 400 bar
NH
90 %

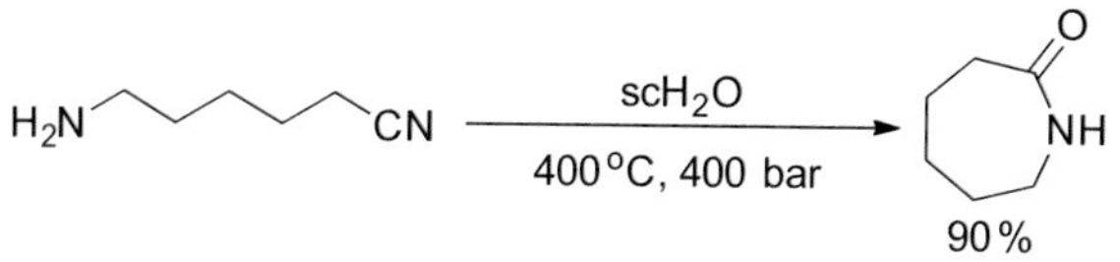

Scheme 4.12 Hydrolysis-cyclization reaction in scH_2O.

- Diels–Alder reaction

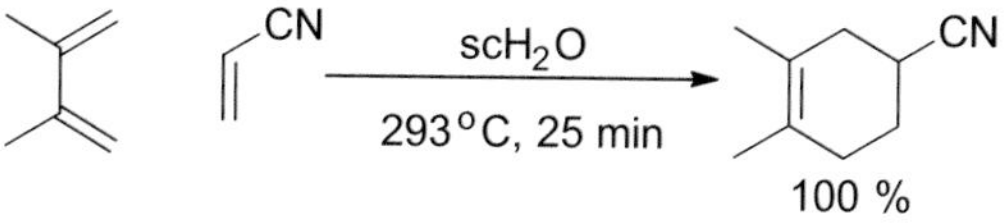

CN
scH_2O
293 °C, 25 min
CN
100 %

Scheme 4.13 Diels–Alder reaction in scH_2O.

- Hydrolysis reaction

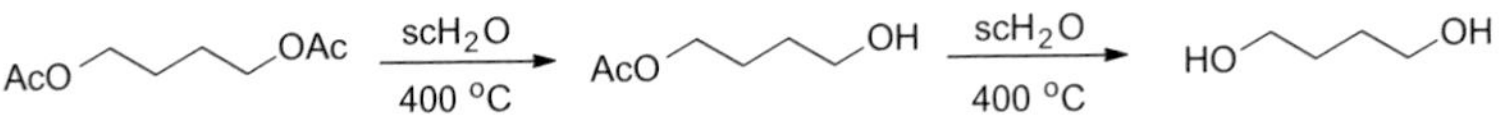

$$\text{AcO}\sim\sim\text{OAc} \xrightarrow[400\ ^{\circ}\text{C}]{\text{scH}_2\text{O}} \text{AcO}\sim\sim\text{OH} \xrightarrow[400\ ^{\circ}\text{C}]{\text{scH}_2\text{O}} \text{HO}\sim\sim\text{OH}$$

Scheme 4.14 Hydrolysis reaction in scH_2O.

4.5.3 Ionic Liquids

4.5.3.1 Introduction to ionic liquids

ILs are defined as organic salts with melting points below 100°C. An IL is usually composed of an organic cation and an inorganic or organic anion. Figure 4.5 represents some of the commonly used cations and anions in ILs. They can be easily prepared using a simple one- or two-step methodology. Many a times, ILs are formed by a simple protonation or quaternization of the central atom involved (Scheme 4.15). However, quite a few times, anion exchange reactions can be carried out to convert one IL into another with a different anion (Scheme 4.16). The synthesis of a few ILs is illustrated next:

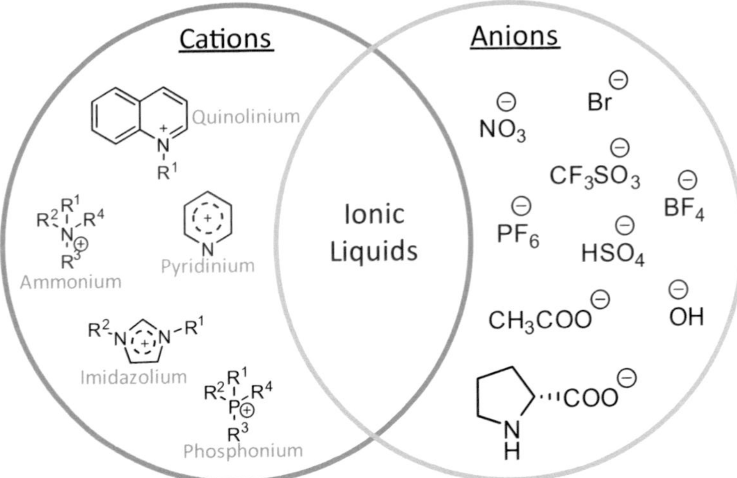

Figure 4.5 Some of the commonly used cations and anions of ILs.

- Synthesis of 1-butyl-3-methylimidazolium chloride IL ([BMIm]Cl)

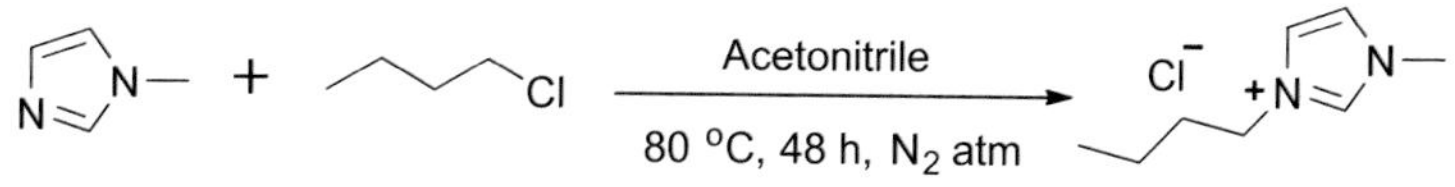

Scheme 4.15 Synthesis of [BMIm]Cl IL.

- Synthesis of an amino acid–based IL

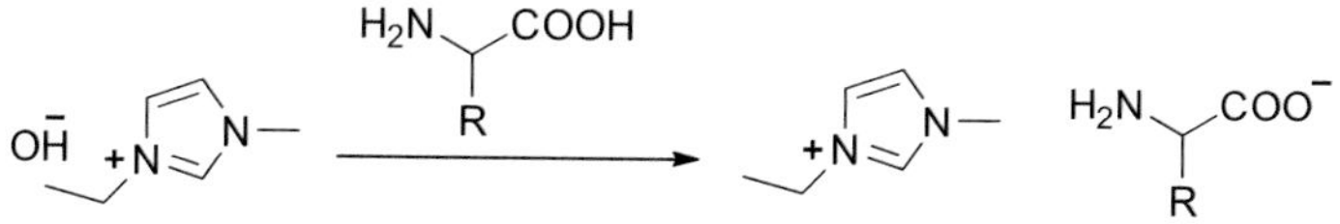

Scheme 4.16 Exchange of an anion in one IL to form another IL.

4.5.3.2 Properties of ionic liquids

ILs are known as "diverse chemicals" in the field of synthetic chemistry. They possess numerous features that designate them as green alternative solvents. A few of the important properties are listed next.

- Low melting point: ILs are composed of large and asymmetric ions; usually one of the two ions is organic in nature. Due to poor packing of their respective ions, they have low melting points, generally below 100°C.
- Low vapor pressure: ILs have low volatility and a nonflammable nature and thus low vapor pressure, due to which they are known as environmentally benign solvents and excellent alternatives to classical volatile organic solvents.
- Tunability: The properties of ILs (density, viscosity, melting point, acidity/basicity, etc.) are highly dependent upon the respective ions constituting the ILs. It is estimated that approximately 10^{18} ILs can be synthesized by combinations of the cations and anions and functionalization of the alkyl chain.
- Thermal stability: Many ILs are stable at temperatures over 300°C. This provides an opportunity to carry out reactions even at high temperatures.
- Polarity: ILs are highly polar, due to which they are often immiscible in organic solvents. This property renders facile separation of products from the IL medium during liquid–liquid extraction processes.

- Reusability: Wherever used as the reaction medium, the IL phase can be recovered and reused.

> Table salt and [BMIm]Cl both are ionic salts, but their melting points are far apart, 800°C and 65°C, respectively. This is because of the difference in their lattice enthalpies. The constitutional ions of ILs have large sizes and conformational flexibility, which leads to exceptionally small lattice enthalpies and a large negative Gibbs free energy of fusion and, thereby, low melting points in comparison to classical inorganic salts.

4.5.3.3 Ionic liquids as solvents

ILs have the unique ability to dissolve transition metal salts and organometallic compounds and hence can be used as excellent solvents for homogeneous metal-catalyzed reactions [9]. ILs may themselves act as catalysts or cocatalysts, depending upon the coordinating properties of the anion. However, in this chapter, only those applications are discussed that involve the role of ILs as solvents.

- Hydrogenation reaction

  Many biphasic reactions have been investigated having an IL phase containing a metal catalyst and an organic phase containing the organic reactants. For example, hydrogenation of 1-pentene was carried out in [BMIm]SbF_6 containing a rhodium-based organometallic complex (Scheme 4.17). Although the reactants had very little solubility in the IL phase, a five times enhanced reaction rate than that of a comparable reaction in acetone was observed.

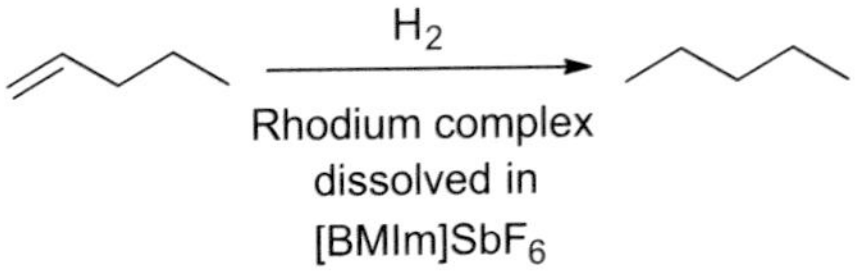

H_2

Rhodium complex dissolved in [BMIm]SbF_6

Scheme 4.17 Hydrogenation reaction in [BMIm]SbF_6.

- Oxidation reaction

  The use of ILs has also been explored in various oxidation reactions, such as epoxidation. The reaction mentioned below presents manganese complex–catalyzed epoxidation of 2,2-dimethylchromene in a mixture of $[BMIm]PF_6$ and dichloromethane (Scheme 4.18). Enhancement in catalytic activity was observed after the addition of IL to the organic solvent medium. In the presence of the IL, 86% conversion was achieved in 2 h, whereas the same conversion was achieved in 6 h in the absence of an IL.

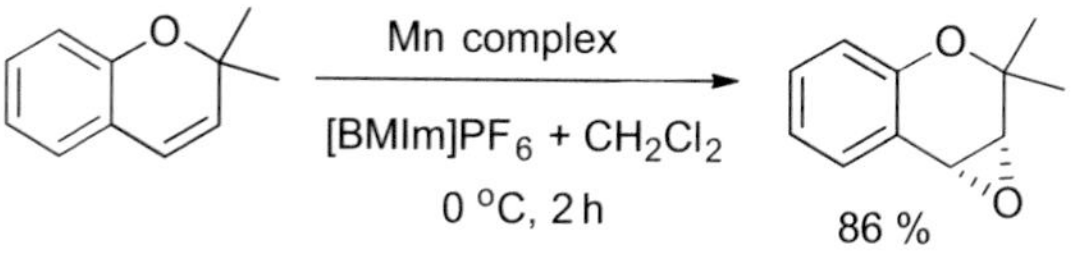

Mn complex

$[BMIm]PF_6 + CH_2Cl_2$

0 °C, 2 h

86 %

Scheme 4.18 Oxidation reaction in $[BMIm]PF_6$.

- Hydroformylation reaction

  Hydroformylation refers to the formation of aldehydes from alkenes. Hydroformylation of 1-octene was performed in $[BMIm]SnCl_3$ in the presence of a platinum complex as the catalyst (Scheme 4.19). Very high activity was reported despite the limited solubility of octene in the IL.

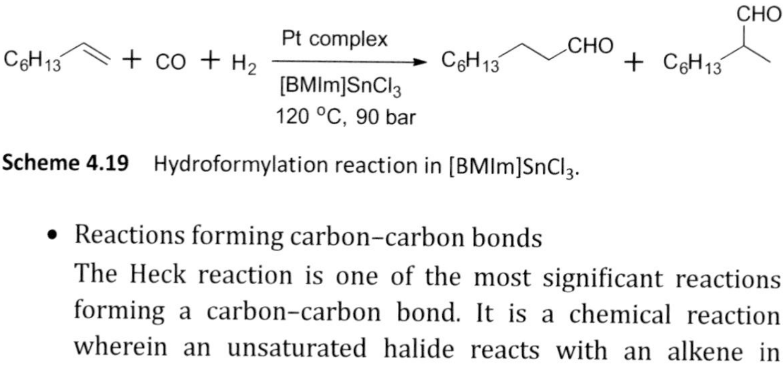

C_6H_{13} + CO + H_2 → (Pt complex; $[BMIm]SnCl_3$; 120 °C, 90 bar) C_6H_{13}–CHO + C_6H_{13}–CH(CHO)–

Scheme 4.19 Hydroformylation reaction in $[BMIm]SnCl_3$.

- Reactions forming carbon–carbon bonds

  The Heck reaction is one of the most significant reactions forming a carbon–carbon bond. It is a chemical reaction wherein an unsaturated halide reacts with an alkene in the presence of a palladium catalyst and a base to form a substituted alkene. Scheme 4.20 depicts the formation of the Heck reaction product in the presence of an IL.

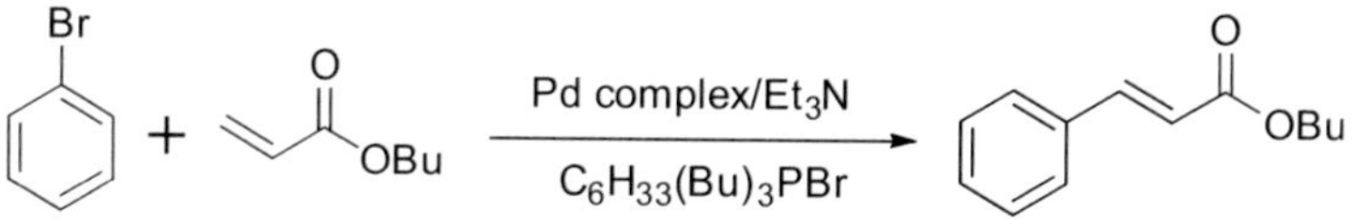

Br + O OBu → (Pd complex/Et_3N; $C_6H_{33}(Bu)_3PBr$) O OBu

Scheme 4.20 Formation of a carbon–carbon in $C_6H_{33}(Bu)_3PBr$.

4.5.4 Polyethylene Glycols

Polyethylene glycols (PEGs), H–(O–CH_2–CH_2)$_n$–OH, are polyether compounds available in a range of molecular weights, varying from 200 to 20,000 units. Their names are usually expressed as a function of molecular weight, for example, PEG with an average molecular weight of 2000 units is named PEG-2000. PEGs with molecular weights <600 are colorless, viscous liquids, whereas those with molecular weights >800 are waxy, white solids. They have also shown stability against acids, bases, and high temperatures. They are also known to be biodegradable and biocompatible and hence are used in many personal care products, such as shampoos. Low-molecular-weight PEGs are soluble in water in nearly all proportions. However, with increasing molecular weight, the solubility decreases. They are also soluble in a variety of organic solvents, for example, dichloromethane, toluene, alcohol, and acetone, but are insoluble with most aliphatic hydrocarbons, including hexane, cyclohexane, and diethyl ether. Due to this reason, such solvents are often used to extract compounds from the polymer phase. PEGs possess hydrophobic methylene groups in their backbones, which leads to an increase in the solubility of organic compounds. Additionally, they bear hydrophilic ether and alcohol groups, which provide solubility to inorganic salts and other reagents. For these reasons, they are used as inert and nonvolatile solvents for various organic reactions [10]. Some of these reactions are discussed next.

- Substitution reaction

  Hydrolysis was carried out in a PEG-300-H_2O medium (Scheme 4.21). The hydrolysis rate constant was found to have increased by nearly three folds in comparison to those for methanol, ethanol, acetone, and acetic acid as cosolvents with H_2O.

$$(CH_3)_3CCl + H_2O \xrightarrow{\text{PEG-300}} (CH_3)_3COH$$

Scheme 4.21 Hydrolysis reaction in PEG-300.

- Reduction reaction

  Carbonyl compounds can be efficiently reduced by $NaBH_4$ in PEG-400, which are otherwise slow in tetrahydrofuran (THF) (Scheme 4.22).

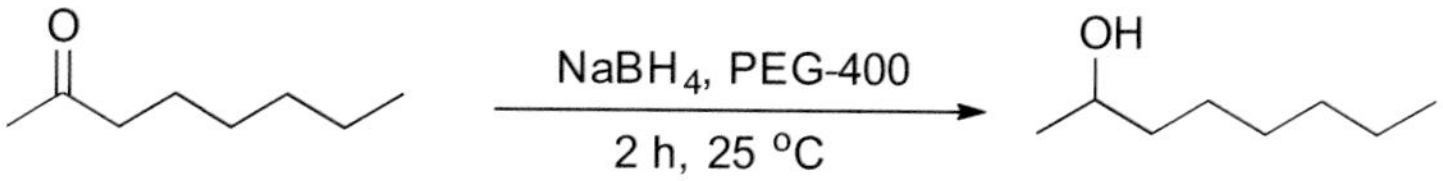

NaBH$_4$, PEG-400
2 h, 25 °C

Scheme 4.22 Reduction of the carbonyl compound in PEG-400.

- Oxidation reaction

  Oxidants such as $K_2Cr_2O_7$ and OsO_4 are soluble in PEGs. Hence, a variety of oxidation reactions have been performed in such solvents without oxidizing the terminal hydroxyl group of the PEG (Scheme 4.23).

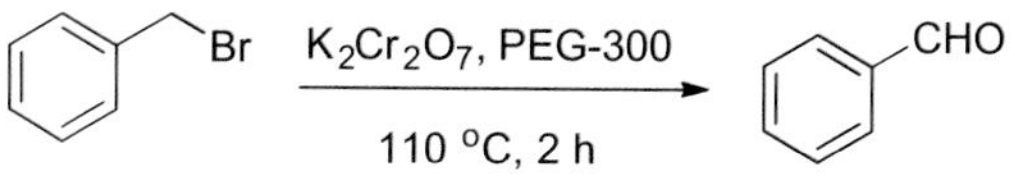

Br $K_2Cr_2O_7$, PEG-300 → CHO
110 °C, 2 h

Scheme 4.23 Oxidation of benzyl bromide in PEG-300.

- Diels–Alder reaction

  In the Diels–Alder reaction involving 2,3-dimethyl-1,3-butadiene and nitrosobenzene in PEG-300, a 3.3-fold increase and a 2.5-fold increase was found in comparison to the same reaction in dichloromethane and ethanol, respectively (Scheme 4.24).

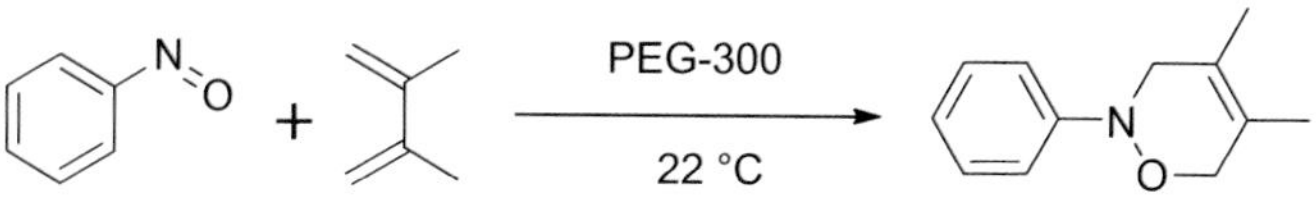

PEG-300
22 °C

Scheme 4.24 Diels–Alder reaction in PEG-300.

- Reactions forming carbon–carbon bonds

  PEG-2000 is used as a reaction medium for palladium-catalyzed reactions forming C–C bonds (Scheme 4.25). The reaction rate and yield were comparable to those with traditionally used nongreen solvents such as dimethyl sulfoxide (DMSO), dimethyl formamide (DMF), and CH_3CN. Besides, due to the immiscibility of PEG with ether, it is used as an extraction medium to remove the organic product from the PEG phase. As a result, both the reaction solvent (PEG) and the catalyst (palladium acetate) could be recycled after the reaction for four consecutive runs without any appreciable compromise in the product yield.

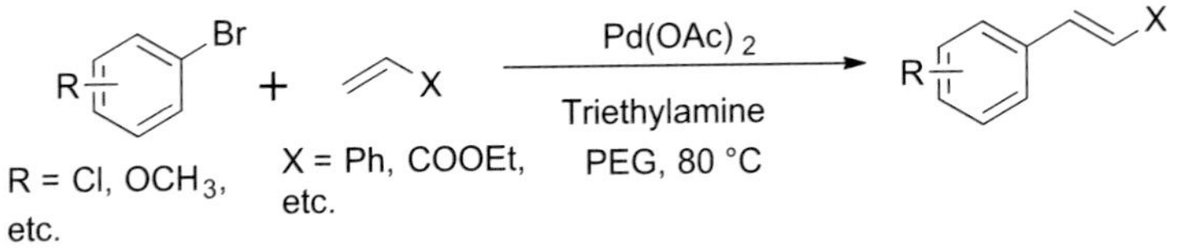

Br
R
+
X
Pd(OAc) 2
Triethylamine
PEG, 80 °C
R
X
R = Cl, OCH3, etc.
X = Ph, COOEt, etc.

Scheme 4.25 Formation of a carbon–carbon bond in PEG-2000.

4.5.5 Organic Carbonates

Organic carbonates represent an interesting class of molecular organic solvents [11]. With respect to polarity, they belong to a class of aprotic highly dipolar solvents, such as DMSO and DMF. However, they show limited or immiscible behavior with water. They offer numerous benefits over traditional solvents, such as wide availability, low cost, low toxicity, and biodegradability. They are also stable under ambient conditions and are not affected by moisture and thus can be stored even under atmospheric conditions. Nevertheless, the boiling points of organic carbonates are usually very high, which makes product isolation difficult because large amounts of VOCs are required for work-up or energy-intensive distillation procedures. Structures of some of the commonly used organic carbonates are shown in Fig. 4.6.

O
MeO OMe
Dimethyl carbonate
O
EtO OEt
Diethyl carbonate
O
O O
Ethylene carbonate
O
O O
Propylene carbonate

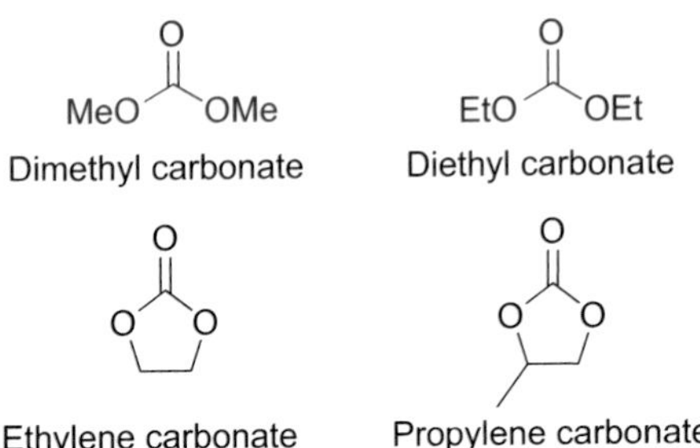

Figure 4.6 Commonly used organic carbonates.

Some examples where organic carbonates serve as the reaction media are described next.

- Oxidation reaction

  Various organic carbonates have been used as solvents in a variety of oxidation reactions. For instance, ethylene carbonate was utilized in the oxidation of alkenes using molecular oxygen as the sole oxidant and $PdCl_2$ as the catalyst (Scheme 4.26). In comparison to commonly used organic solvents, such as DMF, dimethylacetamide (DMA), and *N*-methyl-2-pyrrolidone (NMP), ethylene carbonate showed superior performance under identical reaction conditions.

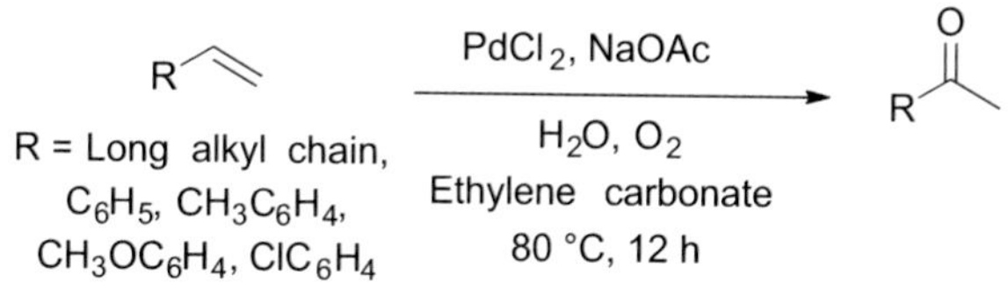

R = Long alkyl chain, C_6H_5, $CH_3C_6H_4$, $CH_3OC_6H_4$, ClC_6H_4

$PdCl_2$, NaOAc; H_2O, O_2; Ethylene carbonate; 80 °C, 12 h

Scheme 4.26 Oxidation of alkenes in ethylene carbonate.

- Reactions forming carbon–carbon bonds

  Sonogashira is one of the most significant reactions forming C–C bonds and utilizes a palladium catalyst and a copper cocatalyst to construct a C–C bond between a terminal alkyne and an aryl or vinyl halide. It has been reported in the literature that the properties of propylene carbonate are similar to that of acetonitrile. Due to this it has been successfully utilized in the copper-free Sonogashira reaction to yield up to 76% of the desired product (Scheme 4.27).

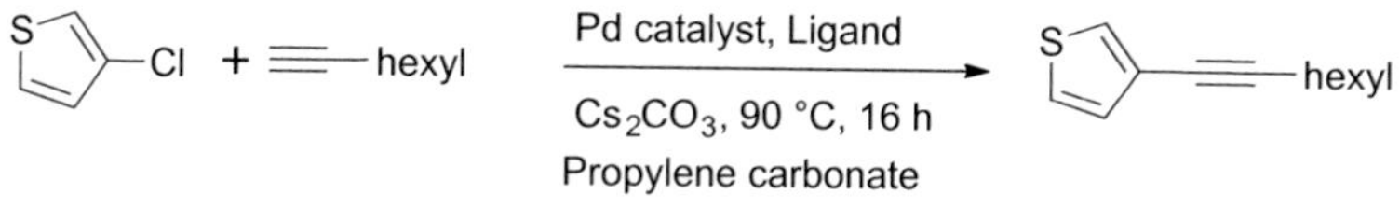

Cl + hexyl → hexyl

Pd catalyst, Ligand; Cs_2CO_3, 90 °C, 16 h; Propylene carbonate

Scheme 4.27 Sonogashira reaction in propylene carbonate.

4.5.6 Solvents Obtained from Renewable Resources

A majority of the current organic solvents are derived from nonrenewable fossil resources, such as petroleum. However, increasing demand for solvents and depleting amounts of such resources have raised considerable concerns with respect to sustainability. In this regard, biomass has come out to be an excellent alternative carbon source, satisfying Principle 7 of green chemistry. As discussed in Chapter 2, biomass feedstock can be utilized to obtain a variety of chemicals. Some of the biomass-derived solvents are listed in Fig. 4.7. Compared to traditional petrochemical-derived solvents, they possess various attractive advantages, such as availability of abundant renewable feedstock (including agricultural, forest, and urban wastes), much lower adverse environmental impact, low toxicity, and high biodegradability [12].

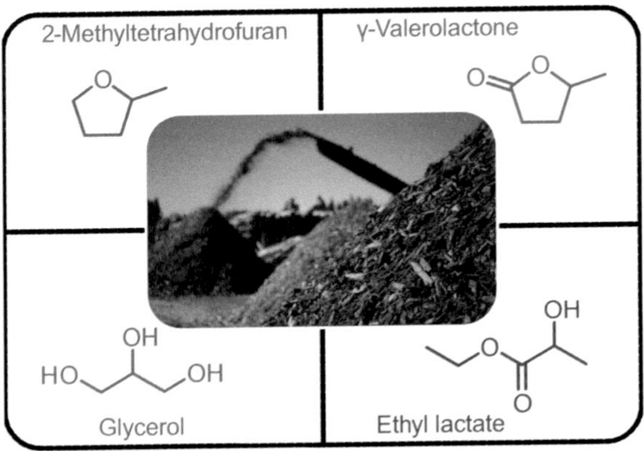

Figure 4.7 Commonly used biomass-based solvents.

Some of the widely used biomass-derived solvents are briefly described next.

4.5.6.1 Glycerol

Glycerol is one of the most interesting biobased solvents [13]. It is obtained in large amounts as a coproduct during the synthesis of biodiesel (Scheme 4.28), due to which it is available at a very low cost. It is nonvolatile, nonflammable, nontoxic, biocompatible, and recyclable. Such features impart "greenness" to its use as an alternative reaction solvent. Glycerol, being a trihydric alcohol, is a polar protic solvent. It is able to dissolve inorganic salts, acids, bases, and various transition metal complexes. Moreover, it is immiscible with many organic solvents, such as hydrocarbons, which allows easy product isolation by simple extraction. Besides its high boiling point, it offers opportunity for carrying out organic reactions at high temperatures, which is otherwise not possible with low-boiling VOCs.

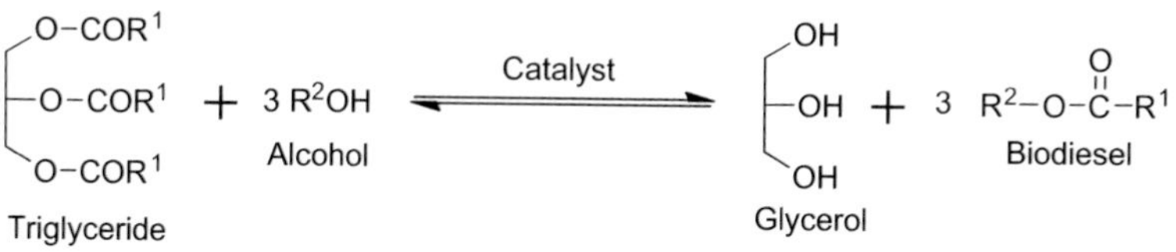

$$\text{Triglyceride} + 3\,R^2OH\ (\text{Alcohol}) \xrightleftharpoons{\text{Catalyst}} \text{Glycerol} + 3\,R^2\text{–O–C(=O)–}R^1\ (\text{Biodiesel})$$

Scheme 4.28 Formation of glycerol as a coproduct during biodiesel synthesis.

Glycerol has successfully been utilized as a solvent in a variety of organic transformations, some of which are discussed next (Schemes 4.29 and 4.30).

- Nucleophilic substitution reaction

$$C_6H_5CH_2Cl + KSCN \xrightarrow[80^{\circ}C,\ 5\ h]{\text{Glycerol}} C_6H_5CH_2NCS + KCl$$

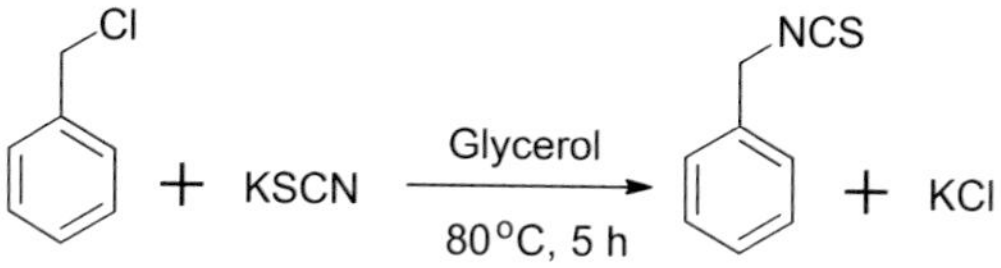

Scheme 4.29 Nucleophilic substitution reaction in glycerol.

- Reduction reaction

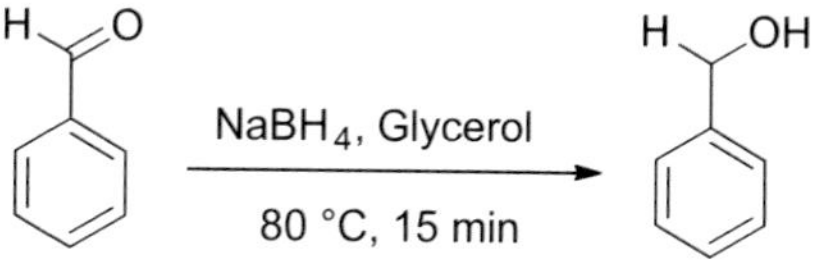

$$C_6H_5CHO \xrightarrow[80\ ^{\circ}C,\ 15\ min]{NaBH_4,\ \text{Glycerol}} C_6H_5CH_2OH$$

Scheme 4.30 Reduction of benzaldehyde to benzyl alcohol in glycerol.

Despite the aforementioned advantages, the use of glycerol is associated with a few shortcomings as well. It is highly viscous and hence results in poor diffusion of the reactants and ultimately in reduced reactivity. The hydroxyl groups in glycerol can react with strong acids and bases, which may lead to the formation of side products. Also, glycerol possesses coordinating properties, which can cause difficulties with transition metal complex catalysts. In spite of these shortcomings, glycerol offers an environmentally viable alternative to current VOCs as a solvent.

4.5.6.2 2-Methyltetrahydrofuran

2-Methyltetrahydrofuran (2-MeTHF) is another promising biomass-derived chemical, a useful alternative to classical THF [14]. It is obtained from furfural and levulinic acid and both, in turn, are obtained from renewable lignocellulosic biomass (Scheme 4.31). Despite resembling THF, it possesses rather different properties, such as a higher boiling point, lower miscibility with water, higher stability, and biodegradability. Due to its environmentally benign nature and unique properties, not only is it being readily used by researchers but it has been widely accepted by industries as well. It can also be degraded abiotically using sunlight and air *via* oxidation and ring opening.

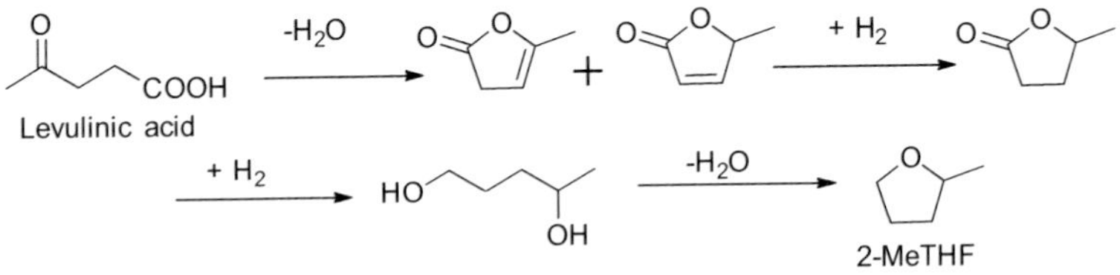

COOH
Levulinic acid
$-H_2O$
$+ H_2$
$+ H_2$
HO
OH
$-H_2O$
2-MeTHF

Scheme 4.31 Synthesis of 2-MeTHF from levulinic acid.

2-MeTHF is being used as a solvent for various reactions (Schemes 4.32 and 4.33).

- Nucleophilic substitution reaction

NO_2
OH
CHO
Br
OEt
Base, 2-MeTHF
70 °C, 30 min

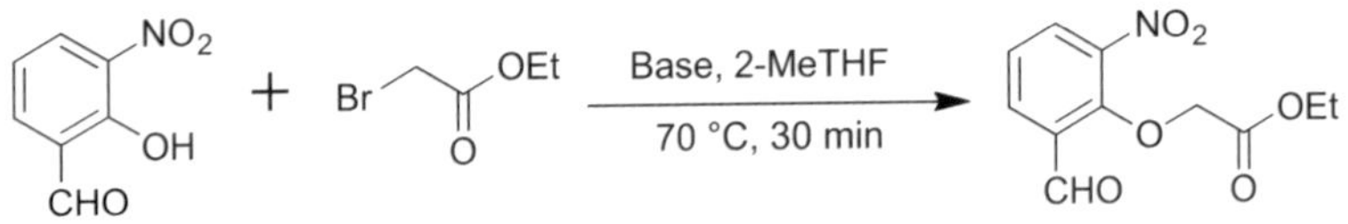

Scheme 4.32 Nucleophilic substitution reaction in 2-MeTHF.

- Oxidation reaction

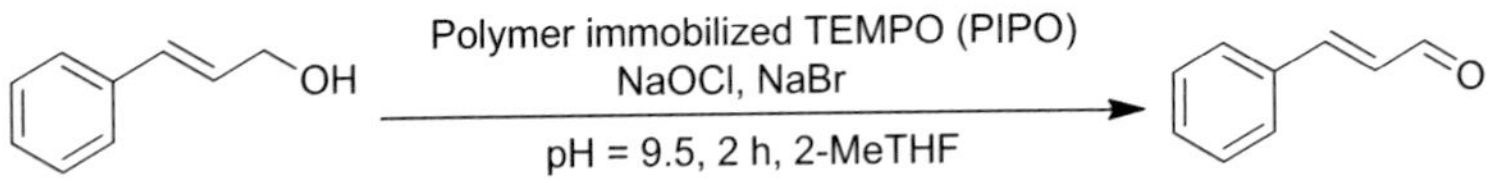

OH
Polymer immobilized TEMPO (PIPO)
NaOCl, NaBr
pH = 9.5, 2 h, 2-MeTHF

Scheme 4.33 Oxidation reaction in 2-MeTHF.

4.5.6.3 Ethyl lactate

Ethyl lactate is a low-cost, environmentally benign compound [15]. It can be produced by the esterification reaction of lactic acid and ethanol (Scheme 4.34), both of which can be generated through the fermentation of biomass. It can exist in either levo (*S*) or dextro (*R*) forms and is industrially produced as a racemic mixture.

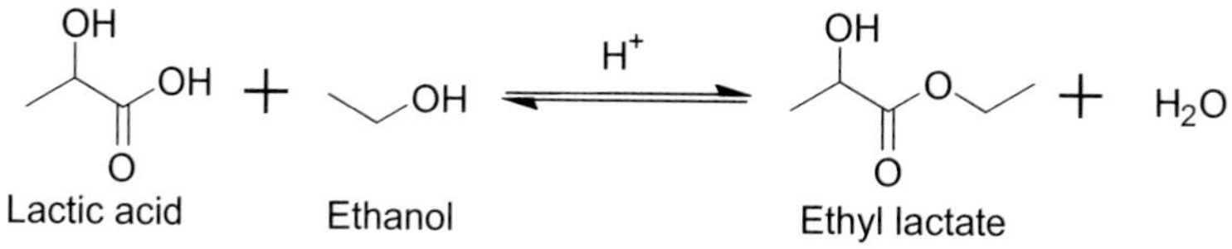

Lactic acid + Ethanol $\xrightleftharpoons{H^+}$ Ethyl lactate + H_2O

Scheme 4.34 Synthesis of ethyl lactate.

It is fully biodegradable, easily recyclable, noncarcinogenic, noncorrosive, and nonozone depleting. It is used in perfumery and as a food additive and cleaning agent. It features various attractive properties, such as high boiling point and moderate polarity, due to which it is being used as a green reaction medium in various organic transformations. Due to the presence of an asymmetric center, it can also be used for asymmetric induction in chemical reactions. Listed next are some examples where ethyl lactate has served as a solvent (Schemes 4.35 and 4.36).

- Reduction reaction

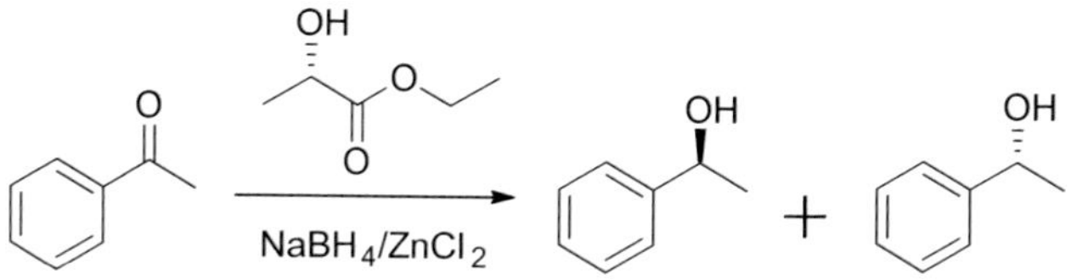

$NaBH_4/ZnCl_2$

Scheme 4.35 Reduction reaction in ethyl lactate.

- Oxime formation

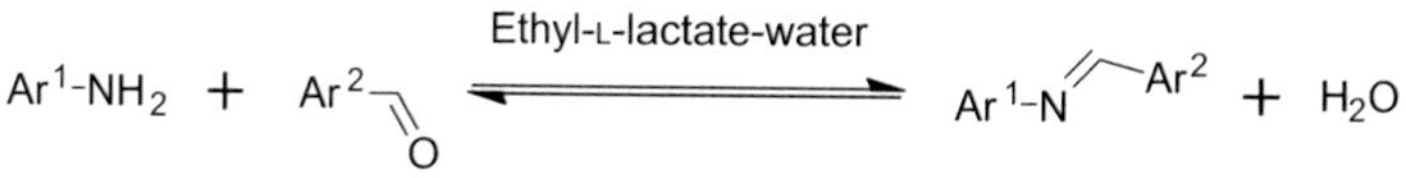

$Ar^1\text{-}NH_2 + Ar^2\text{-}CHO \xrightleftharpoons{\text{Ethyl-L-lactate-water}} Ar^1\text{-}N{=}CH\text{-}Ar^2 + H_2O$

Scheme 4.36 Oxime formation in ethyl lactate.

4.5.6.4 γ-Valerolactone

γ-Valerolactone (GVL) is another biobased chemical produced by the hydrogenative cyclization of levulinic acid (Scheme 4.37) [16]. It possesses a high safety profile due to its high boiling point, good chemical stability, and low toxicity. It has a sweet and herbaceous odor, due to which it is used in perfumes and food additives. Some examples where GVL has been used as a reaction solvent are given here (Schemes 4.38 and 4.39).

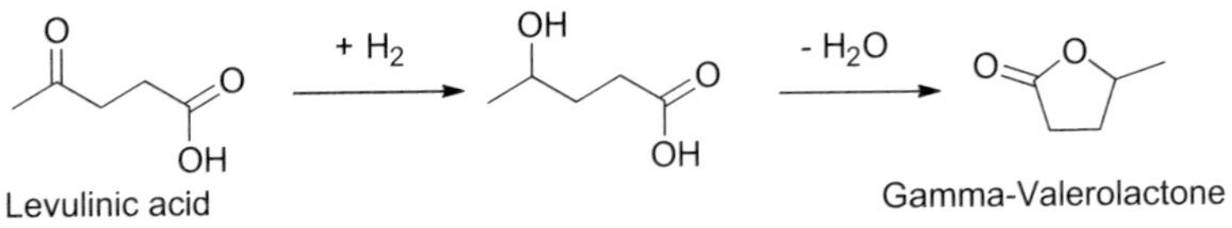

Levulinic acid + H_2 → (OH intermediate) − H_2O → Gamma-Valerolactone

Scheme 4.37 Synthesis of GVL.

- Reactions forming carbon–carbon bonds
  In the present reaction, GVL served as an excellent alternative to the classically used nongreen solvent (DMF).

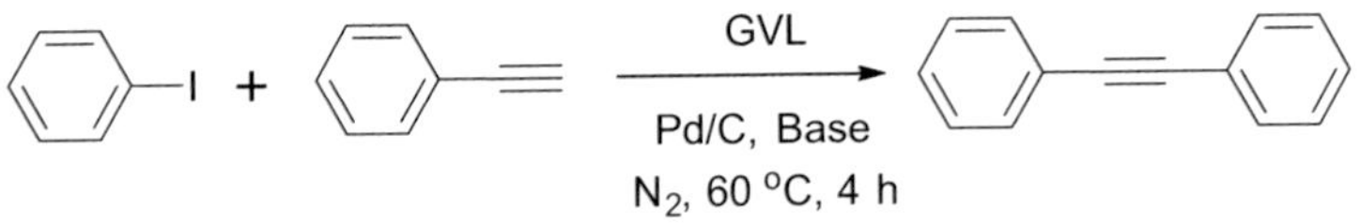

GVL; Pd/C, Base; N_2, 60 °C, 4 h

Scheme 4.38 Formation of a carbon–carbon bond in GVL.

- Aminocarbonylation reaction

PhI + ^{t}Bu−NH_2 + CO → GVL; Pd(OAc)$_2$/PPh$_3$; 1 bar, 50 °C, 72 h

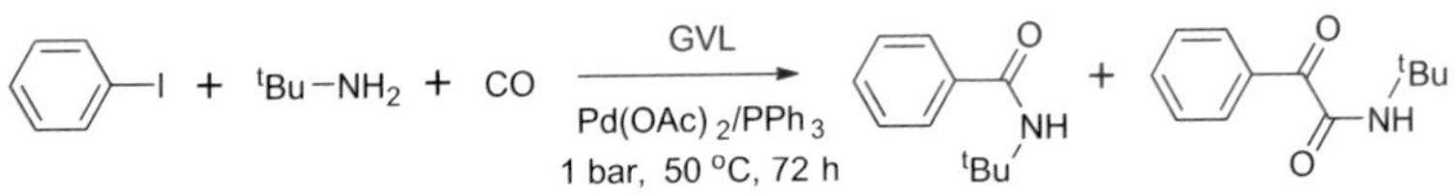

Scheme 4.39 Aminocarbonylation reaction in GVL.

> The carbon of biosolvents comes from sugars, starches, oils, and proteins, which are synthesized by plants by capturing CO_2 from the atmosphere using photosynthesis. The same amount of CO_2, as captured by plants, is released into the atmosphere when a biosolvent is used. Thus, there is no net increase in the greenhouse gas emissions on a life-cycle basis. Therefore, these are considered as carbon-neutral solvents.

4.5.7 Fluorous Biphasic Solvents

4.5.7.1 Introduction to fluorous biphasic solvents

A fluorous biphasic system is defined as a two-phase system that consists of (i) a nonpolar fluorous phase, such as perfluoroalkane, perfluorodialkyl ether, or perfluorotrialkylamine, containing a fluorous-soluble catalyst or reagent and (ii) an inorganic or organic phase, containing the reaction product [17]. Usually, the reagents and catalysts to be used in fluorous phases are modified by attaching fluorocarbon moieties, such as linear or branched perfluoroalkyl chains with a high carbon number, to ligands. These moieties are also known as fluorous ponytails. However, incorporation of such a large number of highly electronegative fluorine atoms could significantly alter the electronic properties and catalytic activities of concerned reagents or catalysts. To circumvent this effect, large, insulating groups (spacers) consisting of methylene groups [$-(CH_2)_x-$] are inserted between the fluorous ponytail and the catalytic center. Some of such fluorous-modified reagents/catalysts are shown in Fig. 4.8.

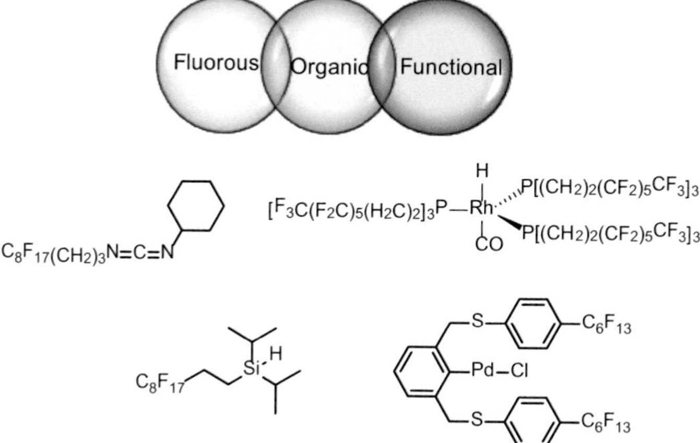

Figure 4.8 Examples representing fluorous-modified reagents and catalysts.

4.5.7.2 Advantages of using fluorous solvents

- A fluorous medium possesses a remarkably high degree of chemical inertness, nonflammability, hydrophobicity, and nontoxicity. It also has a high thermal degradation temperature, higher than that of most reagents and catalysts.

- Fluorinated compounds have limited solubility in commonly used organic solvents at ambient conditions. This allows the formation of a biphasic system in which the reagent or catalyst phase remains separated from the product phase, facilitating easy product isolation and recycling of catalytic reagents.
- The number of phases in the fluorous biphasic system is a function of temperature. The biphasic system may become a single phase at elevated temperatures. For example, a fluorous biphasic system having *n*-hexane (3 mL)–toluene (1 mL)–c-$C_6F_{11}CF_3$ (3 mL) is biphasic at room temperature and monophasic at 36.5°C [18]. Thus, the fluorous biphasic system combines the advantages of a monophasic reaction, that is, ease of mass transfer, at higher temperatures and a biphasic reaction, leading to facile product isolation at room temperature.

> Perfluorinated molecules are synthesized from their hydrocarbon equivalents by fluorination using cobalt trifluoride or electrochemical fluorination. These molecules are then used to synthesize the fluoro-modified reagents and catalysts, without which the concerned reactions cannot take place in a fluorous biphasic system. Thus, the life-cycle analysis of a fluorous solvent represents its nongreen nature. However, its ability to perform efficient separation greatly reduces the amount of solvent required. Because of this, they are considered as green solvents.

4.5.7.3 Fluorous biphasic system as a reaction media

4.5.7.3.1 *Hydroformylation reaction*

Hydroformylation is one of the most important industrial processes and involves the production of terminal aldehydes from olefins, carbon monoxide, and hydrogen using cobalt or rhodium catalysts. Earlier, this reaction was carried out in an aqueous biphasic system, with the catalyst being dissolved in the aqueous phase. Because the reaction was supposed to be carried out in the aqueous phase, poor solubility of olefins in the aqueous phase limited the application of the aqueous biphasic system in the hydroformylation reaction. Later, the hydroformylation of 1-decene was performed in a fluorous biphasic system, with $C_6F_{11}CF_3$ being the fluorous phase and toluene being the other phase. The fluorous-modified catalyst was synthesized in situ using $Rh(CO)_2(acac)$ and $P[CH_2CH_2(CF_2)_5CF_3]_3$. The reaction proceeded at 100°C in an autoclave at 10 bar CO/H_2 (1:1). After reaction completion, the reactor was cooled to room temperature, which converted the one-phase reaction mixture to a biphasic system, with the product being present in the hydrocarbon phase, leaving behind the catalyst in the fluorous phase. The protocol allowed easy product separation and allowed reuse of the fluorous catalyst.

4.5.7.3.2 *Oxidation reaction*

Fluorous solvents are an excellent medium for oxidation reactions due to the high solubility of O_2. Also, perfluoroalkanes are extremely stubborn toward oxidation. An additional advantage of using fluorous solvents in oxidation reactions is that the oxidation products are highly polar, due to which they are less soluble in fluorous solvents, making the separation easier (Fig. 4.9).

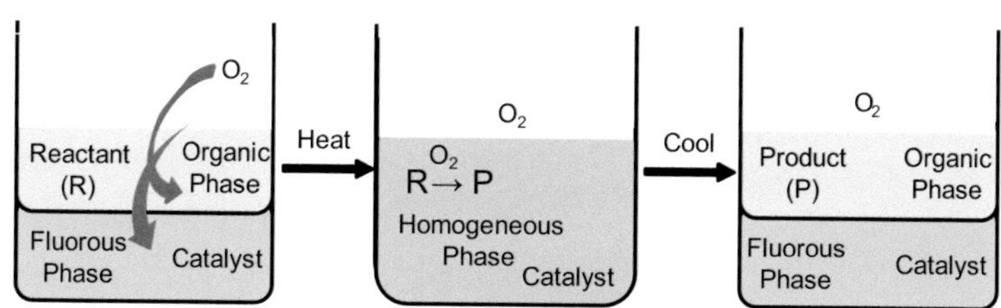

Figure 4.9 Product separation in a fluorous biphasic system.

Scheme 4.40 represents the oxidation of aldehydes into carboxylic acids using a nickel metal catalyst having perfluorinated ligands in a toluene/perfluorinated decalin solvent system. No leaching of the catalyst was reported, and the reaction system was reused for six consecutive cycles.

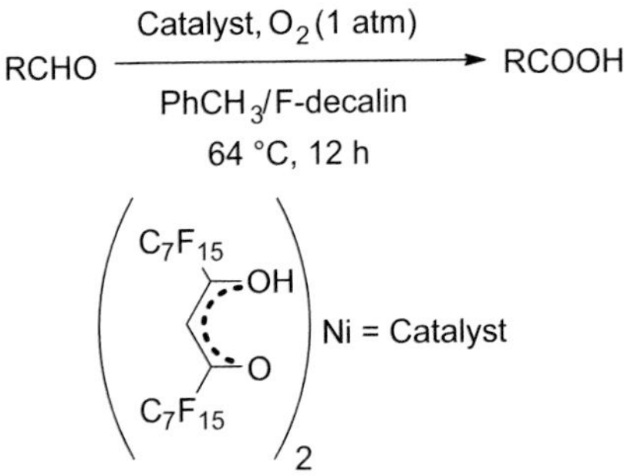

Catalyst, O_2 (1 atm)

RCHO → RCOOH

$PhCH_3$/F-decalin

64 °C, 12 h

C_7F_{15} OH O C_7F_{15} 2 Ni = Catalyst

Scheme 4.40 Oxidation of aldehyde in a fluorous solvent using a metal catalyst having a fluoro-functionalized ligand.

4.6 Solvent-Free Synthesis

Throughout this chapter, the indication has been that synthesis of chemical products requires solvents, either conventional solvents or their greener alternatives. The famous philosopher Aristotle had also said, "No reaction occurs in the absence of solvent." But can reactions occur without any solvent, or do we always need a solvent? In this section, this question will be answered and it will become clear that solvents are not always necessary [19].

4.6.1 When At Least One of the Reactants Is a Liquid

When one of the reactants is a liquid, the reaction may be carried out in a solvent-free condition. The best example of this kind is the solvent-free synthesis of benzoic acid. In this reaction, benzaldehyde reacts with atmospheric oxygen to first form perbenzoic acid, which upon further reaction with benzaldehyde forms benzoic acid (Scheme 4.41). The formation of beautiful crystals of benzoic acid on the walls of a benzaldehyde bottle justifies this reaction.

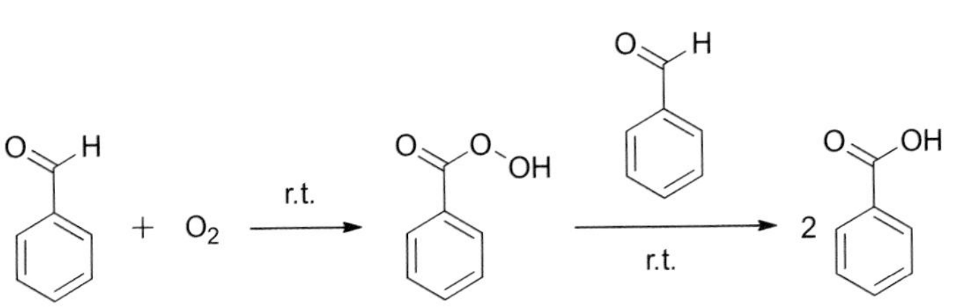

+ O_2 r.t. → r.t. → 2

Scheme 4.41 Conversion of benzaldehyde to benzoic acid under solvent-free conditions.

4.6.2 Gas-Phase Catalytic Reactions

Many industrially significant processes are carried out in the gas phase without using any solvent, for example, the synthesis of methanol using syngas. Typically, methanol production is carried out in two steps:

Step 1: Conversion of feedstock natural gas into syngas (Scheme 4.42)

$$\text{Biomass (dry)} \rightleftharpoons H_2\,(g) + CO\,(g) + CO_2\,(g) + H_2O\,(g) + \text{light hydrocarbons}\,(g) + \text{tar}\,(l) + \text{char}\,(s)$$

Scheme 4.42 Conversion of feedstock natural gas into syngas.

Step 2: Catalytic reaction of syngas (Scheme 4.43)

Usually, a copper-zinc oxide catalyst with aluminum oxide or chromium(III) oxide is used for methanol production.

$$CO + 2\,H_2 \rightleftharpoons CH_3OH$$

$$CO_2 + 3\,H_2 \rightleftharpoons CH_3OH + H_2O$$

Scheme 4.43 Catalytic reaction of syngas.

4.6.3 Solid–Solid Reaction

In a solid–solid synthesis, two reactants are mixed together to form a product under solvent-free conditions, for example, aldol condensation reactions between acetophenone derivatives and benzaldehyde. Aldol condensation is one of most powerful tools for the formation of carbon–carbon bonds, where, in the presence of a dilute acid or a base, two molecules of aldehyde or ketone react together, in an ethanolic solution, to form β-hydroxyaldehyde or β-hydroxyketone (Scheme 4.44). Happily, this reaction can even be performed in a solvent-free condition (Fig. 4.10) [20]. However, it is not necessary that reactions involving solid reactants need to proceed in a solid phase. As mixing of two solids results in the lowering of their melting points, the solid reactants are mixed to give a melt and, thus, the reaction occurs in a liquid albeit solvent-free state.

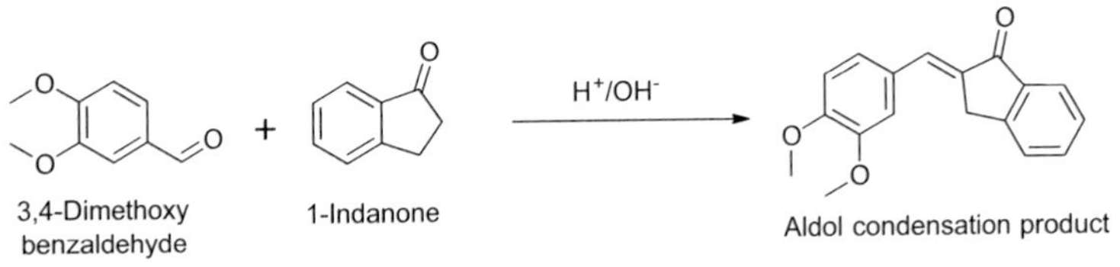

H⁺/OH⁻

3,4-Dimethoxy benzaldehyde

1-Indanone

Aldol condensation product

Scheme 4.44 Aldol condensation reaction.

Figure 4.10 Steps involved in the solvent-free synthesis of the aldol condensation product.

Another example of this class of reactions is the formation of azomethine using *o*-vanillin and *p*-toluidine (Scheme 4.45). The protocol for the synthesis remains the same as above, except that the reaction takes place in the absence of any acid or base.

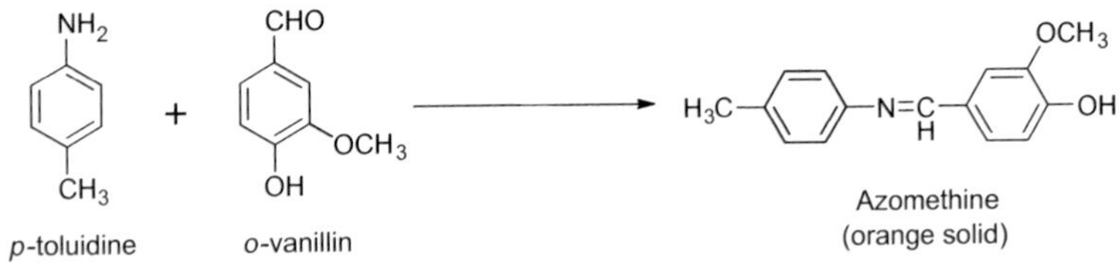

p-toluidine

o-vanillin

Azomethine (orange solid)

Scheme 4.45 Synthesis of azomethine.

Solid–solid reactions seem to be very promising in the field of green chemistry. Scaling up of such reactions at the industrial level can dramatically reduce the cost compared to that of conventional approaches. Ball mills can be used in this regard, in which a ball bearing is placed inside the reaction vessel that is being shaken at high speeds. On a laboratory scale, solid–solid reactions are carried out using a pestle and mortar.

4.6.4 Benefits of Solvent-Free Synthesis

This approach of eliminating the use of solvent during a reaction offers several benefits:

- Solvent-free synthesis eliminates the need for the removal of the solvent from the reaction mixture, thereby minimizing environmental degradation by preventing the immersion of discarded solvents into water bodies.
- Often, a solvent needs to be purified and in some cases even dried before use. A solvent-free protocol reduces the energy consumption.
- Solvent-free reactions lead to simplified work-up procedures.
- Due to the absence of solvents, the reaction can be scaled up easily and safely.
- Besides, they are easy to handle and cheaper to execute.

However, chemists must be cautious when thinking of using solvent-free procedures. Although a reaction can be performed involving neat reactants or reagents, a solvent may be utilized during purifying, isolation, or analysis of the product.

The best solvent is no solvent at all!

4.7 Immobilized Solvents

One of the roles of solvents described in this chapter is as a partner to dissolve either organic substrates or catalysts or both. One such previously discussed solvent is IL. They are excellent media to dissolve transition metal complexes. However, their utility in catalytic processes is still restricted due to some key drawbacks, such as low substrate diffusion due to their high viscosities, generation of large amounts of waste ILs (which are extremely difficult to dispose off), and their expensive nature. These practical inconveniences have given birth to a concept called "immobilization" of solvent, wherein a thin film of an IL is supported on a solid phase. This immobilization process aims to transfer the desired properties of ILs (or metal species containing ILs) onto a heterogeneous solid support material, combining the advantages of both [21]. Although these immobilized solvents appear as solids, the catalytically active species present in the immobilized IL phase still impart the features of homogeneous catalysts, such as high activity and selectivity. Additionally, this technique can bring down the overall cost of a process and may also be acceptable from the environmental viewpoint as it circumvents the aforementioned disadvantages.

Using this concept, Hagiwara and group converted the pores of silica gel into micro reaction vessels by supporting a $Pd(OAc)_2$-containing IL into the pores of silica gel [22]. For this, a suspension of modified silica gel in a solution of $Pd(OAc)_2$ and $[BMIm]BF_4$ (IL) was stirred at room temperature for 4 h. Subsequently, the solid catalyst was obtained by evaporating the solvent. The immobilized Pd-containing IL phase was then utilized in the hydrogenation of cyclohexene (Fig. 4.11). The catalyst is easily separable using unsophisticated filtration/decantation and can be reused up to 10 times.

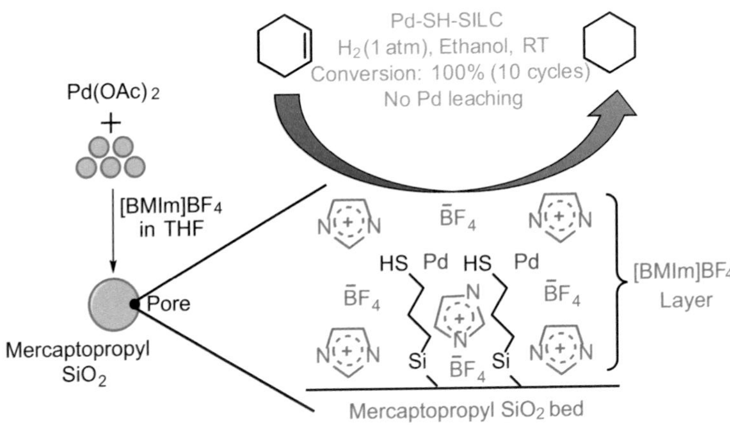

Figure 4.11 Immobilization of Pd-containing IL phase and its utility in the hydrogenation of cyclohexene.

4.8 Learning Outcomes

At the end of this chapter, students will be able to:

- Identify the use of some of the most promising alternative solvents, such as water, SCFs, ILs, PEGs, organic carbonates, biobased solvents, and fluorous biphasic solvents.
- Appreciate the potential benefits of these solvents in the context of sustainability: nonvolatility, economic viability, facile removal, utilization of renewable feedstocks, and environmentally benign nature.

- Learn that every chemical synthesis doesn't necessarily require an organic solvent or rather even a solvent! Many bulk chemicals are manufactured under solvent-free conditions, especially where when one of the reactants is present in the liquid phase.
- Choose an appropriate solvent by considering its EHS aspects and the energy demands during its entire life cycle.
- Acknowledge the use of alternative green solvents in the accomplishment of the 17 Sustainable Development Goals (SDGs). Figure 4.12 highlights how these solvents help in combating the global concerns and will assist in building a better world for people by 2030.

1. NO POVERTY	2. ZERO HUNGER	3. GOOD HEALTH AND WELL BEING	4. QUALITY EDUCATION	5. GENDER EQUALITY
Use of green solvents provides economic benefits	Use of alternative solvents has greatly improved production of crop protecting chemicals and thus, has enhanced food security	Green solvents share excellent environment, health and safety profile	Helps in imbibing correct practices and educates stakeholders from the perspective of sustainability	Works towards a healthier world bridging the gap between gender for their opportunities
6. CLEAN WATER AND SANITATION	**7. AFFORDABLE AND CLEAN ENERGY**	**8. DECENT WORK AND ECONOMIC GROWTH**	**9. INDUSTRY, INNOVATION AND INFRASTRUCTURE**	**10. REDUCED INEQUALITIES**
Use of alternative green solvents as synthetic medium and separating agents can help reduce water pollution	Use of renewable and waste sources to generate affordable and clean green solvents	By 2024, green and bio-solvents market is expected to touch USD 8610 million	Phenomenal green solvents acceptance by industrial fraternity	Continuous development in the area are creating job opportunities for all, irrespective of gender, race or ethnicity
11. SUSTAINABLE CITIES AND COMMUNITIES	**12. RESPONSIBLE CONSUMPTION AND PRODUCTION**	**13. CLIMATE ACTION**	**14. LIFE BELOW WATER**	**15. LIFE ON LAND**
Cities are being made sustainable by taking waste CO_2 from the environment to generate green solvents	Green solvents are synthesized keeping in mind their life-cycle and resource utilization starting from its production, its use, its recycling to its disposal	Unlike VOCs, green solvents are non-volatile, hence, do not contribute towards greenhouse effect	Synthesis of some of the green solvents utilize aquatic biomass and also helps in reducing ocean acidification	Green solvents cause lesser pollution which leads to improved quality of life on land
16. PEACE, JUSTICE AND STRONG INSTITUTIONS	**17. PARTNERSHIPS FOR THE GOALS**	SUSTAINABLE DEVELOPMENT GOALS		
Low dependence on non-renewable resources will discourage conflicts and insecurity among nations	Globally, researchers and industries are collaborating for the development of alternative green solvents			

Figure 4.12 Alternative green solvents taking us a step closer to accomplishing the 17 Sustainable Development Goals.

Despite these solvents having numerous fascinating properties, their use is still in its infancy and is subjected to strict limitations due to a lack of data on the biocompatibility of ILs, high cost of equipment for $scCO_2$, and large synthetic requirements for the use of fluorous biphasic systems, to name a few. When put differently, a universal green solvent doesn't exist, due to which the scientific community is continuously searching for the development of new and sustainable approaches for current solvent innovation so as to widen their use in catalytic and organic processes.

4.9 Problems

1. Discuss the utility of solvents in organic synthesis.
2. Which properties of water highlight its use in organic synthesis?
3. How do you compare the greenness of solvents?
4. Explain using examples how water can act as a better solvent than the conventionally used volatile organic solvents.
5. List the properties of supercritical fluids.
6. Describe the utility of supercritical carbon dioxide.
7. Describe the synthesis of ionic liquids using a suitable example.
8. How do polyethylene glycols act as recyclable reaction media?
9. The use of a compound as a solvent in organic synthesis is based upon four significant factors: (i) greenness, (ii) availability and price, (iii) feasibility, and (iv) necessity. On the basis of these aspects, comment upon the use of glycerol as a solvent.
10. Discuss the advantages of biomass-derived solvents over conventionally used solvents.
11. Explain the utility of organic carbonates as solvents with examples.
12. Write a short note on fluorous biphasic systems.
13. What are fluorous ponytails?
14. Can a reaction proceed under a solvent-deficit condition? Justify the answer using suitable reasoning and examples.
15. "The best solvent is no solvent at all!" Justify the statement.

16. Have you ever seen crystals deposited on the surface of a benzaldehyde bottle? What are these crystals, and how are they formed?
17. How can a solvent be immobilized? Discuss the advantages and give suitable examples.

References

1. Kerton, F. and Marriott, R. (2013). *Alternative Solvents for Green Chemistry*, 2nd Ed. (The Royal Society of Chemistry, UK).
2. Lancaster, M. (2002). *Green Chemistry: An Introductory Text* (The Royal Society of Chemistry, UK).
3. Capello, C., Fischer, U. and Hungerbühler, K. (2007). What is a green solvent? A comprehensive framework for the environmental assessment of solvents. *Green Chem.*, **9**, pp. 927–934.
4. Alfonsi, K., Colberg, J., Dunn, P. J., Fevig, T., Jennings, S., Johnson, T. A., Kleine, H. P., Knight, C., Nagy, M. A., Perry, D. A. and Stefaniak, M. (2008). Green chemistry tools to influence a medicinal chemistry and research chemistry based organization. *Green Chem.*, **10**, pp. 31–36.
5. Byrne, F. P., Jin, S., Paggiola, G., Petchey, T. H. M., Clark, J. H., Farmer, T. J., Hunt, A. J., McElroy, R. and Sherwood, J. (2016). Tools and techniques for solvent selection: green solvent selection guides. *Sustain. Chem. Process.*, **4**, pp. 1–24.
6. Anastas, P. T. and Li, C. J. (2014). *Green Solvents: Reactions in Water*, Volume 5 (Wiley-VCH, Germany).
7. Williams, J. R. and Clifford, A. A. (2000). *Supercritical Fluid Methods and Protocols* (Humana Press, USA).
8. https://www.nottingham.ac.uk/supercritical/scintro.html
9. Wasserscheid, P. and Keim, W. (2000). Ionic liquids - new "solutions" for transition metal catalysis. *Angew. Chem., Int. Ed.*, **39**, pp. 3772–3789.
10. Chen, J., Spear, S. K., Huddleston, J. G. and Rogers, R. D. (2005). Polyethylene glycol and solutions of polyethylene glycol as green reaction media. *Green Chem.*, **7**, pp. 64–82.
11. Schäffner, B., Schäffner, F., Verevkin, S. P. and Börner, A. (2010). Organic carbonates as solvents in synthesis and catalysis. *Chem. Rev.*, **110**, pp. 4554–4581.

12. Santoro, S., Ferlin, F., Luciani, L., Ackermann, L. and Vaccaro, L. (2017). Biomass-derived solvents as effective media for cross-coupling reactions and C–H functionalization processes. *Green Chem.*, **19**, pp. 1601–1612.

13. Wolfson, A., Dlugy, C. and Shotland, Y. (2007). Glycerol as a green solvent for high product yields and selectivities. *Environ. Chem. Lett.*, **5**, pp. 67–71.

14. Pace, V., Hoyos, P., Castoldi, L., Domínguez de María, P. and Alcántara, A. R. (2012). 2-methyltetrahydrofuran (2-MeTHF): a biomass-derived solvent with broad application in organic chemistry. *ChemSusChem*, **5**, pp. 1369–1379.

15. Pereira, C. S. M., Silva, V. M. T. M. and Rodrigues, A. E. (2011). Ethyl lactate as a solvent: properties, applications and production processes: a review. *Green Chem.*, **13**, pp. 2658–2671.

16. Alonso, D. M., Wettstein, S. G. and Dumesic, J. A. (2013). Gamma-valerolactone, a sustainable platform molecule derived from lignocellulosic biomass. *Green Chem.*, **15**, pp. 584–595.

17. Horváth, I. T. (1998). Fluorous biphase chemistry. *Acc. Chem. Res.*, **31**, pp. 641–650.

18. Horváth, I. T. and Rábai, J. (1994). Facile catalyst separation without water: fluorous biphase hydroformylation of olefins. *Science*, **266**, pp. 72–75.

19. Tanaka, K. (2009). *Solvent-Free Organic Synthesis* (Wiley-VCH, Germany).

20. Sharma, R. K., Sidhwani, I. T. and Chaudhuri, M. K. (2012). *Green Chemistry Experiments: A Monograph* (I. K. International, India).

21. Li, H., Bhadury, P. S., Song, B. and Yang, S. (2012). Immobilized functional ionic liquids: efficient, green, and reusable catalysts. *RSC Adv.*, **2**, pp. 12525–12551.

22. Hagiwara, H., Nakamura, T., Hoshi, T. and Suzuki, T. (2011). Palladium-supported ionic liquid catalyst (Pd-SH-SILC) immobilized on mercaptopropyl silica gel as a chemoselective, reusable and heterogeneous catalyst for catalytic hydrogenation. *Green Chem.*, **13**, pp. 1133–1137.

Chapter 5

Greening Energy Sources

Gunjan Arora, Pooja Rana, and Rakesh K. Sharma
Green Chemistry Network Centre, Department of Chemistry,
University of Delhi, Delhi 110007, India
gunjanarora.892@gmail.com

> *The primary objective of science and technology is to create a society where people can have healthy, comfortable and long lives. The crucial thing in science and technology is to develop a new concept that can be applied to actual products and services, and these new products and services will eventually make people happy.*
>
> —Prof. Fujishima

5.1 Introduction

Energy plays a vital role in chemical processes. In order to form a product by a chemical reaction, some of the existing bonds in the reactants must be broken and a new set of bonds must be formed. Energy is required to break bonds and is released during bond formation. On the basis of energy requirements, reactions have

Green Chemistry for Beginners
Edited by Rakesh K. Sharma and Anju Srivastava
Copyright © 2021 Jenny Stanford Publishing Pte. Ltd.
ISBN 978-981-4316-96-5 (Hardcover), 978-1-003-18042-5 (eBook)
www.jennystanford.com

been classified into two categories: exothermic and endothermic. Reactions in which the energy required to break the bonds is greater than the energy released on bond formation are termed as endothermic reactions. In contrast, exothermic reactions result in the net release of energy in the form of heat or light.

Most of the chemical transformations rely on thermal sources of energy that utilize mainly traditional fuels. Commonly, the energy is supplied by a Bunsen burner, a hot plate, chemical reactors, or an oven. However, the energy transferred to the reaction systems using these sources is nonspecific, that is, the supplied energy is not directly absorbed by the targeted molecule/bond undergoing a chemical transformation. In fact, a large amount of the energy is wasted in heating the reaction vessel and/or the solvent system. In this scenario, the use of alternative and more specific sources of energy will be highly appreciated.

Green chemistry strongly promotes the ways and technologies that help in minimizing energy consumption. "Design for energy efficiency" is one of the important principles of green chemistry. However, many times synthetic chemists do not consider temperature and pressure requirements. According to this principle, chemical transformations should be performed at ambient temperature and pressure. Also, the energy requirements of a reaction should be minimized and recognized for their environmental and economic impacts [1]. In Chapter 3, we saw how the use of a catalyst can speed up a reaction, resulting in reduced energy requirement in the overall process. Moreover, some catalysts offer the additional advantages of recoverability and reusability, which help minimize waste. Apart from a catalyst, techniques like the use of microwaves (MWs), photochemical energy, and ultrasonic frequency that tend to improve the energy efficiency of a reaction and also ensure minimum energy wastage and significant cost reduction. The use of such alternative forms of energy is not new, but they are presently becoming popular among manufacturing industries and are viewed as emergent green technologies. In fact, these processes are quite supportive in achieving Sustainable Development Goals (SDGs, Fig. 5.1), especially goal 7, aspiring "to ensure access to affordable, reliable, sustainable and modern energy for all" by 2030 [2]. In this chapter we have focused on some evolving technologies that can lead to improved energy efficiency as well as help in reducing the overall cost of the process.

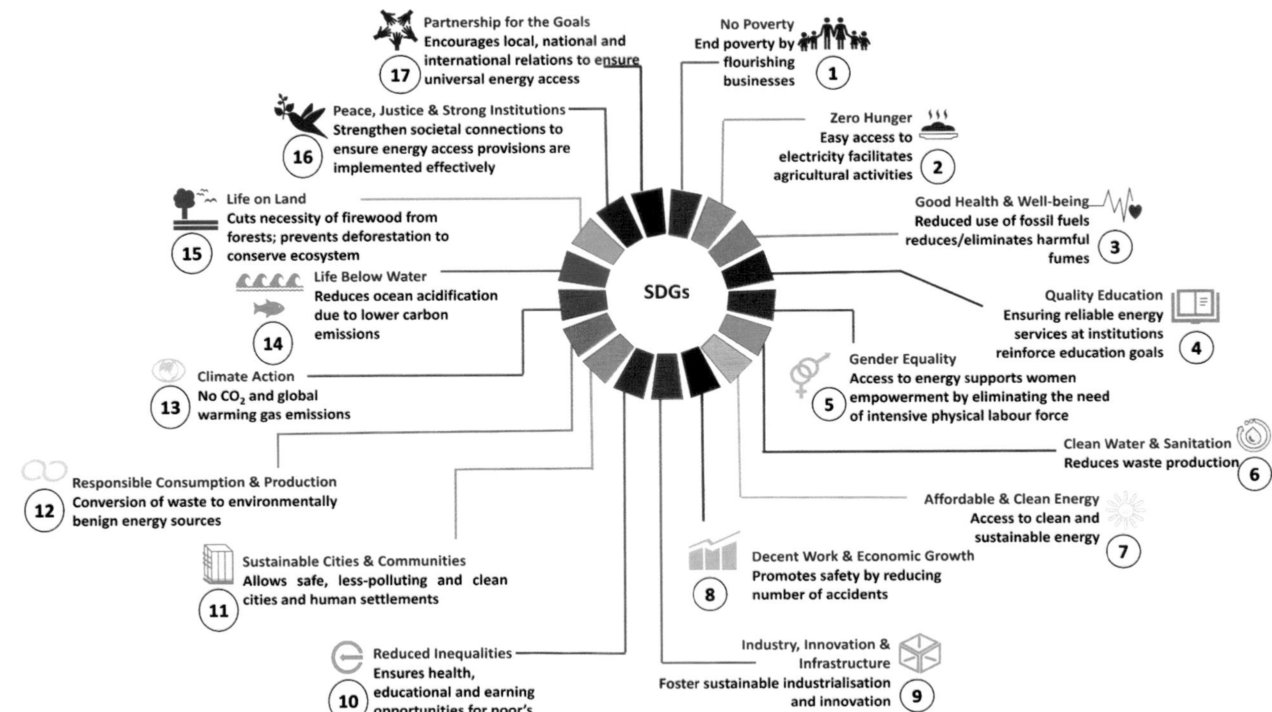

Figure 5.1 Role of alternative energy sources in achieving Sustainable Development Goals.

process highly efficient as it eliminates the need to heat the reaction vessel, allowing energy to be targeted at some specific parts of the reactive system. Thus, MW heating results in significant energy saving.

5.2.1.1.2 *Accelerated reaction rate*

Compared to conventional heating, MW heating may accelerate the rate of some chemical reactions by several times. The rate acceleration has been accredited to superheating of solvents and catalysts by MWs. MW radiations are first absorbed by MW-susceptible components in the reaction mixture and then converted into heat with high efficiency. This results in "superheating" at ambient pressure.

5.2.1.1.3 *Homogeneous heating*

In conventional heating, the walls of the reaction vessel or the reactor heat up first by convection and then by conduction. However, the core of the sample takes much longer to attain the desired temperature. So, there always remains a temperature difference between the walls and the reaction media, which not only results in a loss of energy but also leads to slow and less effective heating. However, in MW-assisted synthesis, energy is directly absorbed by active components and solvent, which results in uniform and effective heating of the reaction mixture.

5.2.1.1.4 *Improved yield and selectivity*

Certain chemical reactions furnish a better yield and selectivity of the targeted product under MW irradiation in comparison to conventional heating. Selective heating under MW irradiation is attributed to the fact that different materials absorb MWs to different extents. Moreover, there is a possibility to increase the temperature more quickly, leading to quite different kinetics.

5.2.1.1.5 *Eco-friendly alternative*

Unlike conventional methods, MW technology is much cleaner and environment friendly. In some chemical reactions, the use of solvents can be reduced or eliminated, offering economic and environment benefits. For instance, in a solventless approach, the reactants adsorbed onto the surface of a solid support like zeolite react at a much faster rate under MW irradiation than conventional heating.

5.2.2 Microwave-Assisted Chemical Reactions

MW-assisted reactions can be broadly classified into two categories, non-solid-state reactions and solid-state reactions.

5.2.2.1 Non-solid-state reactions

Non-solid-state reactions are conducted in some solvent media. It is important to note that reactions should be carried out under MW irradiation only if the solvent, reactants, and products are nonflammable. Otherwise, there is a serious risk of fire or explosion. Hence, in this type of reaction, the selection of an appropriate solvent is crucial. The solvent should have the following properties:

- The solvent to be used as the source of heat must couple effectively under MW radiation. For this, the solvent must possess some dipole moment to absorb the MWs.
- The solvent to be employed must have a boiling point at least 20°C–30°C higher than the reaction temperature.

> *N,N*-dimethyl formamide is considered to be an excellent solvent of choice since it has a high boiling point (160°C) and dipole moment (ε = 36.7). Moreover, it can retain water generated in the reaction, thus eliminating the necessity for water separation.

5.2.2.1.1 *Reactions using water as a solvent*

Water as a solvent is always a prime choice of chemists. It is a readily available, nontoxic, and nonflammable green solvent. However, the wide utility of water as a solvent is hampered due to the low solubility of organic compounds in it. To resolve this problem, various strategies has been introduced, such as utilization of some organic cosolvents, high temperature, and hydrophobic effects. Employment of high temperature is quite beneficial as at high temperatures, water shows some unique properties. The dielectric constant of water at 25°C is 78.5, which decreases to a value of 27.5 at 250°C (similar to acetonitrile at 25°C) and further decreases to

5.2.2.1.2 *Reactions using organic solvents*

Reactions in which one or more reactants are liquid (that can act as a solvent) or reactions in which an organic solvent is used to promote the reaction fall under this category.

- Esterification

  Reaction between benzoic acid and *n*-propanol in the presence of a catalytic amount of sulfuric acid gives a good yield of propylbenzoate upon irradiation with MW for 6 min. (Scheme 5.5).

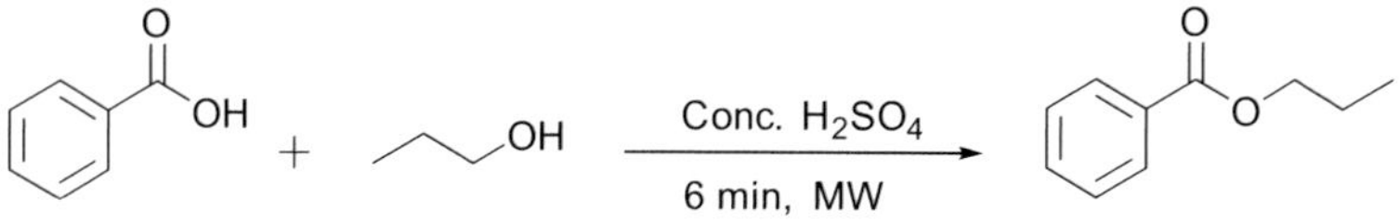

Scheme 5.5 Esterification reaction.

- Fries rearrangement

  Fries rearrangement is defined as a Lewis acid–catalyzed rearrangement reaction of phenolic ester to give hydroxy aryl ketone. A significant rate enhancement has been observed in the synthesis of phenolic ketones under MW irradiation. For MW-mediated Fries rearrangement, a mixture of *p*-cresyl acetate and anhydrous aluminum chloride is heated in dry chlorobenzene for 2 min. to obtain 85% product (Scheme 5.6).

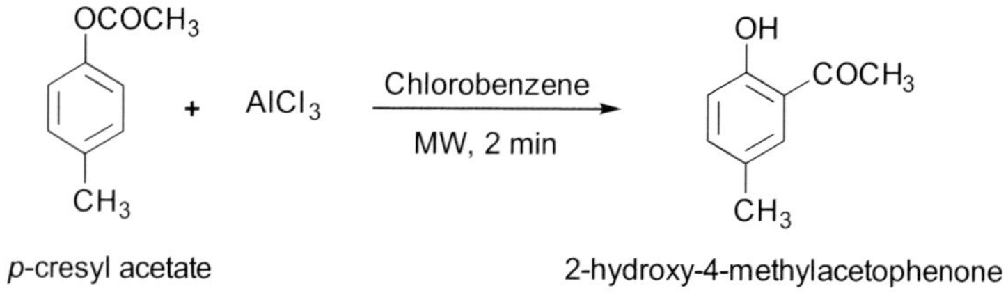

Scheme 5.6 Fries rearrangement.

- Diels–Alder reaction

  The Diels–Alder reaction is defined as the 1,4-addition of an alkene to a conjugated diene, resulting in the formation of a six-membered ring adduct. Use of MW irradiation

(Scheme 5.7) has significantly improved the reaction conditions, and 80% product is obtained within 90 sec., which otherwise require 90 min. under refluxing conditions.

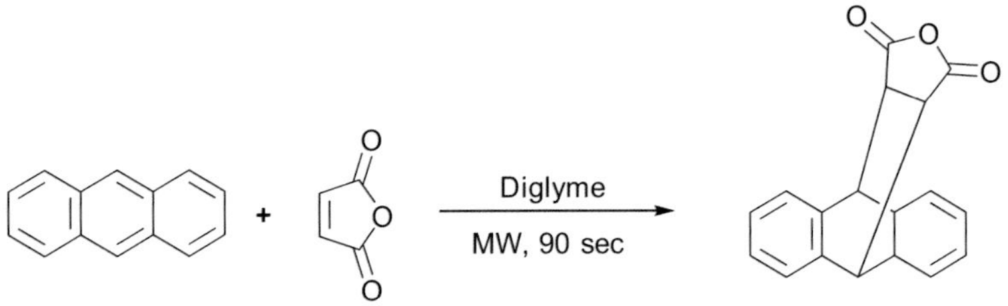

Anthracene + Maleic anhydride —(Diglyme, MW, 90 sec)→ Six-membered adduct

Scheme 5.7 Microwave-mediated Diels–Alder reaction.

5.2.2.2 Solid-state reactions

In solid-state synthesis, reactions are conducted under neat conditions, that is, no solvent is employed. These reactions are very fascinating in MW chemistry as the radiations are directly absorbed by reactants, delivering enhanced energy efficiency. Here, the reactants and a solid support (like silica gel, clay, zeolite, or alumina) are stirred in a suitable solvent. Then, the solvent is evaporated under vacuum. The obtained dried solid support on which reactants have been adsorbed is treated under MW irradiation for carrying out the reaction. Some of the important applications of solid-state synthesis have been highlighted next.

- Deacetylation

  Some reactions require the protection of aldehyde, phenol, and alcohol groups present in the reactants, which is usually done by acetylation. After completion of the reaction, deacetylation of the obtained product needs to be done. Conventionally, deacetylation is conducted under acidic or basic conditions that require a long reaction time and the yields are low. Fortunately, deacetylation under MW irradiation enables better yield of the product, that too in a short amount of time (Scheme 5.8).

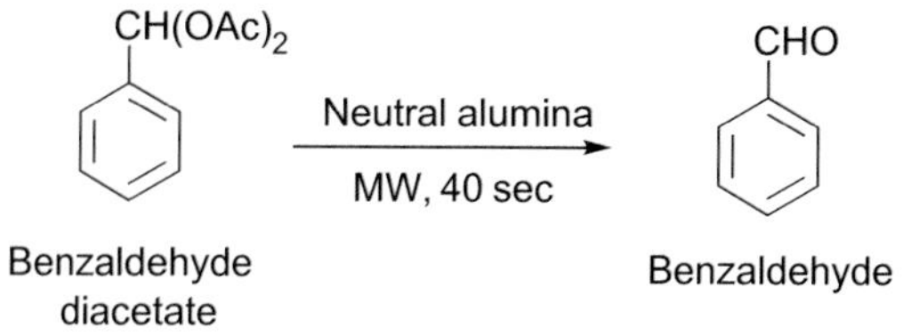

$CH(OAc)_2$ Neutral alumina / MW, 40 sec → CHO

Benzaldehyde diacetate → Benzaldehyde

Scheme 5.8 Deacetylation reaction.

- Synthesis of anhydrides from dicarboxylic acid

  Anhydrides can be obtained from dicarboxylic acids by using acetyl chloride (CH_3COCl), thionyl chloride ($SOCl_2$), and acetic anhydride ($(CH_3CO)_2O$). However, these reagents are corrosive and hazardous to the environment. MW irradiation using montmorillonite-KSF in the presence of isopropenyl acid (which acts as a water scavenger) is a much more rapid, convenient, and cleaner approach (Scheme 5.9).

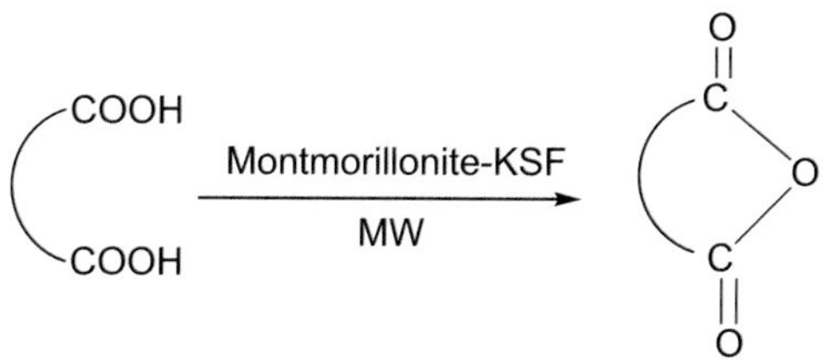

COOH, COOH — Montmorillonite-KSF / MW →

Scheme 5.9 Synthesis of anhydrides from dicarboxylic acid.

- Synthesis of copper phthalocyanine

  Metallophthalocyanine complexes are useful chemicals that find potential applications in textiles, coloring for metal surfaces, photocopiers, and laser printers. Conventional synthesis of copper phthalocyanine requires a mixture of phthalic anhydride, urea, and copper(II) chloride to be heated in the presence of a catalytic amount of ammonium molybdate in high-boiling solvents, such as nitrobenzene. Contrary to this, green synthesis of copper phthalocyanine can be done under solventless conditions using catalyst and MW energy (Scheme 5.10). High selectivity, waste minimization, low cost, and simplicity of process are some of the salient features of this green protocol.

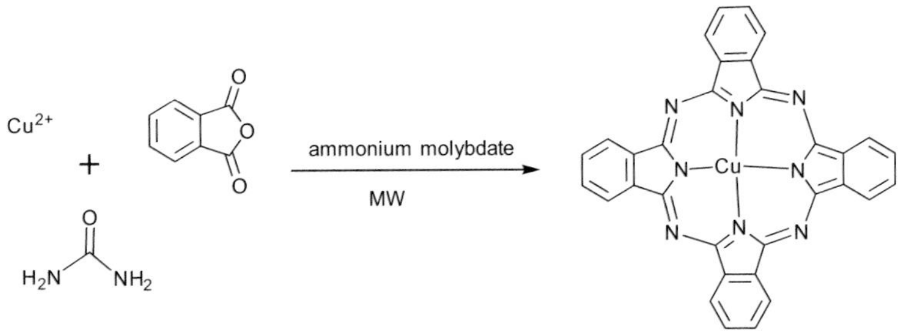

Scheme 5.10 Green synthesis of copper phthalocyanine.

5.2.3 Challenges Faced by Microwave Technology

- Difficulty in scaling up

  The maximum achievable yield using a domestic MW is limited to few grams only. Therefore, there is a large room for improvement in MW-assisted reactions to make this technology highly scalable.
- Restricted applicability

  The application of MW technology is only limited to the materials that absorb MW radiations. It cannot heat materials that are transparent to MW radiations.
- Safety and health hazards

  Usually the MW apparatus is considered to be safe equipment. However, sometimes, uncontrolled reactions may take place that result in undesirable explosive conditions.

  Deep penetration of high-frequency MWs causes various health hazards. Prolonged exposure to MWs might cause degeneration of body tissues, cells, and DNA strands.

5.3 Chemistry Using Ultrasonic Energy

Ultrasound corresponds to the sound waves with frequencies above 20 KHz, which is far beyond the detectable capability of the human ear. In the past few decades, ultrasound has received tremendous attention from the scientist community due to its widespread applications in many fields, such as in medical diagnosis, material testing, cleaning ultrasonic bath, and chemical reactions (Table 5.1) [7].

Table 5.1 Some useful applications of ultrasound

No.	Field	Applications
1.	Medicine	Treatment of muscle strains, ultrasonic imaging, etc.
2.	Biology	Power ultrasound is used in the destruction of cancerous tissues, enhanced chemotherapy, dissolution of blood clots, etc.
3.	Industry	Pigment and solid dispersal in paints, inks, and resins; ultrasonic drying, welding, and drilling; and so on
4.	Sonochemistry	Catalysis, electrochemistry, wastewater treatment, environmental protection, and waste control, etc.
5.	Cleaning	Cleaning of medical instruments, laboratory glassware, computer components, jewelry, etc.
6.	Geology	Eco/Pulse technique to locate oil and mineral deposits, SONAR, etc.

Sonochemistry particularly deals with the study of the effect of ultrasonic sound waves on chemical reactions. Nowadays, sonochemistry has gained acceptance in nearly all branches of chemistry. It is among the most inexpensive advance technologies that help conserve energy and minimize waste. One of the most exciting features of sonochemistry is that it does not require any distinct characteristic of the reaction system, like the presence of a conducting medium, dipole moment, or chromophore. The only requirement is the presence of liquid to generate a cavitation effect. Acoustic cavitation is a phenomenon through which power ultrasound influences the chemical reactivity [8, 9]. It involves the creation, growth, and violent collapse of gas bubbles in a liquid. This phenomenon was introduced by Sir John Thornycroft and Sidney Barby in 1895.

> Several animals can produce ultrasonic emissions. These animals emit sound (calls) into the local surroundings. Due to the presence of various nearby objects, some echoes are produced. These animals capture the returned echoes and use them to locate and identify the objects, which further helps them in navigation and in hunting. This is called "echolocation." Echolocating animals include some mammals and a few birds, for example, oil birds, bats, toothed whales, and dolphins.

5.3.1 How Sonochemistry Works

Sound waves propagate through a fluid via a series of altering compression and rarefaction cycles (Fig. 5.3). At high power, the negative pressure developed during the rarefaction cycle may become strong enough to exceed the attractive binding forces of the molecules of the fluid and tear them apart, forming cavitation bubbles. These microbubbles however collapse in the succeeding compression cycle, generating a huge amount of energy for chemical and mechanical effects. This collapse generates very high local temperatures and pressures, with estimated values of 5000 K and 1000 atmosphere. respectively. So, sonication of a fluid generates high-energy "hot spots" throughout the system [10]. To provide enough time for a bubble collapse, a wave of a suitable frequency and power must be supplied [8, 9, 11]. In medical applications, the compression cycle follows the rarefaction cycle too quickly to allow bubble collapse. Hence, no hot spots are formed, making ultrasounds quite safe to use on the human body.

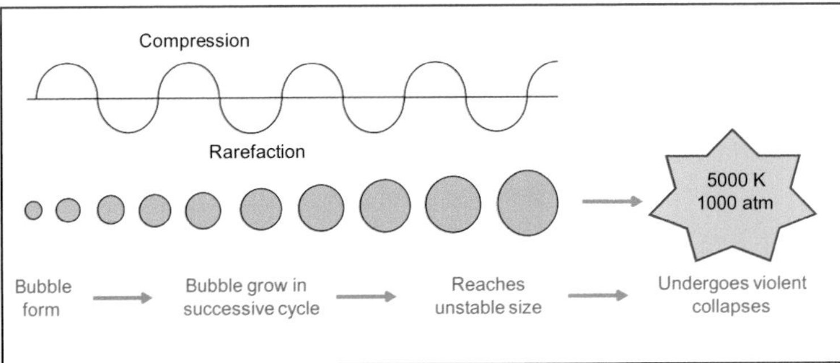

Figure 5.3 Impulsive collapse of bubbles.

The effects of bubble collapsing are felt in two discrete regions: within the bubble and in the immediate vicinity of the bubble. The former results in radical production (homogeneous sonochemistry), while in latter case, a shock wave is produced in the collapse as a result of the release of pressure (heterogeneous sonochemistry) (Fig. 5.4) [8].

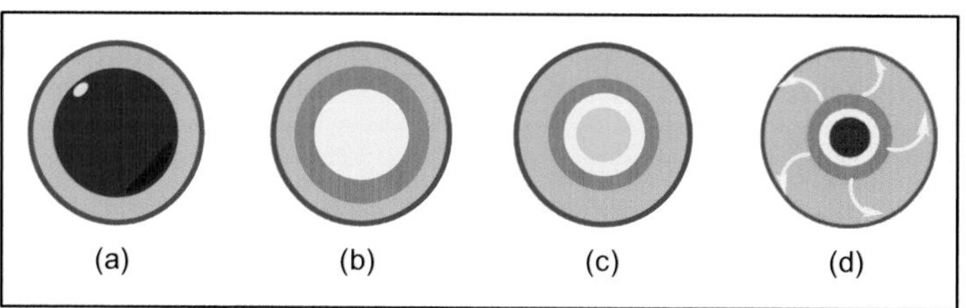

Figure 5.4 Steps during bubble cavitation: (a) collapsing of vapor bubbles, (b, c) intrinsic shock wave, and (d) increase of temperature and light emission.

5.3.2 Factors Affecting the Cavitation Effect

The cavitation effect is dependent on various factors such as the frequency of the ultrasound, solvent, temperature, and pressure [12].

- Ultrasonic frequency
  The sonochemical effect is usually restricted at higher frequencies. At high frequencies, bubbles have less time to grow and collapse. Most sonochemical works are done in the frequency range of 20–50 KHz.
- Effect of temperature
  The cavitation effect is best observed at lower temperatures. This is attributed to the fact that high temperature alters the solvent properties like density, viscosity, and surface tension.
- Effect of external pressure
  Sonic vibrations create an acoustic pressure (P_a) in the fluid system, which is given by Eq. 5.1:

$$P_a = P_A \sin 2\pi f t, \tag{5.1}$$

  where P_A is the hydrostatic pressure, f is the frequency of the sound wave, and t is time. If the external pressure is increased, the acoustic pressure will also increase. This means that the system requires a higher ultrasonic intensity to generate cavitation bubbles.

- Presence of gas

  Presence of dissolved gases facilitate the cavitation effect. This is because dissolved gases serve as the site for nucleation. When these gases are removed from the fluid due to the collapse of cavitation bubbles, instigation of a new cavitation affair becomes relatively difficult. In this situation, purging of gases will facilitate the formation of bubbles.

Success in sonochemistry relies both on the quality of the equipment and expertise with its use to benefit from the full potential offered by ultrasound in the field of chemistry.

—Jean-Louis Luche

5.3.3 Sonochemistry for Efficient Organic Synthesis

As already stated, most physical, chemical, and mechanical effects of ultrasonic frequency arise from cavitation. Ultrasound does not alter the rotational or vibrational states of molecules. The kinetic energy released by cavitational collapse is the only driving force that enables chemical transformations. Sonochemistry is a cleaner and ecofriendly alternative for conducting chemical reactions with improved selectivity and product yield. It is widely used for reducing long reaction times, avoiding high temperatures, and improving the compatibility of functional groups [3].

Some interesting examples of ultrasound-mediated chemical transformation are given next.

- Oxidation reaction

  Potassium permanganate–catalyzed oxidation of alcohols in hexane or benzene is enhanced significantly by ultrasonic irradiation (Scheme 5.11). By mechanical stirring, only 2% yield of the oxidized product was obtained, which upon ultrasonic irradiation improved to 92.8%.

OH → ($KMnO_4$; hexane/benzene, rt, ultrasound) → O

Scheme 5.11 Oxidation of alcohols by ultrasonic irradiation.

- Addition reaction

  The Diels–Alder reaction is accelerated by sonication (Scheme 5.12). It has been proven that the yield and stereoselectivity of cycloaddition reactions are highly enhanced by ultrasonic irradiation in the presence of halogenated solvents. Ultrasound promotes the *in situ* generation of hydrogen halide, which acts as a catalyst.

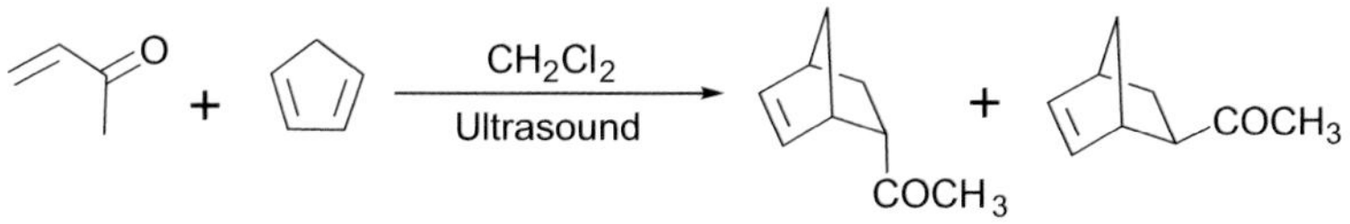

CH_2Cl_2 / Ultrasound; $COCH_3$

Scheme 5.12 Ultrasound-assisted Diels–Alder reaction.

- Substitution reaction

  The alumina-catalyzed reaction of benzyl bromide and potassium cyanide in the presence of toluene under sonication gives a 76% yield of the corresponding substitution product, benzyl cyanide (Scheme 5.13). However, in the absence of ultrasonic frequency, Friedel–Crafts alkylation is preferred over substitution. A different pathway under ultrasonic irradiation is observed since the ultrasonic frequency promotes the contact of cyanide with the surface of alumina, resulting in decreased catalytic ability of alumina toward alkylation. This enhances the possibility of a nucleophilic attack by the cyanide ion at the alumina surface.

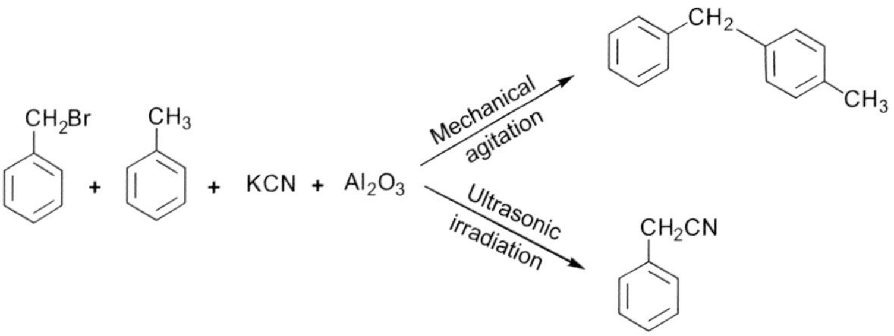

CH_2Br + CH_3 + KCN + Al_2O_3; Mechanical agitation → CH_2, CH_3; Ultrasonic irradiation → CH_2CN

Scheme 5.13 Substitution reactions in the presence of mechanical agitation versus ultrasonic irradiation.

- Simmons–Smith reaction

  The reaction of diiodomethane with zinc in the presence of an alkene to form cyclopropane is known as the Simmons–Smith reaction. The Zn/CH_2I_2 couple generates a carbene intermediate, which on addition with targeted alkene gives a high yield of the cyclopropane derivative (Scheme 5.14).

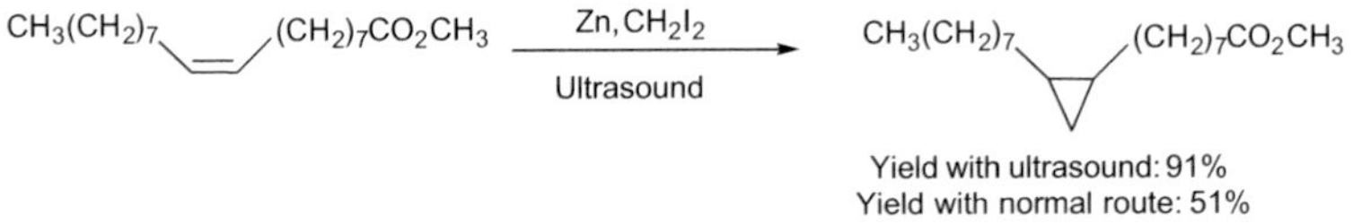

$CH_3(CH_2)_7CH{=}CH(CH_2)_7CO_2CH_3 \xrightarrow[\text{Ultrasound}]{\text{Zn, }CH_2I_2}$ cyclopropane derivative [$CH_3(CH_2)_7$ / $(CH_2)_7CO_2CH_3$]

Yield with ultrasound: 91%
Yield with normal route: 51%

Scheme 5.14 Ultrasound-assisted Simmons–Smith reaction.

- Coupling reaction

  Sonication is highly effective in facilitating the homocoupling of organometallic species formed during the reaction of alkyl, aryl, or vinyl halides with lithium wire in tetrahydrofuran (Scheme 5.15). Interestingly, no reaction is observed in the absence of ultrasound.

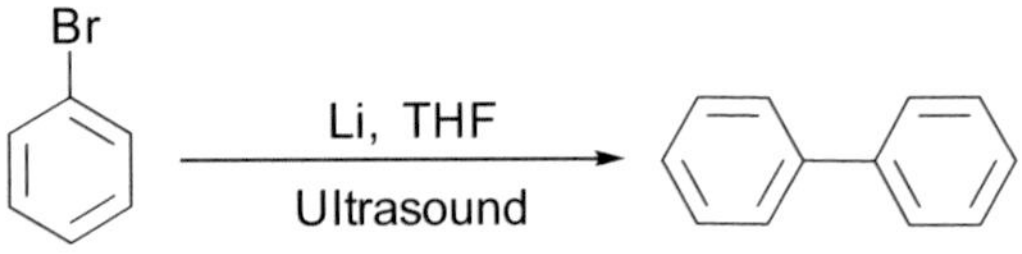

Bromobenzene (Br) $\xrightarrow[\text{Ultrasound}]{\text{Li, THF}}$ biphenyl

Scheme 5.15 Coupling reaction in the presence of ultrasound.

5.3.4 Applications in Wastewater Treatment

Ultrasonic irradiation can also be used in wastewater treatment and effectively destroys the contaminants present in the water. As explained earlier, ultrasonic irradiation leads to the conditions of local hot spots along with the formation of some oxidizing radical species. These two effects combine together to provide the essential elements for water decontamination. The primary radicals produced during the homogeneous sonolysis of water are $HO^\bullet$ and $H^\bullet$. The highly reactive $HO^\bullet$ radical can oxidize most chemical compounds present in the aqueous media, which is primarily accountable for the degradation of organic pollutants in water under ultrasonic irradiation. Therefore, the formation of $HO^•$ is an important aspect in wastewater treatment.

Further, the efficiency of sonochemical treatment strongly relies on the type and nature (hydrophobic or hydrophilic) of the pollutants. The extent of degradation is higher for hydrophobic compounds than for hydrophilic compounds. Hydrophobic compounds generally move toward the collapsing cavity and, hence, experience the maximum chemical and mechanical effects. On the other hand, hydrophilic compounds tend to gather in the liquid bulk, where they react with active radicals. Since the quantum of these radicals is low, hydrophilic compounds have a moderately lower degree of degradation.

5.3.5 Challenges Faced by Sonochemical Processes

Implementing the sonochemical technology on an industrial scale is the biggest challenge ahead of sonochemistry. Currently, there is a severe lack of accessibility to large-scale ultrasonic reactors. However, various industries, including the dairy industry and other food processing industries, are gradually developing methods to introduce sonochemistry into their processing plants.

5.4 Visible Light–Driven Processes: Photochemistry

Photochemistry is the branch of chemistry that deals with the interaction of electromagnetic radiations with matter and results in isomerization and chemical and physical changes. Such reactions are carried out by the absorption of visible (vis, 400–800 nm) or ultraviolet (UV, 200–400 nm) radiation in the electromagnetic spectrum. Various processes, such as photosynthesis and synthesis of vitamin D and carbohydrates, are photocatalyzed and render life in its present form. Inspired by Mother Nature, chemists have successfully utilized light for conducting organic transformations. The term "photocatalysis" is derived from Greek, where "photo" stands for light and "catalysis" means a process to enhance the rate of reaction by the addition of a substance without getting consumed itself during the course of reaction. Photocatalysis deals with the concurrent arrangement of photochemistry and catalysis to accelerate a chemical reaction. The advocates of photochemistry recognized its potential and applied it in various fields, such as solar energy, water purification, and photocatalysis [13].

5.4.1 Classification of Photocatalysts

Photocatalysts are classified into two categories on the basis of their physical state and nature in the reaction mixture.

- Homogeneous photocatalyst: Here, the catalyst and reactants exist in the same phase. Examples include various dyes, such as rose bengal and eosin Y; metal complexes such as $[Ru(bpy)_3]^{2+}$and $Ir(ppy)_3$; and polyacids such as $[Si(W_3O_{10})_4]^{4-}$.
- Heterogeneous photocatalyst: Here, the catalyst and reactants exist in different phases. Examples include metal oxide–based photocatalysts such as TiO_2 and ZnO, metal sulfides such as CdS, and nitrides such as Ta_3N_5 and C_3N_4 [13].

5.4.2 Basic Principle of Photochemistry

The basic principle used to exploit the photochemical technology is the absorption of light by the involved atom or molecule (Grotthuss–Draper law). The Stark–Einstein law suggests that a molecule absorbs a quantum/photon of light for activation [14]. Here, the quantum yield (Eq. 5.2) of the reaction should have a maximum value of 1. In photochemical reactions having quantum yields in thousands, the implication is that the photon initiates the chain reaction.

$$\text{Quantum yield} = \frac{\text{Number of reacted electrons or holes}}{\text{Number of photons absorbed by the photocatalyst}} \quad (5.2)$$

Unlike conventional heating, in homogeneous photocatalytic reactions, the absorption of light promotes an electron from the ground state, or the highest occupied molecular orbital (HOMO), of a molecule to its excited electronic state, or the lowest unoccupied molecular orbital (LUMO). All excitations of electrons from the ground state to the lowest excited electronic states strictly follow the selection rules. Here, the excited electron takes part in the reaction to generate the desired product [13]. In the case of a heterogeneous catalyst, an electron is excited from the valence band (HOMO) to the conduction band (LUMO) on absorption of light energy higher than the bandgap. Consequently, the generation of holes in the valence band takes place as a vacancy of the electrons. These excited states of both homogeneous and heterogeneous photocatalysts are responsible for the electron-initiated chemical transformations (Fig. 5.5) [15].

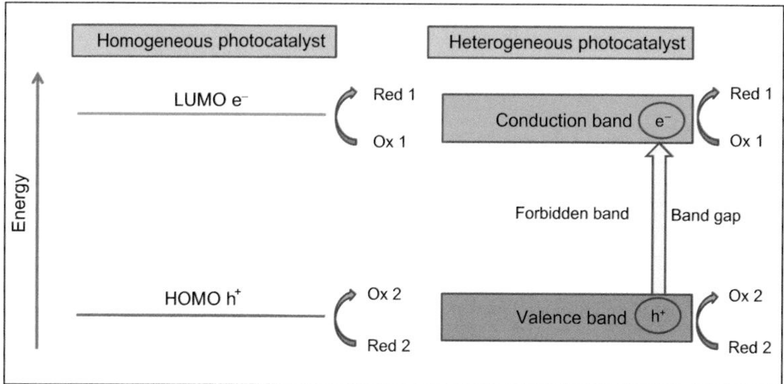

Figure 5.5 Photoexcitation and redox reactions of homogeneous and heterogeneous photocatalysts.

> In 1972, Fujishima and Honda discovered the phenomenon of photocatalytic water splitting under UV light. However, the idea of utilizing light was not accepted in that era due to the belief that water electrolysis is a high-voltage phenomenon and cannot take place at low voltages.

5.4.3 Photocatalytic Organic Transformations

In nature, plants utilize light energy for the photosynthesis of sugars by uptake of CO_2 and H_2O in the presence of chlorophyll. Researchers devoted their attention to the evolutionary photocatalytic science for performing various organic transformations in academics as well as at the industrial level. Some examples of photocatalyzed reactions are mentioned next.

5.4.3.1 Photochemical cycloaddition reactions

Photochemical cycloaddition of olefins is generally utilized for the synthesis of a four-membered ring product. One relevant example is the dimerization of cyclopentenone in dichloromethane to provide a mixture of dimmers upon irradiation with light (Scheme 5.16).

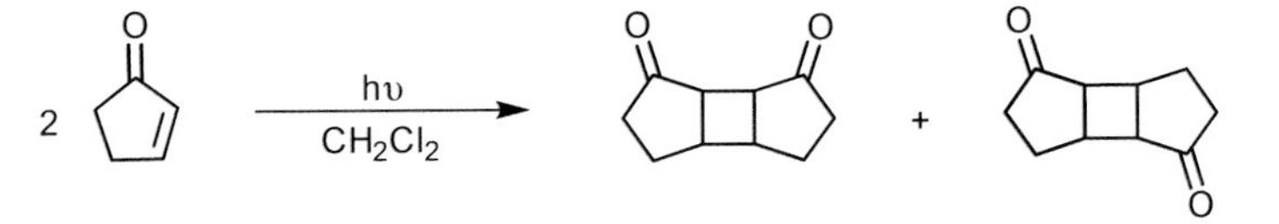

2 Cyclopentenone —hυ, CH_2Cl_2→ Head to head dimer + Head to tail dimer

Scheme 5.16 Photochemical cycloaddition of olefins.

Similarly, photochemical cycloaddition can also take place in an intramolecular fashion, as shown in Scheme 5.17.

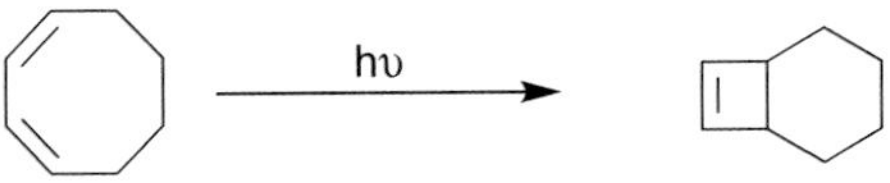

1,3-Cyclooctadiene —hυ→ Bicyclo[4.2.0]oct-7-ene

Scheme 5.17 Intramolecular photochemical cycloaddition reaction.

An interesting application of a photochemical [2+2] cycloaddition reaction is the nonstereospecific Paterno–Büchi reaction (Schemes 5.18 and 5.19). The reaction is named after Emanuele Paternò and George Büchi, who utilized the reaction for the photocatalytic formation of four-membered rings. Here, photocycloaddition of the triplet state of carbonyl compounds with the ground state of olefins leads to the formation of oxetane (oxacyclobutanes).

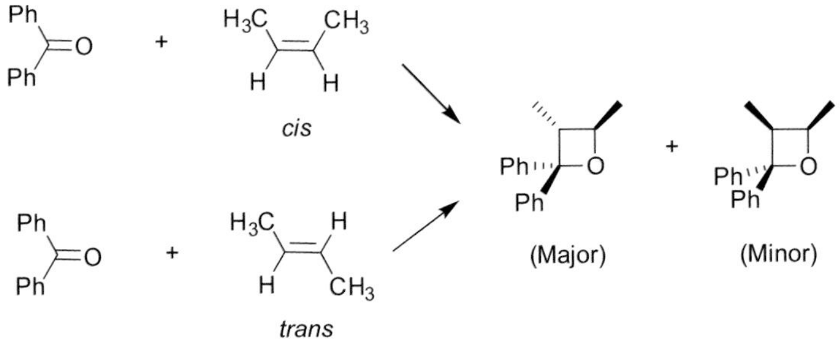

Ph, Ph, O + H_3C, CH_3, H, H (*cis*)

Ph, Ph, O + H_3C, H, H, CH_3 (*trans*)

→ Ph, Ph, O (Major) + Ph, Ph, O (Minor)

Scheme 5.18 Paterno–Büchi reaction.

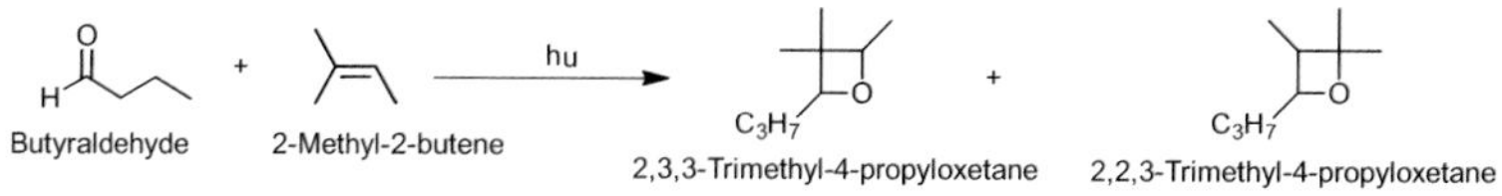

Butyraldehyde + 2-Methyl-2-butene —hu→ 2,3,3-Trimethyl-4-propyloxetane + 2,2,3-Trimethyl-4-propyloxetane

Scheme 5.19 Paterno–Büchi reaction.

5.4.3.2 Photoinduced isomerization

Alkenes are known to exhibit geometrical isomers. Some of the olefins can easily induce geometrical isomerization (*cis-trans*) upon irradiation with light or in the presence of sensitizer/photocatalyst. One example is *trans*-stilbene, which undergoes isomerization to form *cis*-stilbene upon irradiation with UV light (Scheme 5.20). After some time, equilibrium between *cis-trans* isomers is attained, which is known as the photostationary state. After this state is reached, there is no change in the ratio of *cis-trans* even on further exposure to light. The equilibrium favors the formation of the less stable *cis*-stilbene in the ratio of 10:1 (*cis*:*trans*).

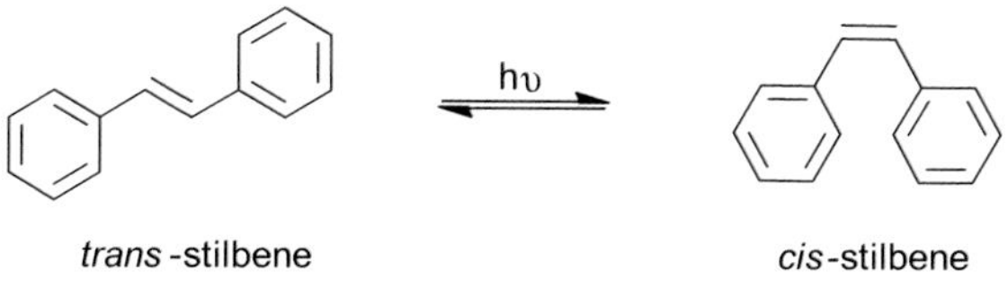

trans-stilbene ⇌(hυ) *cis*-stilbene

Scheme 5.20 Photoinduced isomerization reaction.

5.4.3.3 Photodimerization effect in water

As depicted in Scheme 5.21, photodimerization of stilbenes increased considerably in water in comparison to benzene. Further increase in yield on addition of LiCl is attributed to the increase in the hydrophobic effect.

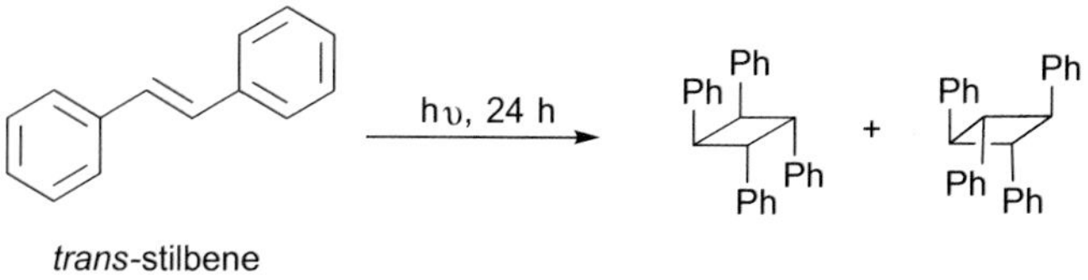

trans-stilbene —hυ, 24 h→ (Ph-substituted cyclobutane) + (Ph-substituted cyclobutane)

Benzene	0%	0%
Water	12%	10%
Water + LiCl	25%	17%

Scheme 5.21 Photodimerization effect.

Similarly, alkyl cinnamate also shows intermolecular dimerization upon irradiation with light (Scheme 5.22).

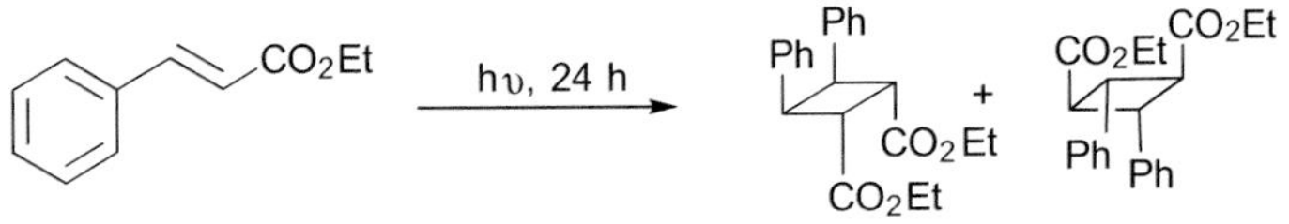

Scheme 5.22 Intermolecular dimerization.

5.4.4 Industrial Applications of Photochemistry

The constant endeavor of pharmaceutical industries is to find new and advanced methods for conducting organic transformations. Though newer methodologies are being discovered frequently, these are often costly and have limitations. Search for readily available and potential methods diverted the attention of industries toward photochemistry. Photochemistry has been utilized in numerous reactions, for example, free radical chlorination, sulfochlorination, sulfoxidation, nitrosation, oxygenation, and Barton reaction; the synthesis of various vitamins, such as vitamin D_3 and hydroxyl derivatives of vitamin D_3; and photoisomerization of vitamin A acetate.

5.4.4.1 Photonitrosation

A Japanese chemical company, Toray, synthesized caprolactum on a large scale by the photochemical method using nitrosyl chloride and cyclohexane as starting materials (Scheme 5.23).

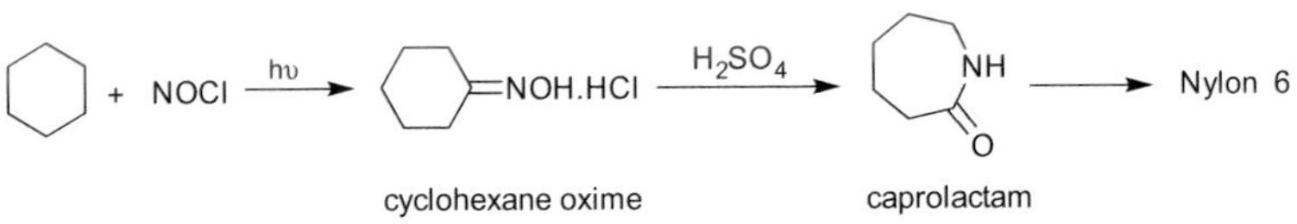

Scheme 5.23 Photonitrosation reaction.

5.4.4.2 Photo-oxygenation

One of the most important examples of photo-oxygenation in industrial chemistry is the synthesis of fragrance by using singlet oxygen as the reactive species. Here, the singlet oxygen is generated by photoexcitation of the ground state of molecular oxygen in the presence of rose bengal (photocatalyst). Then, the reaction of citronellol with hydroperoxide gives a mixture of two isomeric hydroperoxides, which on reduction with sulfite provide the corresponding alcohols. The major product undergoes allylic rearrangement in acidic condition, followed by cyclization, to give rose oxide (Scheme 5.24).

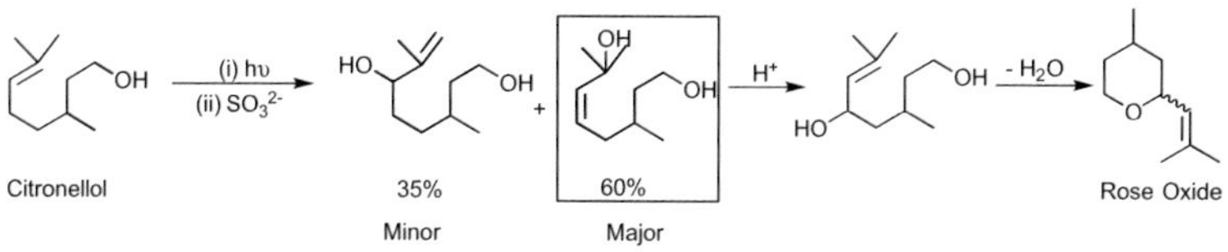

Scheme 5.24 Photo-oxygenation reaction.

5.4.5 Advantages of Photochemistry

Photochemistry has several advantages over the conventional methods. For example:

- Photochemistry is ecofriendly. Photons are clean and sustainable reagents that do not generate any residue.
- It shows high activity. Owing to a more directed energy process, photochemical protocols require mild reaction conditions and offer high activity and selectivity by reducing the possibility of side-product formation.
- It allows targeted selectivity. Some reaction pathways can be easily achieved by photochemical processes and lead to the targeted product that wouldn't have been possible by other routes.
- It can be successfully applied in industrial applications, for example, in pharmaceutical industries for developing various synthetic moieties. This protocol requires less raw material as compared to conventional ones, leading to significant cost reduction, for example, in photo-initiated halogenation reactions [3].

5.4.6 Photocatalytic Degradation of Organic Pollutants

In the 21st century, the rapid growth of technologies has led to a wide spectrum of challenges associated with the contamination of drinking water due to the presence of various harmful and waste materials. A photocatalyst based on the advanced oxidative process (AOP) is seen as a potential candidate for wastewater treatment. It proves its supremacy over other methods (such as coagulation, flocculation, sedimentation, filtration, and disinfection) in terms of successful dye degradation due to its low cost and sustainable nature. AOPs have the tendency to eradicate the potential contaminations (such as sludge and organic content) and reduce color and odor by redox reactions in water [16]. In the past few decades, various photocatalysts, such as organometallic polypyridyl metal complexes (e.g., $[Ru(bpy)_3]Cl_2{\cdot}6H_2O$), organic dyes (e.g., eosin Y and rose bengal), inorganic semiconductors consisting of metal oxides or sulfides (e.g., TiO_2, ZnO, $PbBiO_2Br$, CeO_2, and CdS), graphitic carbon nitride ($g\text{-}C_3N_4$) polymers, photoactive metal-organic frameworks, and covalent organic frameworks, have been developed and studied (Fig. 5.6) for photocatalytic degradation of various dyes [17].

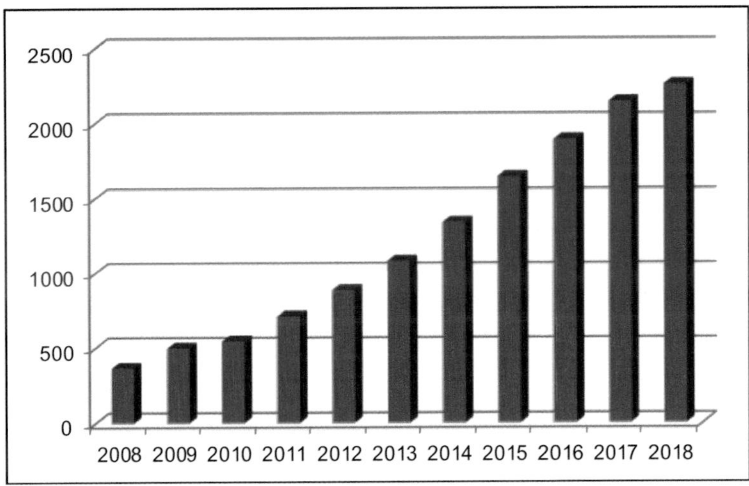

Figure 5.6 Number of publications falling under the category of photocatalytic degradation of dyes (*Source*: Web of Science).

5.4.6.1 Mechanism of photocatalytic degradation of organic pollutants

The literature precedent suggests that photocatalytic degradation involves a series of oxidation and reduction (redox) reactions upon contact with organic pollutants in the presence of UV-vis light that degrade the pollutant molecules present in water and air (Fig. 5.7) [3]. The photocatalytic redox mechanism for dye degradation involves the following fundamental steps:

1. Photoexcitation: Photocatalytic reactions are primarily initiated by photon energy, which promotes electrons from a filled valence band to the empty conduction band of the semiconductor, provided that the energy of the absorbed photon is higher than or equivalent to the bandgap energy of the semiconductor. Here, the excitation process will generate holes in the valence band (h^+(VB)) and electrons in the conduction band.
2. Oxidation process: The photogenerated h^+(VB) react to oxidize donor molecules. Afterward, the adsorbed water on the surface of the photocatalyst is oxidized by the positive holes present in the valence band to produce $OH^\bullet$ radicals (Eq. 5.3).

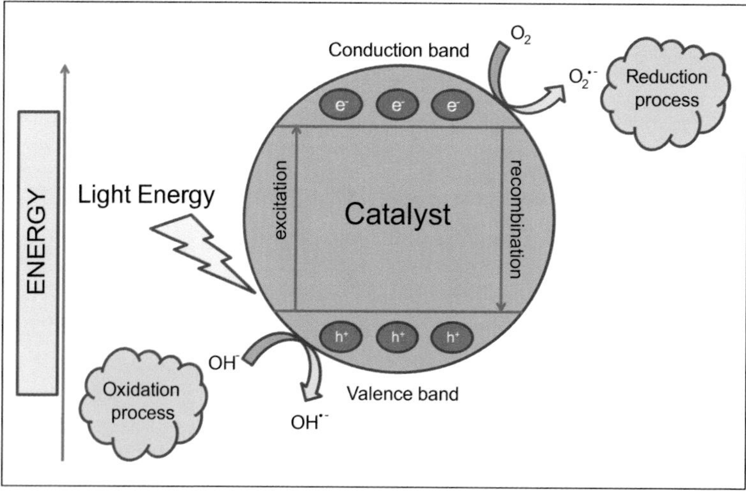

Figure 5.7 Schematic representation of the semiconductor-based photocatalytic mechanism.

$$H_2O(ads) + h^+(VB) \longrightarrow OH^\bullet\,(ads) + H^+(ads) \quad (5.3)$$

These $OH^\bullet$ radicals are extremely strong oxidative decomposing agents and further mineralize the organic pollutants.

3. Reduction process: The electron in the conduction band (e^- (CB)) reacts with the oxygen to generate anionic superoxide radicals ($O_2^{\bullet-}$), as shown in Eq. 5.4.

$$O_2 + e^-(CB) \longrightarrow O_2^{\bullet-}(ads) \quad (5.4)$$

These superoxide radicals not only take part in the oxidation process but also forbid the recombination of holes and electrons within the photocatalytic semiconductor.

4. Protonation of superoxide: Afterward, superoxide radicals ($O_2^{\bullet-}$) get protonated to form a hydroperoxyl radical ($HO_2^\bullet$) and successively form H_2O_2, which further splits into hydroxyl radicals ($HO^\bullet$) and initiates the process of degradation (Eqs. 5.5–5.10) [18].

$$O_2^{\bullet-}(ads) + H^+ \rightleftharpoons HOO^\bullet(ads) \quad (5.5)$$

$$2HOO^\bullet(ads) \longrightarrow H_2O_2(ads) + O_2 \quad (5.6)$$

$$H_2O_2(ads) \longrightarrow 2\,HO^\bullet(ads) \quad (5.7)$$

$$\text{Dye} + HO^\bullet \longrightarrow CO_2 + H_2O\ (\text{dye intermediates}) \quad (5.8)$$

$$\text{Dye} + h^+(VB) \longrightarrow \text{Oxidation products} \quad (5.9)$$

$$\text{Dye} + e^-(CB) \longrightarrow \text{Reduction products} \quad (5.10)$$

5.4.7 Factors Affecting Photocatalytic Degradation

5.4.7.1 Effect of the concentration of organic pollutants

The amount of organic pollutant and its type play a significant role and also influence the degradation rate. On increasing the concentration of the organic pollutant, there is a decrease in the catalytic rate because the active sites get covered, which leads to a decrease in the generation of $OH^\bullet$ radicals on the surface of the catalyst. Moreover, a high concentration of organic pollutants causes a decrease in the path length of the absorbed photons and, therefore, a net decrease in the rate constant.

5.4.7.2 Effect of the catalyst amount

The amount of catalyst used is a noteworthy parameter in the photocatalytic purification of water. In a system, an increased amount of catalyst boosts the number of active sites, which further enhance the generation of $OH^{\bullet}$ and $O_2^{\bullet -}$ radicals for effective photocatalytic degradation. Moreover, the degradation rate is directly proportional to the amount of catalyst (number of active site present). However, increasing the amount of catalyst beyond the optimized value leads to unfavorable results due to the low penetration of light.

5.4.7.3 Effect of pH

The pH of the reaction mixture makes a crucial contribution in photocatalytic dye degradation as it influences the surface charge property of the photocatalyst for the successful adsorption of the dye over the surface and manifests the degradation of organic contamination.

5.4.7.4 Size, structure, and surface area of the photocatalyst

The structure of the catalyst also influences the photocatalytic activity for dye degradation. Modification of the structure of the catalyst induces specific changes in their characteristic properties, such as stability, position of the conduction band, the degree of hydroxylation, and adsorption. Moreover, morphology also has key importance in photocatalytic activity. For example, nanosized materials show more catalytic efficiency due to a large surface area and a small size in comparison to the bulk materials. The smaller size leads to a high surface-to-volume ratio, which boosts the number of active sites and rate of interfacial charge transfer for achieving higher catalytic activities.

5.4.7.5 Effect of the reaction temperature

Various studies suggest that photocatalytic degradation is dependent on temperature. Increasing the temperature to a value above 80°C results in the recombination of the hole-electron. Consequently, desorption of the dye takes place, which decreases the activity of photocatalytic degradation.

5.4.7.6 Effect of the light intensity and wavelength of irradiation

The light intensity has a considerable effect on the degradation rate of the photocatalytic reaction. Better results can be achieved with a UV-vis source rather than solar radiations as reproducibility of the results is not assured. Moreover, the performance of the photocatalyst for dye degradation is influenced by the light intensity and is categorized into three ranges:

- Low-intensity range (0–20 mW/cm^2): The rate of degradation is directly proportional to the intensity of the light.
- Intermediate-intensity range (~25 mW/cm^2): The rate of degradation is directly proportional to the square root of the intensity of the light.
- High-intensity range: The rate is independent of the intensity of the light because a higher radiation facilitates the formation of more electron–hole combinations, causing an unpredictable rate of reaction. However, the rate of reaction is dependent on the wavelength of the light, which alters the electron–hole formation and the recombination rate.

5.4.8 Challenges Faced by Photochemical Synthesis

Photochemistry has been utilized in academic research for decades, although its uptake by industries is still limited due to some significant unresolved and inherent problems. There are some major challenges associated with photochemistry that limit its wide utilization for cleaner chemical manufacturing. These are enlisted here:

- In photoreactors, prevention of fouling on the photochemical window or the reactor wall is a tough challenge because even slight fouling can prevent the passage of light, which may hinder catalytic activity.
- The initiation of electronic transition from the ground state to the excited state requires radiation of a particular wavelength, especially monochromatic. And most light sources are polychromatic and emit a wide range of wavelengths and hence cannot be utilized for excitation. This decreases the energy efficiency and increases the costs.

- Some light sources generate heat and require installation of cooling devices to resolve the heating problem, which further increase the overall cost.
- The thick wall quartzes (as in mercury lamps) used in photoreactors are often expensive and delicate and make the set-up cost higher in comparison to that of the thermal process.
- The reaction setup should be in close proximity of the light source for effective reaction since the power of the transmitted light is inversely proportional to the square of the distance.
- Utilization of UV rays as a radiation source leads to various health hazards as constant exposure to UV rays leads to skin cancer, a weak immunity system, aging, damage to eyes, etc. [4].

5.5 Electrochemistry for Clean Synthesis

Electrochemistry is a branch of physical chemistry that allows the transfer of electrons/charges between electrodes by using an electrolyte as the medium for chemical changes. It involves the relationship between electricity and chemistry and is widely utilized in industries. Typically, the applied voltage between the electrodes is one of the driving factors that affect the rate of reaction. Most commonly, organic electrochemistry involves the oxidation and reduction of organic molecules at the electrodes, conserving both ions and electrons across the whole cell.

In 1834, Michael Faraday carried out an electrochemical reaction and observed the formation of hydrocarbons, but he missed the opportunity as he did not identify the product or state Faraday's two laws of electrolysis. Later, Kolbe successfully used an electrochemical reaction for the synthesis of decarboxylated coupling products by the oxidation of carboxylic acids at the platinum electrode (Eq. 5.11) [19–22].

$$2RCOO^- \longrightarrow R\text{–}R + 2CO_2 + 2e^- \quad (5.11)$$

> In 1791, the Italian physicist Luigi Galvani checked the profound effect of electricity on muscular motion (in frog leg) and marked the birth of electrochemistry by establishing a bridge between chemical reaction and electricity.

Various other studies of electrochemistry are reported in Table 5.2 [21].

Table 5.2 Historical progress of organic electrochemistry

Discovery	Explorers
Electro-organic synthetic transformation	Faraday and Kolbe
Organic electrode processes	Tafel and Haber
Polarography at dropping mercury research	Heyrovsky
Polarographic research	Lingane, Kolthoff, and Laitenen
Potentiostat marketing computer era	Hickling

5.5.1 Basis of Electrochemical Synthesis

Usually, electrocatalysis involves the use of a heterogeneous catalyst for initiating any reaction on the surface of the electrode by adsorption or desorption processes. A good electrocatalyst exhibits high catalytic activity by showing moderate interaction between the reactants and catalyst (Sabatier principle) [23]. A weak adsorption leads to weak binding of the reactants to the catalyst, and hence no reaction will occur. However, strong interaction between the catalyst and reactants makes the desorption process tough, which decreases the catalytic activity. The electrochemical pathway of organic reactions is dependent on the experimental conditions. The basic step of electrocatalysis is the conversion of substrates (Sub) into reactive intermediates and subsequently into the targeted product. Initially, an electron is removed from the cathodic electrode's surface (S_{ox}) and consequently, transmitted to the LUMO of the organic compound to generate a reduced intermediate species. In a similar manner, an electron is removed from the HOMO of the molecule as a nucleophile by the anodic electrode (Fig. 5.8) [24]. The resultant unstable intermediate generated during both the processes undergoes a reaction according to the following electrolytic conditions:

- Nucleophilic substitution occurs when the leaving group is replaced by a nucleophile via electrode oxidation.

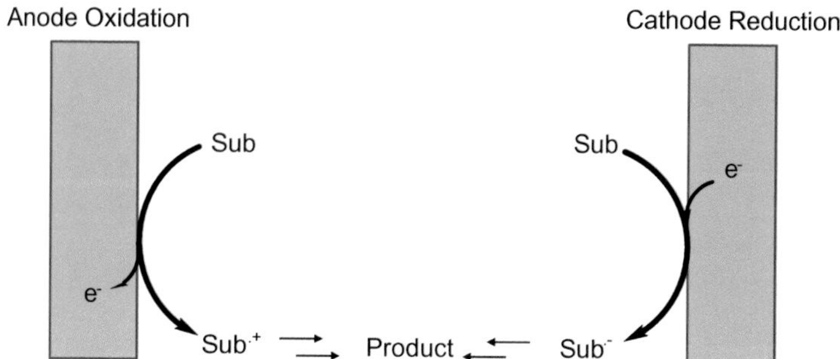

Figure 5.8 Schematic representation of the mechanism of an electrochemical cell.

- Similarly, electrophilic substitution occurs on the replacement of the leaving group by an electrophile at the cathodic electrode.
- Dimerization occurs in the presence of excess of radicals in the electrolytic solution.

Additionally, electroauxiliaries (nonsolvent nucleophiles) are utilized to promote an electron to achieve high selectivity with low energy consumption [25].

5.5.2 Types of Organic Electrochemical Synthesis

5.5.2.1 Anodic oxidative processes

In this process, a compound gets oxidized *via* abstraction of an electron from it by the anode. Here, the process is dependent upon various factors, such as the solvent, the pH of the reaction medium, and the oxidation potential.

5.5.2.2 Cathodic reductive processes

Reduction of organic compounds is achieved *via* the electron provided by the cathode. Generally, acidic conditions are most favorable to enhance the rate of reduction [26].

5.5.2.3 Paired organic electrosynthesis

Here, reactions on both cathode and anode electrodes show a cooperative effect and lead to the formation of the targeted product.

5.5.3 Examples of Electrochemical Synthesis

Nowadays, electrocatalysis has applications in several branches of science, including various fields, such as organic electrosynthesis, material chemistry, catalysis, biochemistry, medicinal chemistry, and environmental chemistry. The diverse commercial applications of electrochemical technology are known at the academic and industrial levels, such as in recycling, water purification and effluent treatment, corrosion prevention electroplating, electrochemical synthesis, analysis, sensors and monitors, and energy storage and generation. Organic electronics applications include transistors, photodiodes, solar cells, light-emitting diodes, and lasers and can be applied on various surfaces, such as papers, plastic foils, and metals. The potential synthetic utility of electrochemistry in chemical reactions is discussed next.

5.5.3.1 Anodic oxidations

The anodic oxidation reaction of neutral substrates generates a reactive radical intermediate that can further undergo various types of reactions, such as elimination, fragmentation, trapping reaction, and electron transfer reaction, to form a second reactive intermediate [25].

Electroauxiliaries have also been used for the synthesis of alkylated amines. Initially, pyrrolidine-based carbamates were utilized to synthesize *N*-acyliminium ion's cationic pools in the absence of a nucleophile, which were further directed to build a small library of alkylated amines (Scheme 5.25) [26].

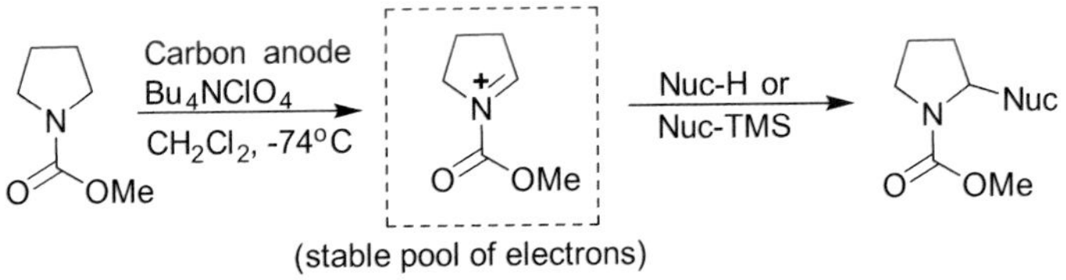

Carbon anode
Bu_4NClO_4
CH_2Cl_2, -74°C
N
O OMe
N
O OMe
Nuc-H or
Nuc-TMS
N Nuc
O OMe
(stable pool of electrons)

Scheme 5.25 The anodic oxidation reaction for the synthesis of alkylated amines.

In a similar manner, a stable pool of oxonium ions was synthesized by employment of electrochemical reactions and subsequent addition of nucleophiles to the carbon bearing the oxygen to obtain the targeted compound (Scheme 5.26).

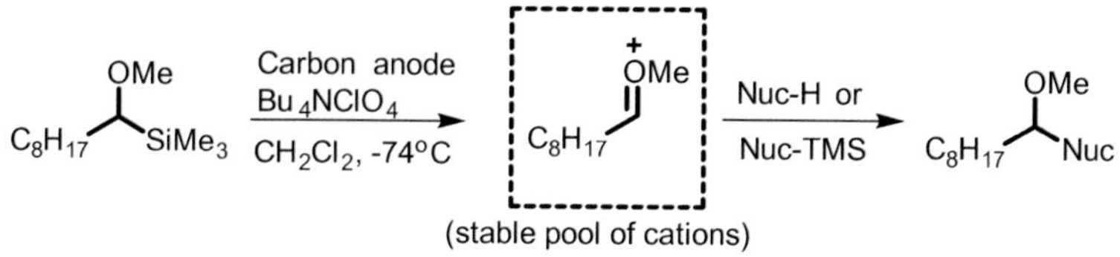

$C_8H_{17}CH(OMe)SiMe_3$ $\xrightarrow[CH_2Cl_2,\ -74^\circ C]{\text{Carbon anode},\ Bu_4NClO_4}$ $[C_8H_{17}CH{=}\overset{+}{O}Me]$ $\xrightarrow[\text{Nuc-TMS}]{\text{Nuc-H or}}$ $C_8H_{17}CH(OMe)Nuc$

(stable pool of cations)

Scheme 5.26 The anodic oxidation reaction for the synthesis of a stable pool of oxonium ions.

5.5.3.2 Cathodic reductions

Adiponitrile is an important moiety used for the synthesis of hexamethylenediamine and adipic acid (to a small extent), which are the core materials for the production of nylon-6,6. Over 1 million tons of adiponitrile per annum is commercially synthesized by the electroreductive coupling of acrylonitrile, followed by protonation, reduction, Michael addition, and a final protonation step (Scheme 5.27) [25].

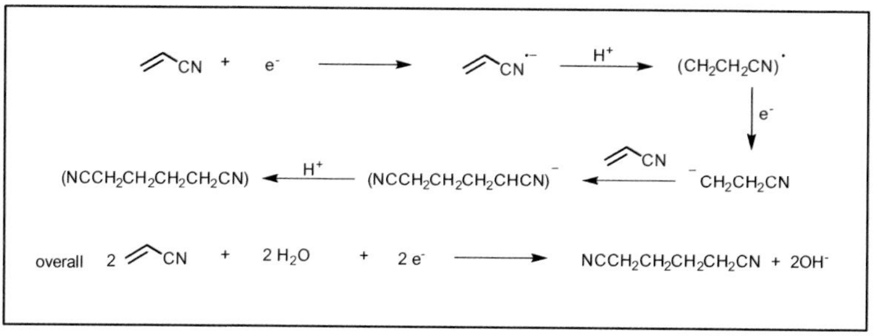

$CH_2{=}CHCN + e^- \longrightarrow CH_2{=}CHCN^{\bullet -} \xrightarrow{H^+} (CH_2CH_2CN)^{\bullet}$

$(CH_2CH_2CN)^{\bullet} \xrightarrow{e^-} {}^-CH_2CH_2CN$

${}^-CH_2CH_2CN \xrightarrow{CH_2=CHCN} (NCCH_2CH_2CH_2CHCN)^- \xrightarrow{H^+} (NCCH_2CH_2CH_2CH_2CN)$

overall $2\,CH_2{=}CHCN + 2\,H_2O + 2\,e^- \longrightarrow NCCH_2CH_2CH_2CH_2CN + 2OH^-$

Scheme 5.27 Electrochemical adiponitrile synthesis.

Electrochemical synthesis of 3-bromothiophene proves its potential over the conventional synthesis (Scheme 5.28). During the conventional synthesis method, excess of zinc and acetic acid are used as a reducing agent, which generates a lot of waste in the form of zinc bromide and waste acid, while in the case of electrocatalysis, there is no requirement of metal-based reducing agents and bromide is recovered in the form of bromine at the anode.

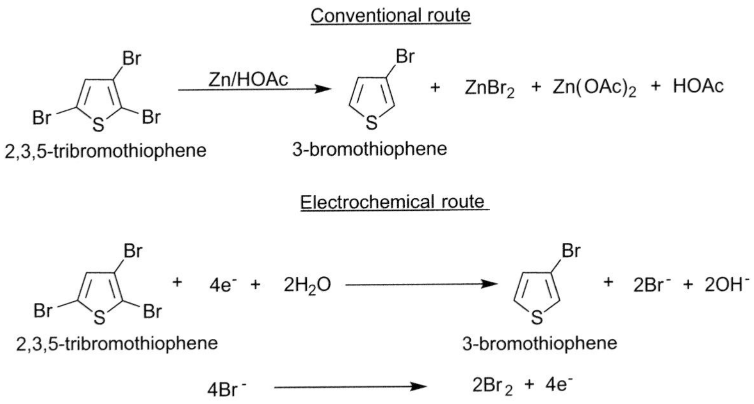

Conventional route

2,3,5-tribromothiophene $\xrightarrow{Zn/HOAc}$ 3-bromothiophene + $ZnBr_2$ + $Zn(OAc)_2$ + HOAc

Electrochemical route

2,3,5-tribromothiophene + $4e^-$ + $2H_2O$ ⟶ 3-bromothiophene + $2Br^-$ + $2OH^-$

$4Br^-$ ⟶ $2Br_2$ + $4e^-$

Scheme 5.28 Conventional and electrochemical synthesis of 3-bromothiophene.

In the electrocatalysis process, additional sodium bromide is used as the electrolyte and dioxane is employed as the cosolvent to increase the solubility of the starting material [3].

5.5.3.3 Paired organic electrosynthesis

Cyanoacetic acid and its derivatives are known as important synthons for their applications in pharmaceutical industries. Conventionally, cyanoacetic acid was synthesized by the reaction of chloroacetic acid and alkaline cyanides, which are toxic and environmentally hazardous substrates. Alternatively, it can be easily synthesized by the electrochemical process, as shown in Scheme 5.29 [27].

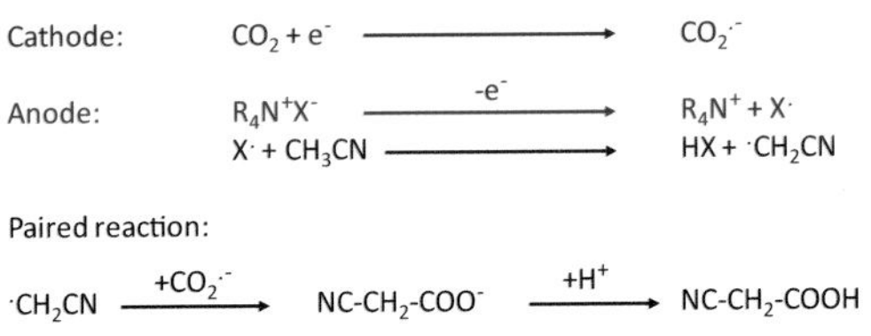

Cathode: $CO_2 + e^- \longrightarrow CO_2^{\cdot -}$

Anode: $R_4N^+X^- \xrightarrow{-e^-} R_4N^+ + X^{\cdot}$

$X^{\cdot} + CH_3CN \longrightarrow HX + {}^{\cdot}CH_2CN$

Paired reaction:

${}^{\cdot}CH_2CN \xrightarrow{+CO_2^{\cdot -}} NC\text{-}CH_2\text{-}COO^- \xrightarrow{+H^+} NC\text{-}CH_2\text{-}COOH$

Scheme 5.29 Paired organic electrosynthesis of cyanoacetic acid.

In the process shown in Scheme 5.29, cyanoacetic acid is formed by cathodic reduction of CO_2 and anodic oxidation of tetraalkylammonium salt anion in acetonitrile.

Another example of paired organic electrosynthesis is the formation of 2,6-dimethyl-4-arylpyridine-3,5-dicarbonitrile in a two-compartment cell using platinum electrodes and acetonitrile as the solvent (Scheme 5.30) [28].

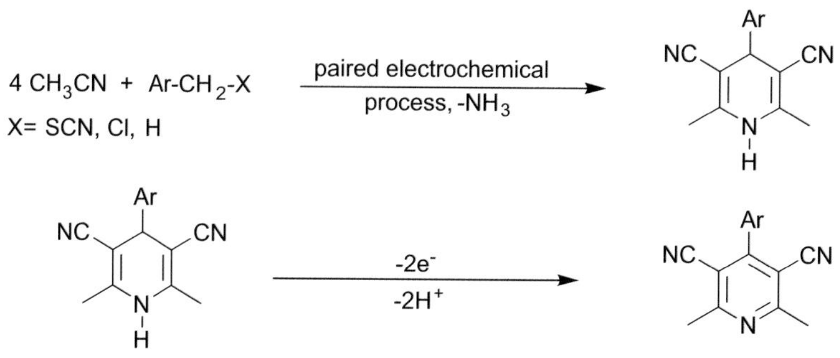

4 CH_3CN + Ar-CH_2-X —paired electrochemical process, -NH_3→ [dihydropyridine product]
X= SCN, Cl, H

[dihydropyridine] —$-2e^-$, $-2H^+$→ [pyridine product]

Scheme 5.30 Paired organic electrosynthesis of 2,6-dimethyl-4-arylpyridine-3,5-dicarbonitrile.

5.5.4 Advantages of Electrochemical Synthesis

The diverse advantages of the environmentally friendly electrochemistry can be summarized as follows [29]:

- Electrons are clean reagents.
- Milder reaction conditions are required as the reactions are generally operated at ambient temperature and pressure due to a tunable redox potential (voltage).
- It displays a high functional group tolerance, for example, for halogens, azo compounds, and salts.
- Stoichiometric reagents are replaced by an electric current.
- The energy consumption is low.
- Conductive nonvolatile reaction media (water, ionic liquid, supercritical fluids, etc.) are required, and the process prevents the release of toxic and expensive solvents into the atmosphere.
- The protocol for synthesis, separations, and pollution control is selective and convenient.
- The electrode works as a heterogeneous catalytic system.
- The process is cost effective due to simple equipment and operation.

- The process is atom efficient as the reagents are replaced by electrons.
- Waste is minimized.
- The process enables efficient recovery and recycling of metal ions.

5.5.5 Challenges Faced by Electrochemical Synthesis

The use of electrochemical protocols is limited in industries because of the mechanical processes involved, the engineering problems, and the setup cost. Hence, various factors need to be considered to carry out cost-effective electrochemical operations, such as current, voltage, choice of electrode, resistance to passivation and corrosion, and energy consumption. Moreover, the production of explosive gases, such as hydrogen and oxygen, needs to be minimized. Also, the incorporation of electrochemistry in the field of synthetic organic chemistry remains an unexplored area.

5.6 Future Outlook

"Today's youth is tomorrow's leader to live the dream of a green and sustainable environment." To achieve this dream, one should focus on reducing the consumption of energy.

The use of various emerging sources of energy is seen as an optimistic step in improving chemical synthesis. Especially in the case of bulk chemical synthesis, the development of a much more energy-efficient process is required to provide significant benefits and help abandon older energy-intensive processes. The future is likely to encompass and exploit a much wider range of applications of various energy-efficient processes, and more studies of a combination of various energy sources will be conducted.

5.7 Learning Outcomes

At the end of this chapter, students will be able to:

- Unlock the potential of various clean and efficient sources of energy for conducting chemical transformations in a much sustainable way

- Appreciate the beauty of alternative energy technologies in reducing energy wastage by specifically targeting the chemical bond/molecule undergoing reaction
- Optimize energy-efficient and atom-economic conditions
- Choose one or more green components (solvent, reagents, catalyst, etc.) for a successful reaction
- Develop more sustainable and green processes that help to reduce overall waste generation and process cost
- Adapt the emerging alternate sources of energy in laboratories as well as in manufacturing industries

5.8 Problems

1. What is the need to accelerate chemical reactions with heat? Briefly describe one green alternative source of energy.
2. Explain the basic principle behind microwave heating.
3. What makes microwave technology superior to conventional heating?
4. What do you mean by cavitation effect? Discuss various factors affecting cavitation effect.
5. Compare the products formed in the alumina-catalyzed reaction of benzyl bromide and potassium cyanide in the presence of toluene under conventional heating with ultrasonic irradiation.
6. Briefly discuss the mechanism of and factors responsible for photocatalytic dye degradation.
7. Discuss the basic principles of photo- and electrochemistry.
8. What is paired organic electrocatalysis? Give one example.
9. Highlight the challenges associated with the utilization of photocatalysis on an industrial scale.

References

1. Anastas, P. T. and Warner, J. C. (2000). *Green Chemistry: Theory and Practice* (Oxford University Press, Oxford).
2. McCollum, D., Gomez, E. L., Riahi, K. and Parkinson, S. (2017). *A Guide to SDG Interactions: From Science to Implementation*, eds. Griggs,

D. J., Nilsson, M., Stevance, A.-S. and McCollum, D., SDG7: ensure access to affordable, reliable, sustainable and modern energy for all (International Council for Science, Paris) pp. 127–173.

3. Lancaster, M. (2002). *Green Chemistry: An Introductory Text*, 3rd Ed. (The Royal Society of Chemistry, Cambridge).
4. Stankiewicz, A. and Stefanidis, G. (2016). *Alternative Energy Sources for Green Chemistry* (The Royal Society of Chemistry, Cambridge, UK).
5. Schmink, J. R. and Leadbeater, N. E. (2010). *Microwave Heating as a Tool for Sustainable Chemistry*, ed. Leadbeater, N. E., Chapter 1, Microwave heating as a tool for sustainable chemistry: an introduction (CRC Press, Boca Raton, FL).
6. Loupy, A. and Varma, R. S. (2006). Microwave effects in organic synthesis: mechanistic and reaction medium considerations. *Chim. Oggi*, **24**, pp. 36–39.
7. Mason, T. J. (1990). *Sonochemistry: The Uses of Ultrasound in Chemistry*, ed. Mason, T. J., Chapter 1, A general introduction to sonochemistry (The Royal Society of Chemistry, Cambridge, UK) pp 1–8.
8. Alarcon-Rojo, A. D., Carrillo-Lopez, L. M., Reyes-Villagrana, R., Huerta-Jiménez, M. and Garcia-Galicia, I. A. (2019). Ultrasound and meat quality: a review. *Ultrason. Sonochem.*, **55**, pp. 369–382.
9. Cintas, P. and Luche, J.-L. (1999). Green chemistry. The sonochemical approach. *Green Chem.*, **1**, pp. 115–125.
10. Ashokkumar, M. (2011). The characterization of acoustic cavitation bubbles: an overview. *Ultrason. Sonochem.*, **18**, pp. 864–872.
11. Ashokkumar, M. and Mason, T. J. (2000). Sonochemistry, *Kirk-Othmer Encylcopedia of Chemical Technology* (John Wiley and Sons, New York).
12. Luo, J., Fang, Z., Smith, R. L. and Qi, X. (2015). *Production of Biofuels and Chemicals with Ultrasound*, eds. Fang, Z., Smith, R. L. and Qi, X., Chapter 1, Fundamentals of acoustic cavitation in sonochemistry (Springer, Dordrecht) pp. 3–33.
13. Stephenson, C. R. J., Yoon, T. P. and MacMillan, D. W. (2018). *Visible Light Photocatalysis in Organic Chemistry* (Wiley-VCH, Weinheim).
14. Arumainayagam, C. R., Garrod, R. T., Boyer, M. C., Hay, A. K., Bao, S. T., Campbell, J. S., Wang, J., Nowak, C. M., Arumainayagam, M. R. and Hodge, P. J. (2019). Extraterrestrial prebiotic molecules: photochemistry vs. radiation chemistry of interstellar ices. *Chem. Soc. Rev.*, **48**, pp. 2293–2314.

15. Peter, L. M. (2016). *Photocatalysis: Fundamentals and Perspectives*, eds. Schneider, J., Bahnemann, D. Ye, J., Puma, G. L. and Dionysiou, D. D., Chapter 1, Photoelectrochemistry: from basic principles to photocatalysis (The Royal Society of Chemistry, UK) pp. 3–28.

16. Viswanathan, B. (2018). Photocatalytic degradation of dyes: an overview. *Curr. Catal.*, **7**, pp. 99–121.

17. Pichat, P. (2013). *Photocatalysis and Water Purification: From Fundamentals to Recent Applications* (Wiley-VCH, Weinheim).

18. Ajmal, A., Majeed, I., Malik, R. N., Idriss, H. and Nadeem, M. A. (2014). Principles and mechanisms of photocatalytic dye degradation on TiO_2 based photocatalysts: a comparative overview. *RSC Adv.*, **4**, pp. 37003–37026.

19. Grimshaw, J. (2000). *Electrochemical Reactions and Mechanisms in Organic Chemistry* (Elsevier, Amsterdam).

20. Holzhäuser, F. J., Creusen, G., Moos, G., Dahmen, M., König, A., Artz, J., Palkovits, S. and Palkovits, R. (2019). Electrochemical cross-coupling of biogenic di-acids for sustainable fuel production, *Green Chem.*, **21**, pp. 2334–2344.

21. Surucu, O. and Abaci, S. (2017). Organic electrochemistry: basics and applications. *Int. J. Biosen. Bioelectron.*, **3**, pp. 270–271.

22. Vijh, A. and Conway, B. (1967). Electrode kinetic aspects of the Kolbe reaction. *Chem. Rev.*, **67**, pp. 623–664.

23. Pletcher, D. (1999). *Guide to Electrochemical Technology for Synthesis, Separation, and Pollution Control* (Electrosynthesis Company Inc, Lancaster, NY).

24. Cardoso, D. S. P., Sljukić, B., Santos, D. M. F. and Sequeira, C. A. C. (2017). Organic electrosynthesis: from laboratorial practice to industrial applications. *Org. Process Res. Dev.*, **21**, pp. 1213–1226.

25. Little, R. D. and Moeller, K. D. (2002). Organic electrochemistry as a tool for synthesis. *Electrochem. Soc. Interface*, **11**, pp. 36–42.

26. Yoshida, J.-i., Suga, S., Suzuki, S., Kinomura, N., Yamamoto, A. and Fujiwara, K. (1999). Direct oxidative carbon-carbon bond formation using the "cation pool" method. 1. Generation of iminium cation pools and their reaction with carbon nucleophiles. *J. Am. Chem. Soc.*, **121**, pp. 9546–9549.

27. Batanero, B., Barba, F., Sánchez-Sánchez, C. M. and Aldaz, A. (2004). Paired electrosynthesis of cyanoacetic acid. *J. Org. Chem.*, **69**, pp. 2423–2426.

28. Batanero, B., Barba, F. and Martín, A. (2002). Preparation of 2,6-dimethyl-4-arylpyridine-3,5-dicarbonitrile: a paired electrosynthesis. *J. Org. Chem.*, **67**, pp. 2369–2371.

29. Savinell, R. F., Ota, K.-i. and Kreysa, G. (2014). *Encyclopedia of Applied Electrochemistry* (Springer, New York).

Chapter 6

Implementation of Green Chemistry: Real-World Case Studies

Sriparna Dutta,[a] **Manavi Yadav,**[a,b] **and Rakesh K. Sharma**[a]
[a]*Green Chemistry Network Centre, Department of Chemistry, University of Delhi, Delhi 110007, India*
[b]*Department of Chemistry, Hindu College, University of Delhi, Delhi 110007, India*
sriparna.duttagcncnew@gmail.com

The most sustainable way is to not make things. The second most sustainable way is to make something very useful, to solve a problem that hasn't been solved.

—Thomas Sigsgaard

6.1 Introduction

The concept of green chemistry presents an overarching approach toward the promotion of sustainable development and, thus, opens up scope for multifaceted research. The 12 principles provide basic guidelines that work as a framework of actions to develop

Green Chemistry for Beginners
Edited by Rakesh K. Sharma and Anju Srivastava
Copyright © 2021 Jenny Stanford Publishing Pte. Ltd.
ISBN 978-981-4316-96-5 (Hardcover), 978-1-003-18042-5 (eBook)
www.jennystanford.com

environmentally benign chemical products and processes. In recent years, a resurgence has been observed in the adoption of green chemistry principles to develop safer alternative pathways in chemical syntheses that can be applied on a commercial scale [1]. In this respect, the ideal synthesis is expected to have the following characteristics: it should employ inexpensive and easily available substrates, use energy efficiently with less number of steps, phase out toxic reagents, show high selectivity toward the desired product, use renewable sources instead of fossil fuels, and reduce the risks associated with the overall process. Furthermore, the final product must be easily managed throughout its life cycle. Following this agenda, industries continually work to improve products and processes in addition to their ecological performances [2].

Greening of industries is found to be imperative not only for sustainable growth but also for alleviating poverty. However, in doing so, pharmaceutical industries experience certain challenges in terms of redesigning the drug so that it is readily degradable as this might alter its overall function [3]. In spite of several problems, some well-known pharmaceutical companies have embraced green chemistry in practice for drug designing and manufacturing. Besides, there are several other companies and organizations that have incorporated this discipline, including Hewlett-Packard, Johnson & Johnson, Nike, Amgen, DuPont, Eastman Chemical, Codexis, BASF, Pfizer, and Bayer Material Science. Additionally, numerous efforts are made in both industrial and academic setup to implement greener alternatives [4].

In this chapter, a comparison between traditional routes and alternative greener routes for synthesizing industrially significant compounds has been made. This is followed by some real-world case studies where the entity won the Presidential Green Chemistry Challenge Award.

6.2 Synthesis of Valuable Compounds: Greener Protocols

The implementation of green chemistry principles in several industries has resulted in the discovery of numerous chemicals and blockbuster drugs that have reduced the environmental footprint.

Indeed, significant improvement has been realized in terms of reaction conditions and reduction in solvent/reagent use while manufacturing chemicals. Ranging from feedstocks to processing and product synthesis, several attempts have been made to produce chemicals that are benign to human health and environment [5]. This section explains various approaches adopted for green syntheses of chemicals with industrial significance.

6.2.1 Synthesis of Adipic Acid and Catechol

An enormous amount of adipic acid is required per year as feedstock for the synthesis of nylon, lubricants, and plasticizers. Conventionally, adipic acid was derived from benzene. Similarly, hydroquinone and catechol were also prepared from benzene. However, knowing the carcinogenic nature of benzene and the nonrenewable source from which benzene is obtained, the process was not considered to be environmentally friendly. Besides, N_2O was also obtained as one of the by-products during the course of reaction, which is responsible for the depletion of the ozone layer (Schemes 6.1 and 6.2).

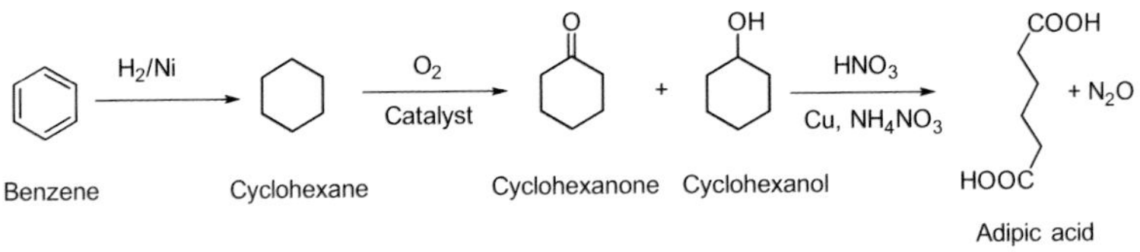

Benzene —(H_2/Ni)→ Cyclohexane —(O_2, Catalyst)→ Cyclohexanone + Cyclohexanol —(HNO_3, Cu, NH_4NO_3)→ Adipic acid (HOOC…COOH) + N_2O

Scheme 6.1 Synthesis of adipic acid from benzene.

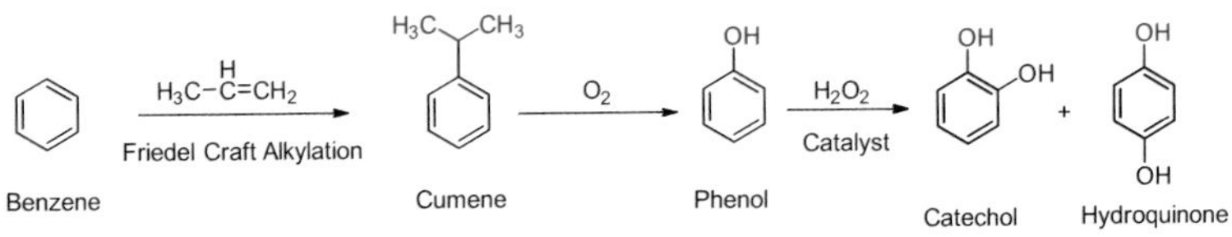

Benzene —($H_3C-CH=CH_2$, Friedel Craft Alkylation)→ Cumene —(O_2)→ Phenol —(H_2O_2, Catalyst)→ Catechol + Hydroquinone

Scheme 6.2 Synthesis of hydroquinone and catechol from benzene.

As an alternative route, Frost and Draths prepared adipic acid using glucose as a starting material and genetically engineered bacteria [6]. Usually, all bacteria use a natural metabolic route known as the shikimic acid pathway (responsible for the natural synthesis of many aromatic compounds) to convert glucose into important metabolites. Frost and Draths utilized the gene-splicing method to produce modified *Escherichia coli*, which could further produce *cis,cis*-muconic acid rather than undergoing the usual shikimic acid pathway. Then, this muconic acid was converted into adipic acid by normal catalytic hydrogenation (Scheme 6.3). With this manipulation of the shikimic acid pathway, Frost demonstrated the synthesis of several compounds with glucose as a starting material, including catechol and hydroquinone, all of which are essential chemicals required in larger amounts. Even though this method is still not adapted to an industrial scale, it promises to minimize the use of toxic reagents and solvents and, thus, presents a greener route. For this work, Frost and Draths earned the Presidential Green Chemistry Challenge Award in 1998.

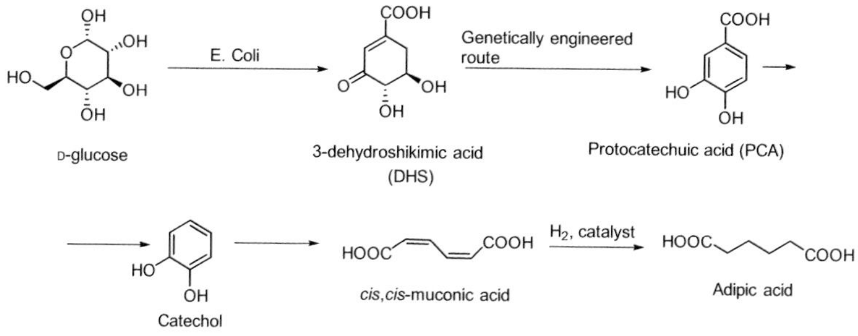

D-glucose → (E. Coli) → 3-dehydroshikimic acid (DHS) → (Genetically engineered route) → Protocatechuic acid (PCA) → Catechol → *cis,cis*-muconic acid → (H_2, catalyst) → Adipic acid

Scheme 6.3 Synthesis of adipic acid and catechol from glucose.

6.2.2 Synthesis of Styrene

Another commodity used in large quantities every year is polystyrene made from monomer styrene. Industrially, styrene is manufactured from benzene (carcinogen) via Friedel–Crafts alkylation, followed by dehydrogenation (Scheme 6.4). As a greener alternative, Chapman developed a method that employs a single step in the conversion of mixed xylenes (noncarcinogen) into styrene (reaction not available) [7].

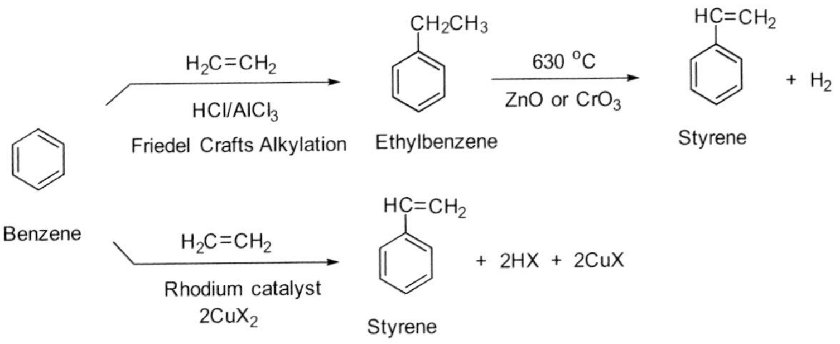

Scheme 6.4 Two routes for styrene synthesis.

6.2.3 Synthesis of Citral

In comparison to other types of reactions, oxidation reactions are performed with stoichiometric inorganic (or organic) oxidants like periodate, manganese dioxide, and potassium permanganate. Therefore, these reactions demand greener alternatives employing clean oxidants such as oxygen or hydrogen peroxide for synthesizing fine chemicals. Citral is one of the valuable chemicals used as an intermediate for producing fragrances and as building blocks for vitamins A and E. It is obtained from β-pinene employing chlorine as an oxidant. Moreover, this method is lengthy, involving five steps, and the yields are generally low (Scheme 6.5).

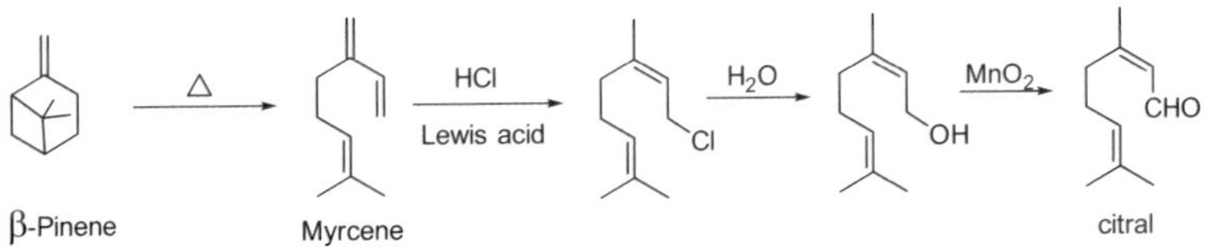

Scheme 6.5 Citral synthesis from β-pinene.

In a new route used by BASF, cheap starting materials were employed, such as formaldehyde and isobutene, to form 3-methylbut-3-en-1-ol (isoprenol). A part of this is oxidized to an aldehyde while another part is isomerized, and the obtained products are reacted together to form citral with a yield of 95%. This route is an atom-efficient and low-salt process (Scheme 6.6) [8].

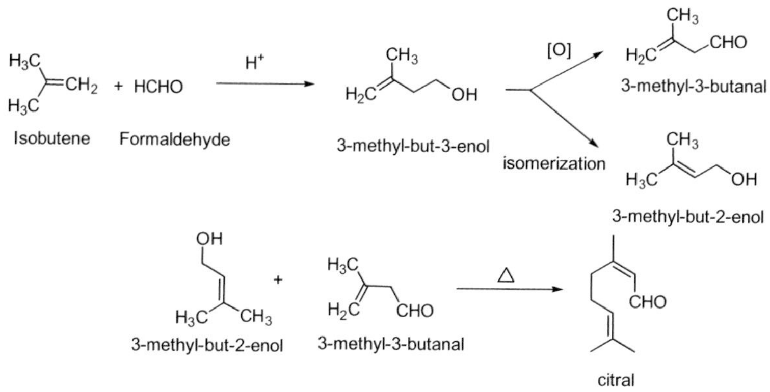

Scheme 6.6 Citral synthesis by a greener route.

6.2.4 Synthesis of Disodium Iminodiacetate

Disodium iminodiacetate (DSIDA), a key intermediate in the manufacture of Monsanto's Roundup(r) herbicide, was conventionally produced by the Strecker process, a well-known method employing ammonia, hydrogen cyanide, formaldehyde, and hydrochloric acid. However, this method had several drawbacks, such as handling difficulties for HCN, generation of highly unstable intermediates, and toxic nature of the overall process, with a lot of waste that required special treatment before disposal (Scheme 6.7).

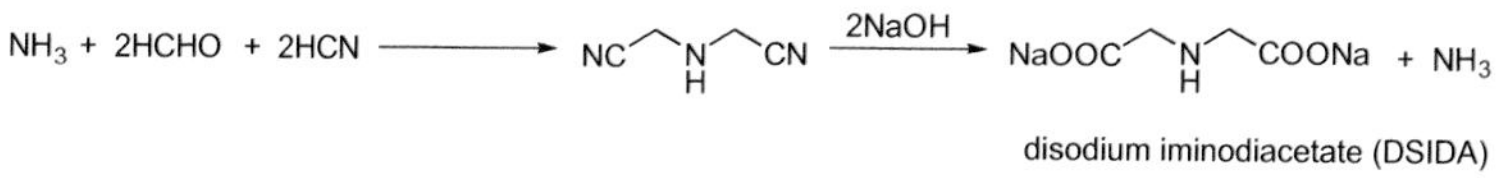

$$NH_3 + 2HCHO + 2HCN \longrightarrow NC\text{-}CH_2\text{-}NH\text{-}CH_2\text{-}CN \xrightarrow{2NaOH} NaOOC\text{-}CH_2\text{-}NH\text{-}CH_2\text{-}COONa + NH_3$$

disodium iminodiacetate (DSIDA)

Scheme 6.7 Strecker synthesis of DSIDA.

In view of this, Monsanto developed an alternate process employing a copper catalyst for the dehydrogenation of diethanolamine. This process is much safer to operate and does not pose any risk of a runaway reaction due to the endothermic dehydrogenation reaction. Besides, this method prevents the use of cyanide and formaldehyde, has a fewer number of steps, gives a high yield, and does not involve any purification difficulties. In fact, the catalyst can be readily filtered off from the product stream, and no waste is obtained (Scheme 6.8) [9].

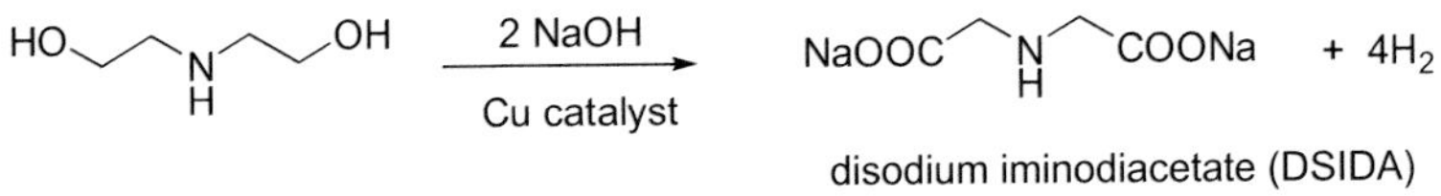

$$HOCH_2CH_2NHCH_2CH_2OH \xrightarrow[\text{Cu catalyst}]{\text{2 NaOH}} NaOOCCH_2NHCH_2COONa + 4H_2$$

disodium iminodiacetate (DSIDA)

Scheme 6.8 Alternate process using a copper catalyst for synthesizing DSIDA.

6.2.5 Synthesis of Acetaldehyde

Commercially, acetaldehyde was obtained by the catalytic oxidation of ethanol or by the hydration of acetylene through a Markovnikov-selective oxymercuration, followed by the oxidation of the resultant alcohol (Scheme 6.9). However, nowadays, the Wacker Chemie oxidation process is the most convenient method for synthesizing acetaldehyde. This process employs oxidation with dioxygen in the presence of a palladium catalyst (Scheme 6.10). Overall, the synthesis is carried out in a single step and requires only a catalytic amount of palladium instead of a stoichiometric amount of toxic mercury deployed in oxymercuration reaction [10].

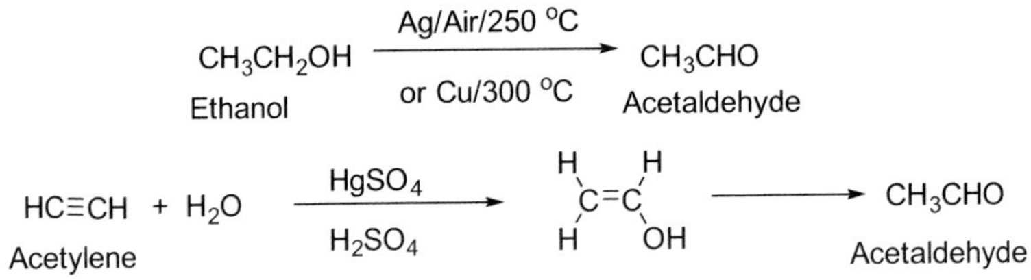

$$\underset{\text{Ethanol}}{CH_3CH_2OH} \xrightarrow[\text{or Cu/300 °C}]{\text{Ag/Air/250 °C}} \underset{\text{Acetaldehyde}}{CH_3CHO}$$

$$\underset{\text{Acetylene}}{HC{\equiv}CH} + H_2O \xrightarrow[H_2SO_4]{HgSO_4} H_2C{=}CH(OH) \longrightarrow \underset{\text{Acetaldehyde}}{CH_3CHO}$$

Scheme 6.9 Conventional syntheses of acetaldehyde.

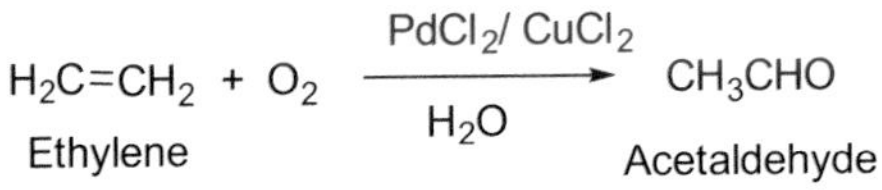

$$\underset{\text{Ethylene}}{H_2C{=}CH_2} + O_2 \xrightarrow[H_2O]{PdCl_2/\ CuCl_2} \underset{\text{Acetaldehyde}}{CH_3CHO}$$

Scheme 6.10 Wacker's oxidation process.

6.2.6 Synthesis of Urethane

Polyurethane is an important class of polymers exploited for numerous commercial purposes. Previously, polyurethane was produced using phosgene gas, which is highly toxic and has detrimental effects (Scheme 6.11).

$$RNH_2 + COCl_2 \longrightarrow RNCO + 2HCl \xrightarrow{R'OH} RNHCOOR'$$

Amine Phosgene — Isocyanate — Urethane

R,R′ = alkyl groups

Scheme 6.11 Urethane synthesis from phosgene.

The Monsanto Company developed a method that completely eliminates the use of phosgene for synthesizing polyurethanes and their isocyanate precursors (Scheme 6.12) [11–14].

$$RNH_2 + CO_2 \longrightarrow RNCO + H_2O \xrightarrow{R'OH} RNHCOOR'$$

Amine Carbon dioxide — Isocyanate — Urethane

R,R′ = alkyl groups

Scheme 6.12 Urethane synthesis without using phosgene.

6.2.7 Selective Methylation of Active Methylene Using Dimethyl Carbonate

Traditionally, methylation reactions used methyl halides or methyl sulfates. However, their toxic nature and uncontrolled multiple alkylations during the reaction imposed several environmental problems and, thus, this method was discontinued. A new method was developed by Tundo, where methylation of active methylene compounds was performed using dimethyl carbonate. In this, aryl acetonitrile was reacted with dimethyl carbonate at 180°C–220°C in the presence of a base K_2CO_3 to selectively give 2-arylpropionitrile (Scheme 6.13) [15–18].

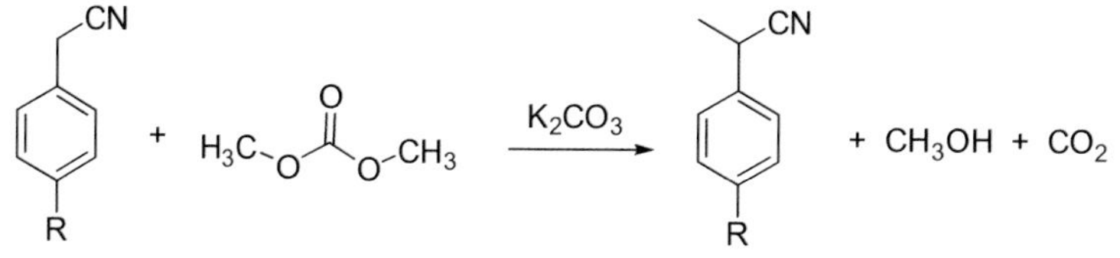

CN
R
+ $H_3C{-}O{-}C(=O){-}O{-}CH_3$ $\xrightarrow{K_2CO_3}$
CN
R
+ CH_3OH + CO_2

Scheme 6.13 Selective methylation of active methylene using dimethyl carbonate.

6.3 Some Real-World Cases: Green Chemistry Efforts Honored

Undeniably, green chemistry has played an indispensable role in resolving many of the pressing environmental issues mankind has been facing since the progressive globalized era began. Though it has been a creative challenge to put the 12 principles of green chemistry pioneered by Paul T. Anastas and John C. Warner into action, incredible innovations have been possible and many of these completely replaced the long traditional cumbersome practices in academic as well as industrial research. These principles have provided deep insight into designing novel chemicals and processes while also enabling the revitalization of the existing ones. The results have been marvelously marked by industries, especially wherein the introduction of the concepts of waste prevention, atom economy, less hazardous chemical synthesis, use of catalysts, renewable feedstocks, etc., has greened the product manufacturing processes, ultimately also resulting in an economic boom apart from adding to the environmental credentials.

To honor the tremendous efforts and accomplishments, in 1995, the Presidential Green Chemistry Challenge Awards were established by President Clinton and are conferred every year at the National Academy of Sciences in Washington DC just before the commencement of the annual meeting of ACS GCI Green Chemistry and Engineering Conference. These awards are promoted by the Environmental Protection Agency (EPA), US, in collaboration with many chemical and pharmaceutical industries, and the basic intention behind conferring these awards is the promotion of better practices and innovations in academia and industry for significant improvement in human health and environment. Princeton University's Paul J. Chirik, a 2016 award recipient, rightly said, "The Green Chemistry Challenge Awards highlight the importance of sustainable chemistry and its impact across a range of disciplines."

Until now, 118 award-winning technologies have been recognized and these are primarily divided into the following five categories:

- Academic
- Small business
- Greener synthetic pathways

- Greener reaction conditions
- Designing greener chemicals

In this chapter, we shall highlight some of these, which will enable the readers to understand how the principles of green chemistry can be put to action and the large-scale benefits of doing so. The award-winning technologies have also become the best examples of real-world case studies of green chemistry.

6.3.1 Development of NatureWorks™ PLA: An Efficient, Green Synthesis of a Biodegradable and Widely Applicable Plastic Made from Corn (A Renewable Resource)

Presidential Green Chemistry Challenge: 2002 Greener Reaction Conditions Award

Plastics constitute an important class of materials that are widely employed in modern society. Almost all aspects of daily life deal with plastic in some form or the other, including clothing, medical equipment, food packaging, building materials, compact disks, sports equipment, and automobile parts. Some of the versatile features of plastics have been depicted in Fig. 6.1.

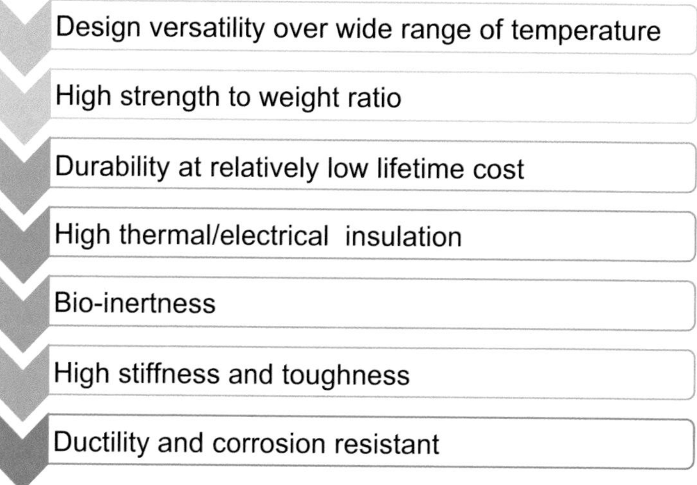

Figure 6.1 Versatile features of plastics.

Plastics can be made from any feedstock that contains carbon and hydrogen, but usually the most preferred feedstock for manufacturing plastics is crude oil, a nonrenewable resource that is rapidly depleting. The resistance of plastics to everyday wear and tear has made it a major environmental hazard. For instance, plastic utensils do not require the same care as glass or ceramic utensils and when these plastic utensils are discarded, they collect in landfills, resulting in ecological concerns. Even the reuse and recycle practices do not help in the reduction of plastic waste as many plastics are "single use," such as the plastic used in diapers and food wraps, while the recycling process (thermally or mechanically) often degrades the physical properties of plastics. Downward recycling still remains an option, but that requires a fresh monomer to create a new product.

6.3.1.1 Greener alternative: Polylactic acid

In the search for greener alternatives to plastics, polylactic acid (PLA) was discovered, which is a polymer derived from lactic acid (LA) found naturally in living organisms and is obtained by anaerobic fermentation of glucose. PLA is an example of a thermoplastic polymer that is fully compostable and can be recycled back (closed-loop recycling) to give the starting material, with which new virgin PLA is made. The physical properties of PLA are comparable or superior to those of most commonly used plastics, such as polyethylene and polystyrene. Figure 6.2 illustrates some of the properties of PLA. These features make PLA a promising candidate for use in items like fibers, food containers, and films. Due to its biocompatible nature, it has also been used in the manufacture of medical objects, like surgical staples and bone screws. However, the high cost of PLA synthesis hampers its use and has resulted only in advanced manufacturing techniques [19, 20].

6.3.1.2 NatureWorks LLC: Synthesizing PLA from corn

The commercial production of PLA began in November 2001, by NatureWorks LLC (formerly Cargill Dow LLC) in Blair, Nebraska, US. They developed a method for synthesizing PLA from the starch harvested from corn, and this work received the Presidential Green Chemistry Challenge Award in 2002. The process majorly involves three steps: (i) conversion of starch to glucose, (ii) fermentation of glucose to LA, (iii) polymerization of LA to PLA (Fig. 6.3) [21].

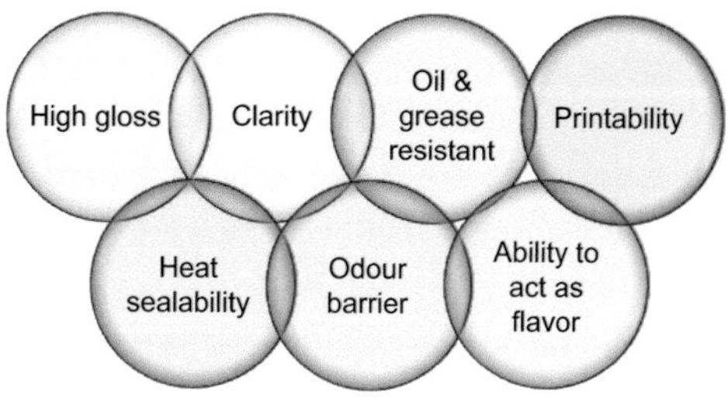

Figure 6.2 Physical properties of PLA.

The various stages involved in the production of PLA are discussed next.

Stage 1: Corn to glucose
Starch is separated from corn by a mechanical process called the wet milling method, after which this corn-derived starch is enzymatically reacted with water to cleave the starch glycosidic linkage to form glucose.

Stage 2: Glucose to LA
At this stage, glucose is converted into LA by fermentation. This process first generates pyruvate, which is further converted into LA (enantiomerically enriched with 99.5% in L form) (Fig. 6.4).

Stage 3: LA to PLA
LA is then polymerized first into a low-molecular-weight "prepolymer" by solventless condensation polymerization, which is then catalytically broken down into a cyclic dimer called lactide with three stereoisomers: L-lactide, D-lactide, and *meso*-lactide (Fig. 6.5). Collectively, all three lactides are monomers for the polymerization reaction, where these lactide subunits are attached to each other through a catalytic solventless ring-opening polymerization reaction to yield a high-molecular-weight polymer. It is possible to tailor the properties of the final product according to its intended use by varying the concentration of D-lactide.

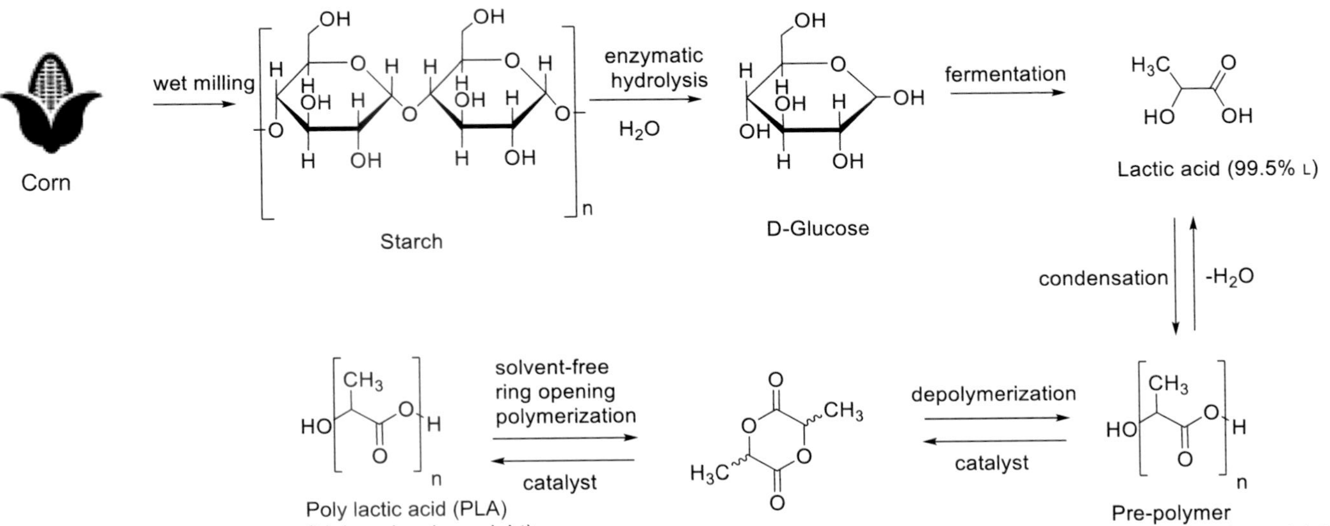

Figure 6.3 NatureWorks LLC production of PLA from corn.

What are fats and oils?

Fats and oils are tricarboxylic esters of glycerol and are called triglycerides. Those having carbon–carbon single bonds are saturated ones, and unsaturated fats and oils have carbon–carbon double bonds.

Why are unsaturated fats/oils hydrogenated?

Unsaturated oils that are found in nature are partially hydrogenated to increase their shelf life and to raise their melting points. As in the case of alkenes, C=C easily undergoes addition and substitution reactions in comparison to the C–C bonds in alkanes. The same is true for the double bonds present in unsaturated fats and oils. Thus, the tendency to go rancid is higher for unsaturated fats in contrast to saturated fats. Exposure of oils to oxygen results in their degradation, especially at higher temperatures, which is usually experienced during cooking. The mechanism of autoxidation of oils is represented in Fig. 6.7. To slow down this autoxidation process, hydrogenation is performed, which removes many double bonds, thereby lowering the reactivity of fats/oils and extending their shelf life. Autoxidation in fatty acid chain follows this order: three C=C > two C=C > one C=C > zero C=C.

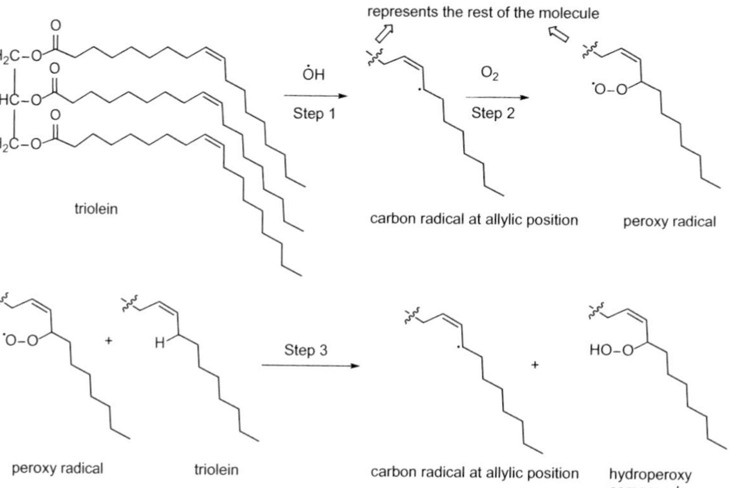

Figure 6.7 Autoxidation of unsaturated triglyceride.

The autoxidation process begins with the attack of the hydroxyl radical (reactive oxygen species), which abstracts hydrogen from the allylic position (the most reactive site) of the oil, generating a carbon radical. This carbon radical further reacts with oxygen to give a peroxy radical, which abstracts hydrogen from another oil molecule to form hydroperoxy compound and a carbon radical. The generated carbon radical can further attack an oxygen molecule, resulting in a free radical chain reaction, and the primary products from autoxidation (hydroperoxy products) can then react with alkenes, carboxylic acids, aldehydes, and ketones.

Additionally, hydrogenation is done to raise the melting points of triglycerides. In general, triglycerides having fewer double bonds have higher melting points. Besides increasing the melting point, hydrogenation also tends to isomerize the double bonds of triglycerides from *cis* to *trans*. Among *cis*- and *trans*-isomers, *cis* has a lower melting point due to its kinked shape (bent structure), while *trans* has a more closely packed structure (Fig. 6.8). Also, a *trans*-isomer is thermodynamically more stable than a *cis*-isomer. Thus, hydrogenation leads to higher melting points by reducing the number of double bonds and isomerizing the remaining double bonds to *trans*. However, this process has a major shortcoming: the trans fats increase the level of low-density lipoprotein cholesterol in the human body, which can increase coronary heart disease. It also contributes to obesity and high blood pressure.

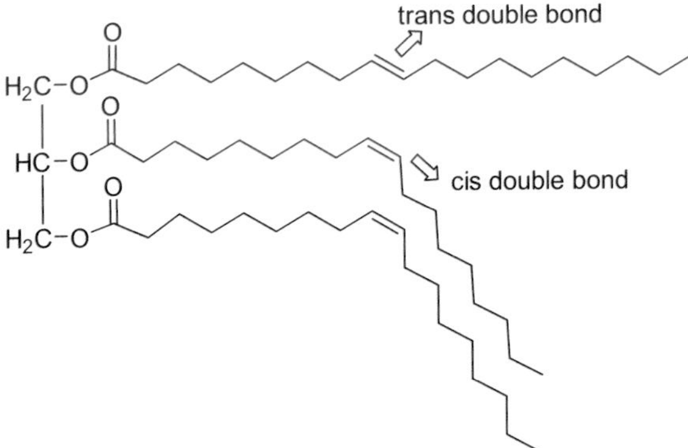

Figure 6.8 *Cis*- and *trans*-molecule in a triglyceride.

Fats are solid triglycerides, whereas oils are liquid triglyceride. Warm-blooded animals have higher quantities of saturated triglycerides and these triglycerides are solid at room temperature but liquid at body temperature. Plants and cold-blooded animals have a high degree of unsaturated triglycerides, which remain in liquid state even at lower temperature. For example, tristearin, a saturated triglyceride, is fat with a melting point of 72°C and triolein, an unsaturated triglyceride, is an oil with a melting point of –4°C.

6.3.2.1 Green chemistry: Interesterification of oils

To avoid the formation of trans fats during the processing of fats/oils, scientists came up with interesterification, a process that has been known for many years. In this process, random rearrangement of fatty acids in triglyceride molecules occurs. For example, if two triglyceride molecules have different fatty acid chains represented by R^1 and R^2, a mixture of the product is obtained. In this case, eight different triglycerides are obtained, with three different fatty acid chains in each molecule (Fig. 6.9). The melting point of the product mixture is in between those of the two starting materials, and no isomerization to trans fats takes place.

The chemical interesterification (CIE) process of a triglyceride is carried out in the presence of a catalyst like sodium methoxide. However, this method suffers from several drawbacks:

- The toxicity exhibited by sodium methoxide can cause severe risk to the workers.
- Sodium methoxide is also sensitive toward peroxides, water, and free fatty acids present in the starting material. Thus, the oil must be pretreated to eliminate these impurities before CIE is conducted.
- Also, if any colored impurity is present, the oil must be treated with bleaching clay.
- The final products are first neutralized with acid and then washed with water, which results in a significant amount of wastewater with a high biological oxygen demand. Also, this

wastewater has to be treated properly before being disposed of. This adds to the total cost of the overall process.

- CIE is usually accompanied by side-reactions that result in loss of up to 5.8% of oils.
- It is an expensive process in comparison to hydrogenation.

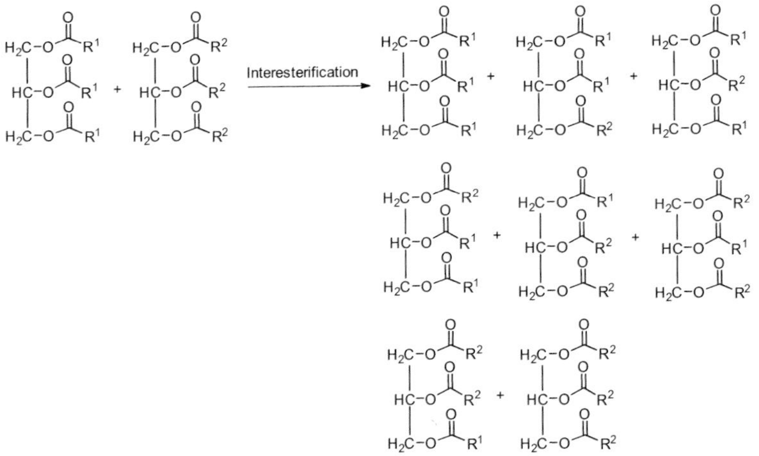

Figure 6.9 Enzymatic interesterification of triglyceride.

6.3.2.2 Enzymatic interesterification of oils

Interesterification can also be performed using enzymatic catalysis. For this, Novozymes developed an improved lipase enzyme (also known as Lipozyme TL) and Archer Daniels Midland Company (ADM) utilized this enzyme to catalyze the interesterification reaction. In 2005, Novozymes and ADM shared a Presidential Green Chemistry Challenge Award [23]. Further, grafting of this enzyme on silica particles helped in their easy removal from oil by filtration and when this grafting is done with improved enzyme, the cost of enzymatic interesterification (EIE) is reduced by more than 50%. Even though both CIE and EIE can produce oils with desirable properties, EIE offers several environmental benefits over CIE. These advantages are illustrated in Fig. 6.10.

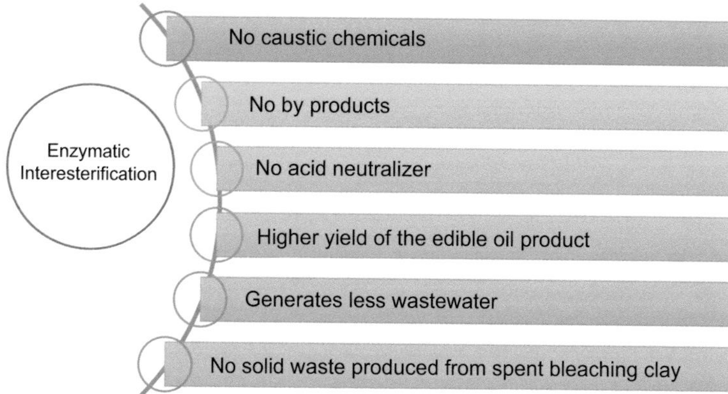

Figure 6.10 Advantages of EIE over CIE.

6.3.3 Development of Fully Recyclable Carpet: Cradle-to-Cradle Carpeting

Presidential Green Chemistry Challenge: 2003 Designing Greener Chemicals Award

In most countries, a lot of area is covered with carpeting and the bulk carpeting is produced from synthetic polymers, which are not readily degradable and when discarded take up a lot of landfill space for centuries. Besides, a huge amount of crude oil is required for making raw materials in order to synthesize polymers. With the continually decreasing landfill space and declining supply of crude oil, it has become imperative to decrease the amount of carpeting that is disposed of each year or in other words, increase recycling of carpets [24].

Moreover, there are several environmental concerns associated with the toxicity of various materials employed to make carpeting. There are two major parts of carpeting, one is the backing, which is usually done with polyvinyl chloride (PVC), and the other part is the face fiber. PVC is manufactured by the addition polymerization of vinyl chloride (VC). A high level of exposure of VC in air for a short span can cause headaches, dizziness, and drowsiness, while in the case of long-term exposure, it can cause damage to the liver or even cancer. Apart from this, plasticizers are added to enhance the flexibility of PVC. These are oily compounds that are neither covalently linked nor ionically bonded to PVC and thus tend to seep out gradually from the polymer. Phthalates are commonly used plasticizers, but some of them can result in liver damage, severe reproductive effects, and cancer [25]. In addition, when these carpets are incinerated, they give rise to toxic fumes that contain furans, dioxins, and hydrochloric acid [26]. Also, PVC-backed carpets are not easily recycled as PVC contaminates the fiber, which prevents its depolymerization and the recycling process requires a huge amount of energy.

6.3.3.1 Green chemistry: Production of a cradle-to-cradle carpet

Shaw industries developed an EcoWorx carpeting with a polyolefin (PO) backing and a nylon face [27]. For this work, Shaw industries won the Presidential Green Chemistry Challenge Award in 2003. The POs are synthesized from ethylene or substituted ethylene by an addition polymerization reaction. No plasticizers are added, but fillers are used to provide bulk to carpeting. Initially, calcium carbonate was used as a filler, which was very soon replaced by fly ash, a waste product from flue gases. This carpeting has numerous environmental benefits over the PVC-backed carpeting. These advantages are as follows:

- The monomers used to synthesize PO are less toxic in comparison to VC.
- During the incineration of carpeting, dioxins are not produced as there is no chlorine in PO.
- The problem of leaching of plasticizers is completely eliminated as there is no such addition.
- Due to the lower density and thinner profile of PO, more carpeting can be shipped with reduced energy consumption and at a lower cost.
- Fly ash is used as a filler, which not only solves the problem of landfills but also eliminates the environmental burden of mining and processing during the production of calcium carbonate (previously used filler).
- Most importantly, this is an example of closed-loop recycling, which allows the use of the carpeting multiple times and create fresh carpeting each time. Thus, it is also termed as "cradle-to-cradle carpeting."

6.3.3.2 Recycling of PO-backed nylon-faced carpeting

In the recycling process, a used carpet is first ground to form smaller granules, followed by the separation of PO from nylon by elutriation, a process in which an air stream is passed through the granules and separation of lighter nylon particles from heavier ones occurs. The left PO is then used to manufacture new backing, while nylon is converted back to caprolactam (monomer). The monomer is again polymerized to form new nylon fiber (Fig. 6.11).

6.3.4 Design and Development of Environmentally Safe Marine Antifoulant

Presidential Green Chemistry Challenge: 1996 Designing Greener Chemicals Award

Fouling is a very common phenomenon that results from the unwanted growth of marine plants and animals on the submerged area of the ship. This is usually caused by slime-forming microorganisms, such as algae, barnacles, bacteria, and diatoms. Due to the slime buildup on the bottom of the ship (hull) below the waterline, a hydrodynamic drag is experienced by the ship, along with decreased ship speed, increased fuel consumption, and high cleaning costs. The estimated total annual global cost caused by fouling is approximately US$3 billion, out of which a major chunk is due to increased fuel consumption. Moreover, increased consumption of fossil fuels contributes to global warming, pollution, and acid rain.

To overcome this fouling, the hulls of the ships were treated with antifoulants, an essential pesticide that is mixed with paints and then applied to the hull. The most commonly used antifoulants are organotin biocides, such as tributyl tin oxide (TBTO) (Fig. 6.12), which when mixed with paints leaches gradually from the hull and kills the microorganisms. Even though TBTO was found to be very effective as an antifoulant and has saved shipping industries billions of dollars, it poses numerous environmental problems, two of which are described here [28]:

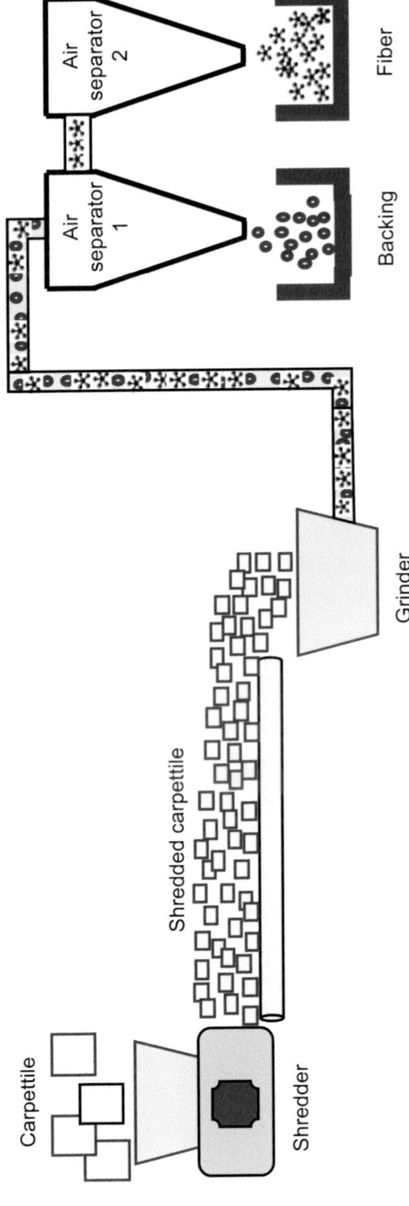

Figure 6.11 Recycling process of EcoWorx Carpeting.

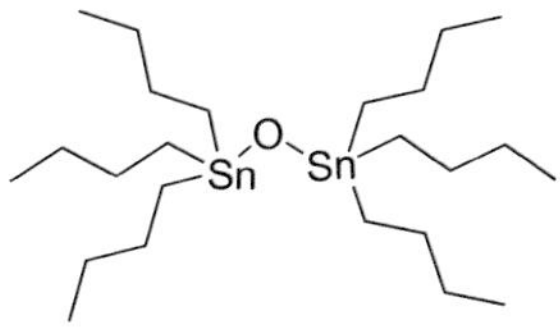

Figure 6.12 Structure of TBTO.

- Organotin compounds are found to have long half-lives. For example, TBTO has a half-life of more than six months in seawater. Thus, it is highly persistent and can harm several organisms by, for example, decreasing reproductivity, causing imposex, increasing shell thickness, and leading to acute toxicity.
- Bioaccumulation of organotin compounds in marine organisms is another major problem and affects the immune system of fishes, snails, turtles, and oysters.

Due to these concerns, the use of organotin compounds, especially TBTO, has been now discontinued and efforts are directed toward the development of environmentally benign antifoulants. An ideal marine antifoulant must address the following requirements:

- For commercial applicability, it must prevent fouling from a broad range of organisms.
- It should degrade rapidly in the surrounding environment.
- It should rapidly partition into sediments, thereby limiting bioavailability to nontarget organisms.
- It should be toxic only to the target organisms and for the rest of the marine organisms, it should have minimal or no toxicity.

6.3.4.1 Development of SeaNine™ 211: Environmentally safe antifoulant

Rohm and Hass Company (now a subsidiary of the Dow Chemical Company) developed several suitable compounds from 3-isothiazolones, out of which 4,5-dichloro-2-*n*-octyl-4-isothiazolin-3-one (DCOI) was found to be the ultimate choice as a safe marine antifoulant for commercial purposes. DCOI is commonly marketed as SeaNine™ 211 (Fig. 6.13). The advantages associated with DCOI are as follows:

- DCOI leaches out quickly from the paint after application to the hull and kills the foulants. It is toxic to algae at the parts-per-billion level and to barnacles to the parts-per-million level.
- Its half-life is less than an hour. So, it biodegrades into nontoxic compounds (Scheme 6.15).
- Due to its fast degradation, it does not persist and the levels of DCOI is always below the toxicity level.
- It displays limited water solubility and gets adsorbed by the soil sediments. Therefore, it does not bioaccumulate in seawater and its bioconcentration in marine lives is very low.

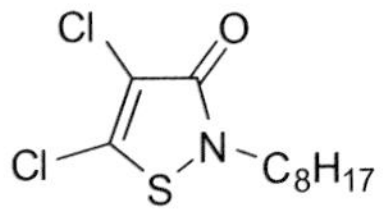

Figure 6.13 Structure of DCOI.

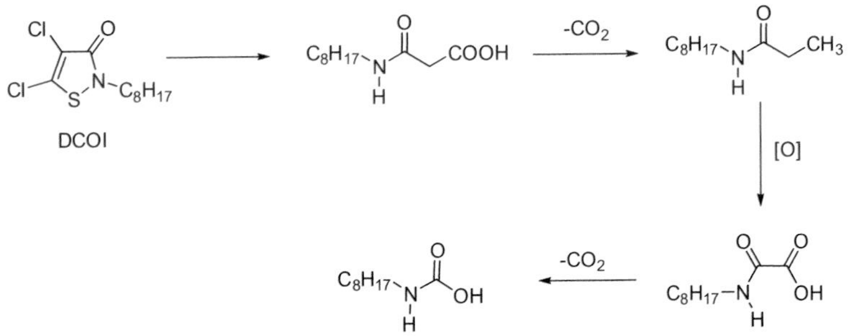

Scheme 6.15 Metabolic products of DCOI.

Due to the effectiveness of DCOI, it is now widely employed as a safe antifoulant instead of organotin compounds. For this work, Rohm and Haas Company won the Presidential Green Chemistry Challenge Award in 1996. Table 6.1 compares the properties of organotin compounds with that of SeaNine™ [29].

Table 6.1 Comparison of organotin compounds with SeaNine™ as a marine antifoulant

No.	Organotin compounds	SeaNine™
1.	They degrade slowly under aerobic and anaerobic conditions.	It degrades rapidly in the environment with the help of microorganisms.
2.	Their half-lives range from 6 to 9 months.	Its half-life in aerobic and anaerobic microcosm studies is less than 1 h.
3.	TBTO degrades into dibutyltin species, which are toxic and persistent in the environment.	The degradation products of SeaNine™ antifoulant are compounds with incredibly reduced toxicity.
4.	Bioaccumulation of TBTO was considerable, with bioconcentration factors as high as 10,000.	The SeaNine™ biocide showed essentially no bioaccumulation.
5.	TBTO showed a wide range of effects on the growth, development, and reproduction of marine species, even at levels as low as 2 parts per trillion (ppt).	The SeaNine™ antifoulant showed no chronic or reproductive toxicity to marine species.
6.	The maximum predicted concentrations of TBTO were as high as 345 ppt, which is far above the UK Environmental Standard in seawater of 2 ppt.	Computer modeling predicted maximum concentrations of SeaNine™ of up to 10 ppt, far below the maximum acceptable environmental concentration of 630 ppt.

6.3.5 Designing Rightfit™ Pigments to Replace Toxic Organic and Inorganic Pigments

Presidential Green Chemistry Challenge: 2004 Designing Greener Chemicals Award

Colors are the most beautiful aspect of life. Dyes and pigments are substances that are added to impart color to clothes, paints, plastics, ceramics, prints, etc., as they possess the phenomenal ability to absorb light in the visible region of the electromagnetic spectrum.

Pigments, in particular, are the colorants, mostly inorganic, that are generally insoluble in the media to which they impart color. Although dyes and pigments have similar chromophores, yet remarkably pigments are larger in size and do not possess solubilizing groups, such as SO_3H, attached to them. Organic pigments normally impart more intense colors than the inorganic pigments. Inorganic pigments often contain heavy metals, which are a class of inorganic pollutants causing a detrimental effect on both human health and environment. These are now designated as persistent environmental toxins because unlike other contaminants, such as pharmaceuticals and dyes, they cannot be easily rendered harmless by chemical or biological remediation processes.

6.3.5.1 Green chemistry innovation

In the year 2004, Engelhard Rightfit™, now known as BASF Corporation, won the Green Chemicals Award for designing Rightfit™ pigments that contain divalent calcium, strontium, or barium, replacing heavy metals such as hexavalent chromium, cadmium, and lead employed commonly in pigments that impart beautiful yellow, orange, and red colors [22]. This is one of the best steps taken so far to ensure the removal of toxicity through green chemistry, as one cannot imagine how much hazardous these inorganic species could be.

6.3.5.2 Issues underlying and resolved

As per the US EPA and the International Agency for Research on Cancer, heavy metals have been classified as human carcinogens. Table 6.2 shows the heavy metals used in pigments and the associated toxicities. It is also significant to understand mechanistically how these metals exhibit toxicity.

The innovation by Engelhard has proven successful in replacing heavy metal-based pigments in markets, saving cost as well as adding to the environmental credentials. The name "Rightfit" thus associated with the pigments is to signify that they have (i) the right environmental impact, (ii) the right color space, (iii) the right performance characteristics, and (iv) the right cost-to-performance value. Many of these pigments have got approval from the US Food and Drug Administration and the Canadian Health Protection Branch for indirect food contact.

Table 6.2 Heavy metal toxicity and mechanism

Heavy metal	Toxicity	Mechanism
Cadmium	At low levels, it damages the kidneys and affects the bones, leading to fractures. On long-term exposure, it may cause cancer and toxicity of organ systems, such as skeletal, urinary, reproductive, cardiovascular, central and peripheral nervous, and respiratory systems.	(i) Cd(II) strongly binds to sulfhydryl groups of cysteine residues of enzymes, for example, carbonic anhydrase, dipeptidase, and carboxy peptidase, affecting the active confirmation of biomolecules due to the strong binding. (ii) Cadmium is similar to zinc. Therefore, Cd(II) can displace Zn(II) in many zinc-containing enzymes. Besides, Cd competes with calcium for binding sites.
Lead	It interferes with a number of body functions, primarily affecting the central nervous, hematopoietic, hepatic, and renal systems, producing serious disorders.	(i) Similar to Hg(II) and Cd(II), lead also inhibits SH enzyme but less strongly. (ii) Primarily, lead exhibits toxicity by interfering with heme synthesis as it inhibits several of the key enzymes involved in the overall process of heme synthesis.
Chromium	Cr(VI) acts as a carcinogen on inhalation.	(i) After Cr(VI) enters the cells, it eventually gets reduced to Cr(III), which is toxic because it induces oxidative stress during the reduction, while the Cr intermediates reacts with protein and DNA. (ii) Cr(III) shows the propensity to form adducts with DNA, which may lead to mutations.

Source: Ref. [30].

Different types of Rightfit™ pigments can be produced in the following ways [31, 32]:

- By altering the substituent present on the benzene ring

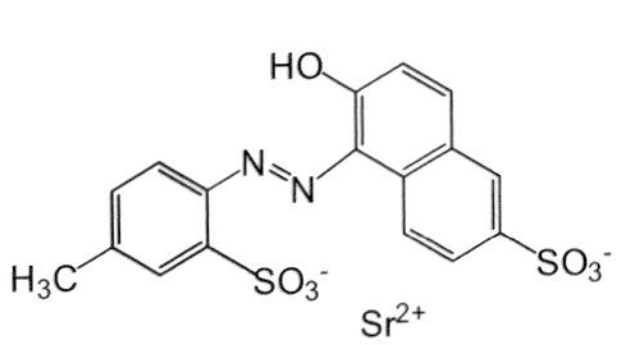

Medium Red

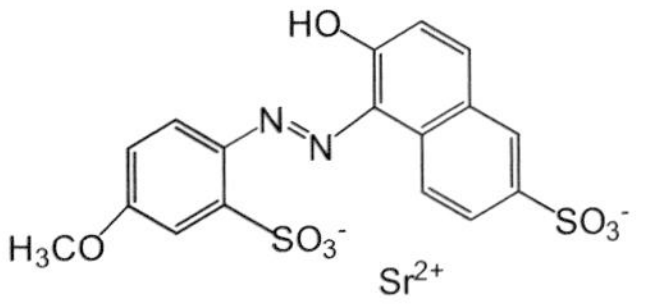

Very Blue Shade of Red

- By protonating the acid functionality

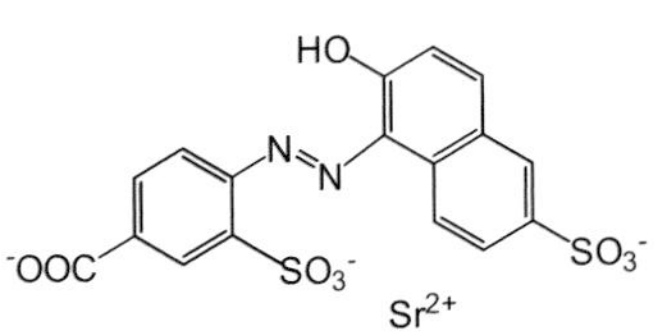

Medium Red

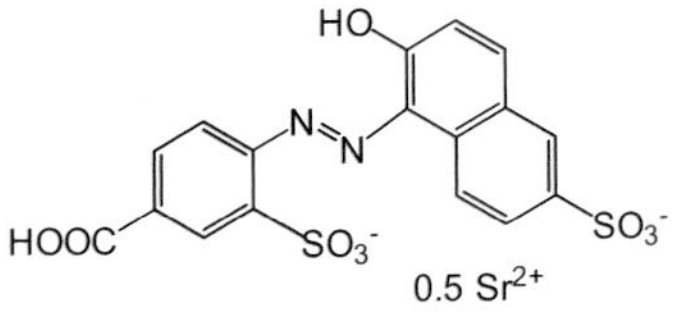

Brilliant Orange

- By changing the metal present

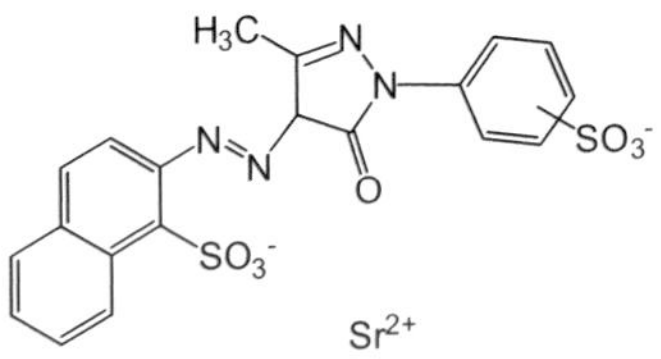

Reddish Yellow

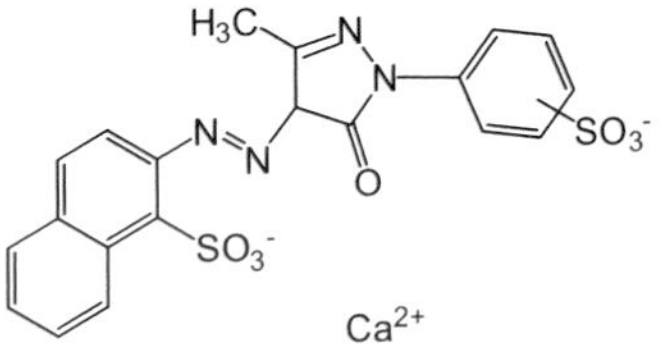

Very Reddish Yellow

The benefits of using Rightfit™ pigments have been depicted in Fig. 6.14.

6.3.5.3 How the Rightfit™ pigments can be synthesized

Here is an example to demonstrate how the Rightfit™ pigments that are commonly sold in Europe and Asia are obtained (Scheme 6.16) [32].

The synthesis demonstrated here is of a Rightfit™ red pigment that is accomplished in one pot and is a classic example of a diazotization reaction. The various green chemistry principles involved in the synthesis are waste prevention, synthesis of a less hazardous chemical, designing of safer products, and use of safer solvents and auxiliaries.

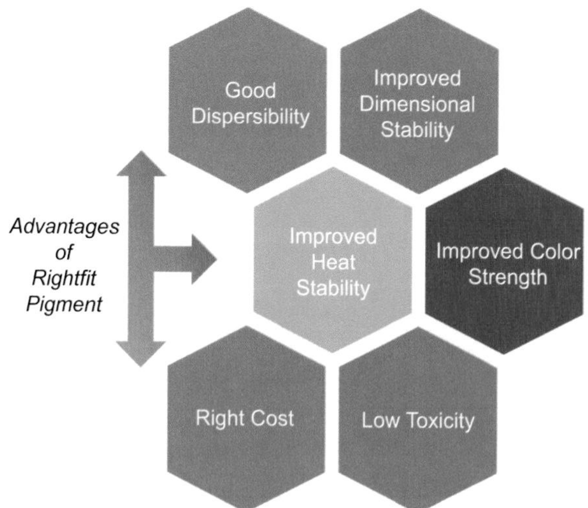

Figure 6.14 Benefits of using Rightfit™ pigments.

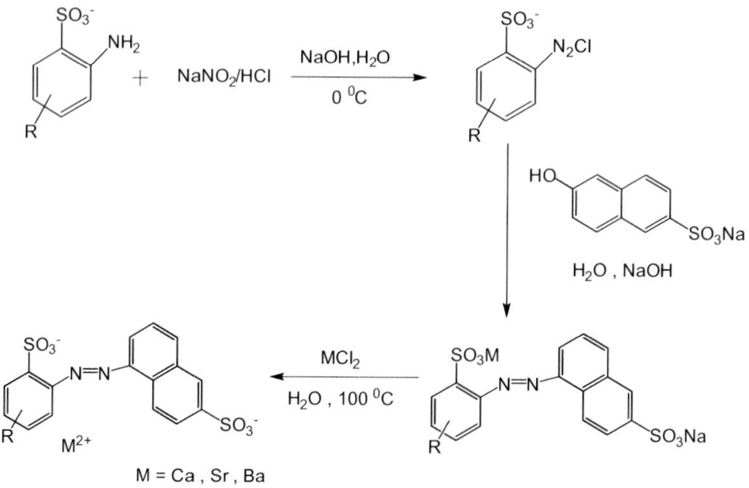

Scheme 6.16 Synthesis of Rightfit™ pigments of various hues.

6.3.6 Design and Application of Surfactants for CO_2 Replacing Smog-Producing and Ozone-Depleting Solvents for Precision Cleaning and Service Industry

Presidential Green Chemistry Challenge: 1997 Academic Award

It has been estimated that worldwide, about 15 billion kilograms of organic solvents are utilized by industries for various purposes, including precision cleaning in optics, electroplating, medical device fabrication, and garment cleaning. Most of the solvents, such as isopropyl alcohol, xylene, toluene, perchloroethylene, and chlorofluorocarbons (CFCs), are volatile organic compounds (VOCs) that are used for the removal of nonpolar impurities, like dirt, stain, greases, and oil, because of their ability to dissolve them. In addition, they easily get vaporized from items at room temperature, thus avoiding the need to heat them for removal of the cleaning solvent. The major issue of using VOCs is they cause environmental pollution and photochemical smog. Smog is created when VOCs mix with sunlight and nitrogen oxides, which are the by-products of fuel combustion, and then get transformed into ground-level ozone, nitric acid, and partially oxidized organic compounds. Exposure to smog can aggravate problems related to asthma and cause other respiratory ailments. As per literature reports, long-term exposure to this can increase the chances of cancer. Besides, many of the halogenated industrial solvents, such as chlorofluorocarbons (CFCs) and hydrochlorofluorocarbons (HCFCs), are ozone-depleting substances. When they are released into the atmosphere, they rise up through the troposphere and reach the stratosphere, where they get photochemically decomposed by the high-energy ultraviolet radiation emitted by the sun, which results in the generation of chlorine radicals. These radicals then work as a catalyst to deplete the ozone layer by abstracting an oxygen atom from the ozone molecule. The so-formed ClO radical then undergoes reaction with the oxygen atom to form an oxygen molecule, regenerating a chlorine radical. Simultaneously, due to the lack of oxygen atom in the lower stratosphere but the abundance of ozone molecules, there is another more prevalent pathway, wherein two ClO radicals react to give the dimer ClOOCl, which undergoes photodissociation to regenerate a Cl radical. These steps repeatedly occur and result in the loss of ozone. The consequence of depletion of the ozone layer is severe; it allows more of sun's UV-B rays to reach the earth's surface, which has serious repercussions, as per evidence, in the form of increased skin cancer and cataracts among human beings.

In view of the harmful effects of CFCs and under the 1990 Clean Air Act Amendment, CFCs can no longer be produced in the United States and instead HCFCs are recommended. Although HCFCs are far more superior, they are still ozone depleting and hence not a suitable long-term replacement.

6.3.6.1 CO_2 as a greener alternative

CO_2 has emerged as an interesting alternative green solvent (Fig. 6.15).

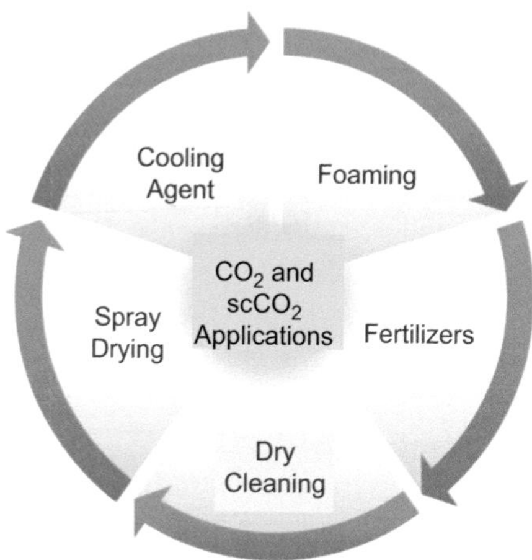

Figure 6.15 Uses of CO_2 and supercritical CO_2 ($scCO_2$) in diverse applications.

Some of the fascinating benefits of CO_2 are as follows:

- It is cheap and readily available and can be recovered as a by-product from ammonia and natural gas wells.
- It is nontoxic, nonflammable, and chemically benign.
- It does not in any way contribute toward smog formation or depletion of the ozone layer.

- It has a low heat of vaporization and hence requires less heat than water or other organic solvents.
- It can be recycled easily to the gaseous state, leaving impurities behind.

However, CO_2 is a greenhouse gas by virtue of the fact that it can reflect back the infrared radiations toward the earth (behaving like a blanket). Thus, it is responsible for global warming. To reduce the problems associated with the global warming caused by CO_2, the gaseous form of this compound can be recovered from ammonia manufacture and gas wells and then this recovered CO_2 can be used as a solvent. This can be converted into liquid CO_2 or supercritical CO_2 ($scCO_2$) [33].

Let us understand how $scCO_2$ can be obtained. As a general principle, we know that a gas can be converted into a liquid by an increase in the pressure exerted on it. However, if a substance is subjected to temperatures above its supercritical temperature (T_c = temperature above which a substance cannot exist in its distinct liquid phase despite the pressure applied) and pressures above its supercritical pressure (P_c = pressure at which a substance can no longer be in a gaseous state), it would generate a supercritical fluid. For CO_2, T_c = 31°C and P_c = 72.8 atm (phase diagram of CO_2 has already been depicted in Chapter 4). What happens in the case of a supercritical fluid is that as a result of the application of high pressure, individual molecules are pressed so close to each other that they start behaving as a liquid. However, the high temperature keeps them excited to a point that intermolecular forces cannot hold them together as a liquid. Thus, supercritical fluids possess densities close to the liquid state and viscosity close to the gaseous state.

6.3.6.2 Benefits of $scCO_2$

It has the ability to dissolve aliphatic hydrocarbons, halocarbons, aldehydes, and ketones. Because of its capability to dissolve caffeine, it has been used as a solvent to decaffeinate coffee. However, it has not yet gained prominence as a replacement for many of the industrial solvents because of its low solubility in common industrial materials, such as heavy oils and waxes, flux, solder residue, greases, proteins, and heavy metals. This problem can be resolved by using surfactants that can enhance the solubility of a lot of substances in liquid and $scCO_2$.

6.3.6.3 How does a surfactant work?

A surfactant works as per the basic concept of "like dissolves like," which means polar solutes would dissolve in polar solvents and nonpolar solutes tend to dissolve in nonpolar solvents. A surfactant has two ends; while one end has polarity similar to the particle to be emulsified, the other end has polarity similar to the solvent. To put it in simple words, the nonpolar end of a surfactant molecule would face intermolecular attraction to a nonpolar particle (i.e., the solute), while the polar end would face attraction to the polar solvent. Thus, the surfactant molecules assemble themselves into spherical structures called "micelles." The surfactant thus stabilizes the nonpolar solute particles in polar solvents, resulting in their dissolution. Soaps and detergents work in a similar manner.

6.3.6.4 Green chemistry innovation

Professor Joseph M. DeSimone, from the University of North Carolina and North Carolina State University, won the Presidential Green Chemistry Challenge Award under the academic category in 1997 for developing new detergents that allowed CO_2 to be used as a solvent in various industrial applications, replacing smog-producing and ozone-depleting hazardous substances for precision cleaning and service industries [34]. He discovered that polymers possessing a carbon chain backbone with the majority or all of the carbon atoms attached to fluorine (which are also called as fluoropolymers) are soluble in liquid CO_2 or $scCO_2$. The solubility may be primarily attributed to the weak van der Waals forces existing between the CO_2 molecules and likewise weak van der Waals forces existing between the fluorocarbon tails of the copolymer.

DeSimone created block copolymers in which some of the segments are soluble in CO_2 while the other segments are not. Thus, a surfactant for CO_2 could consist of a CO_2-phobic block containing a main chain of polymer and a CO_2-philic block containing fluoropolymer grafted onto the main chain (Fig. 6.16). Furthermore, depending on the type of material to be dissolved, CO_2-phobic segments could be made lipophilic (attracted to fats, greases, or oils) or hydrophobic (attracted to water) parts.

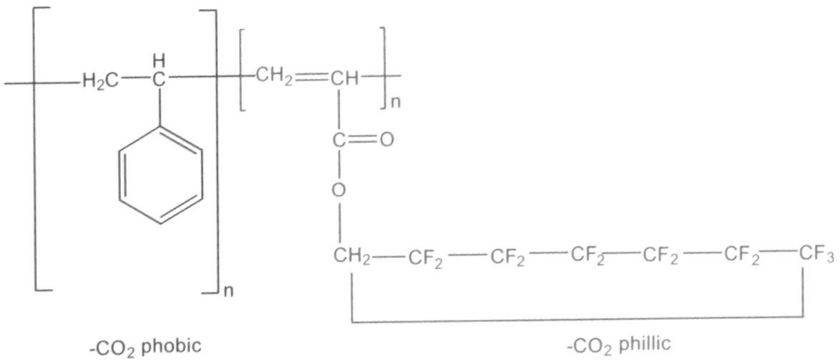

Figure 6.16 Surfactant for the use of CO_2/scCO_2.

DeSimone synthesized a block copolymer having polystyrene blocks that are insoluble in CO_2 and poly(1,1-dihydroperfluorooctylacrylate block) and graft segments. The portion of the graft segment soluble in CO_2 has been depicted in the Fig. 6.16.

6.3.6.5 Mechanism of action

When this particular copolymer is placed in scCO_2, it is arranged into a micellar structure. The CO_2-soluble segments surround the CO_2-insoluble segment. The use of a surfactant enhances the solubility of common industrial materials, such as waxes, greases, and oils. Thus, one can readily use CO_2 as a solvent for industrial cleaning with the use of a copolymer surfactant comprising one segment soluble in CO_2 and another segment with an affinity toward dirt, wax, grease, etc. Thus, the CO_2-insoluble substances would get entrapped in the micellar structure and get carried away/removed by scCO_2. This is an award-winning technology that has shown the promising potential to replace several smog-producing and ozone-depleting substances.

6.3.7 Green Synthesis of Ibuprofen by BHC

Presidential Green Chemistry Challenge: 1997 Greener Synthetic Pathways Award

The pharmaceutical drug industry has witnessed a net economic boom since the discovery of "analgesics"—drugs that are painkillers. The US sales have almost quadrupled since the discovery of analgesics such as aspirin, acetaminophen, and ibuprofen. Ibuprofen, in particular, belongs to a nonsteroidal anti-inflammatory group of drugs widely utilized for reducing inflammation and swelling and is an active ingredient of Motrin, Advil, Nuprin, and Medripen.

The traditional methodology for the synthesis of ibuprofen involves six steps that result in the generation of a large amount of unwanted by-products as waste, and the atom economy is quite low (approximately 40%) (Scheme 6.17) [35]. Enormous amounts of waste are generated as not all the reactants are incorporated into the final product. This procedure was developed by Boots, a company in England, in the year 1960 and remained the method of choice for quite long. Although using this method, millions of pounds of ibuprofen could be synthesized over the past four decades, the flip side of the coin is that it has led to the generation of millions of pounds of unwanted, unutilized, and unrecycled by-products. If we have a look at the synthetic route, we find that 60% of the starting materials go to waste.

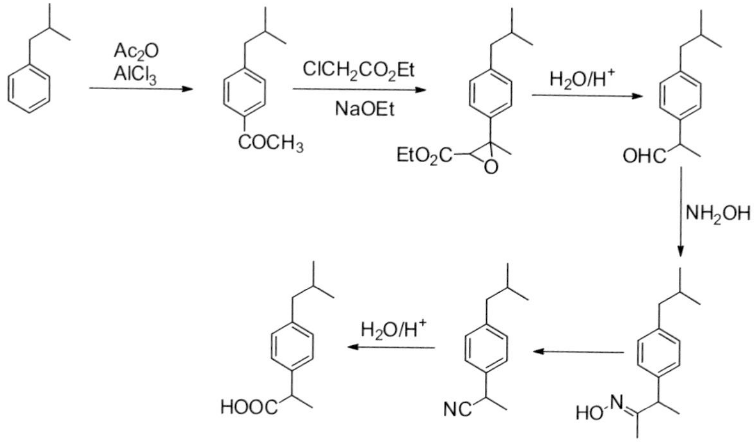

Ac_2O, $AlCl_3$; $COCH_3$; $ClCH_2CO_2Et$, NaOEt; EtO_2C; H_2O/H^+; OHC; NH_2OH; HO–N; NC; H_2O/H^+; HOOC

Scheme 6.17 Old route for the synthesis of ibuprofen.

6.3.7.1 Green chemistry innovation

During the mid-eighties, companies started realizing the potential financial rewards the synthesis of this drug could reap and, consequently, Hoechst Celanese Corporation entered into a joint venture with Boots and the resulting BHC Company decided to develop a new green synthetic route for ibuprofen production and subsequently market the product [36]. The efforts finally culminated into a new modified greener route that drastically improved the atom economy to 77%. Compared to the old methodology that employed a six-step process, this protocol required only three steps, which resulted in lesser amount of waste generation (Scheme 6.18). The BHC strategy was based on the use of catalysts such as hydrogen fluoride (HF), Raney Ni, and Pd to reduce waste and increase atom economy and efficiency. There are also environmental benefits associated with the green synthesis. The very first step in each of the routes renders the same product by acylation of isobutyl benzene. The older route utilizes stoichiometric amounts of aluminum trichloride, which generates a huge amount of aluminum trichloride hydrate as waste that usually reaches landfills. In contrast, the greener route employs HF as a catalyst that is recovered and recycled repeatedly. It is worth noting that the catalysts Raney Ni and Pd in Steps 2 and 3 are also recovered and reused. Besides, this green synthesis offers the advantage of greater throughput that translates into the ability to generate larger quantities of ibuprofen in less time and at less expenditure, resulting in tremendous economic benefits to the company. In October 1992, this synthesis was put into practice on an industrial scale. Because of this great innovation, BHC received the Green Chemistry Challenge Award in 1997.

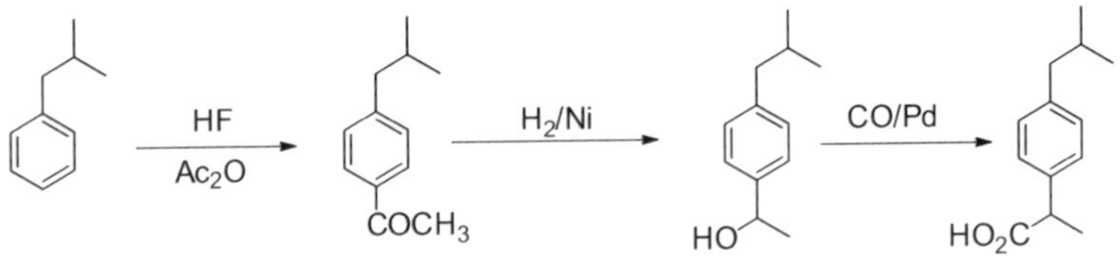

HF, Ac_2O → $COCH_3$ → H_2/Ni → HO → CO/Pd → HO_2C

Scheme 6.18 Greener route for the synthesis of ibuprofen.

6.3.8 TAML Oxidant Activators: General Activation of Hydrogen Peroxide for Green Oxidation Technologies

Presidential Green Chemistry Challenge: 1999 Academic Award

As per the estimates, the per-year consumption of paper pulp from the US paper and pulp industries is 600 lb. The conventional technique to generate paper pulp relies on the use of raw wood, which is subjected to a chemical or mechanical delignification process. Now, wood is composed of polysaccharides (about 70%), which are a combination of cellulose and hemicellulose, used to make paper, and lignin (30%), which primarily works as a glue to hold the polysaccharide fibers together. While making paper, most of the lignin is removed because its presence leads to a brown discoloration. Also, even if white lignin is present, it may eventually cause the paper to yellow over time. The majority of the world's paper mills use the Kraft process to remove lignin from the raw wood starting material. During this process, the logs are chipped into smaller pieces and then placed in a bath of sodium hydroxide and sodium sulfide, which results in the depolymerization of lignin. As the breakdown of lignin occurs, the wood chips lose their rigidity to form a slurry-like solution of polysaccharides (pulp). The treatment with NaOH and NaS causes about 80%–90% decomposition of lignin. At this step, if the pulp were used to produce paper, it would certainly cause the brown discoloration. To obtain bright-white paper, it is important to remove the excess lignin through a bleaching process. The excess lignin may be removed by reacting the pulp with elemental chlorine. Although this method is useful for generating white paper, there are serious environmental and health repercussions associated with it. What might happen during this process is chlorine reacts with lignin and causes the chlorination of the aromatic rings of lignin by an aromatic substitution reaction. The chlorinated organic compounds are in part dioxins, such as 2,3,7,8-tetrachlorodibenzo-*p*-dioxin and furans, which may cause cancer, learning and behavioral problems in offsprings, deficiency in the immune system, and lower testosterone levels in males. Besides, the dioxins and furans might also undergo bioaccumulation and eventually biomagnification (because of their low polarity, they are more soluble in fatty acids of animals than water). As a result of the combined effects of bioaccumulation and biomagnification, contaminant levels in fish can shoot up to levels that are 100,000 times greater than those of the surrounding environment. Though dioxins are released only in small quantities during paper manufacture, they achieve higher concentration in foods because of biomagnification. Ultimately, humans are exposed to dioxins and organochlorine compounds from food such as beef, dairy products, and fish. This can be highly dangerous, even fatal. In view of the harmful effects, the EPA imposed a ban on the use of chlorine in paper bleaching in the year 2001. Many paper manufacturers substituted Cl with chlorine dioxide (also called chlorine-free paper), which reduced the amount of dioxins and other organochlorine compounds drastically (by 91%). However, health risks still remained because the elimination was not 100%. Other chlorine-free bleaching methodologies have also been adopted and H_2O_2, ozone, and elemental oxygen have been employed as delignifying and bleaching agents. However, they have not gained widespread acceptance (Fig. 6.17).

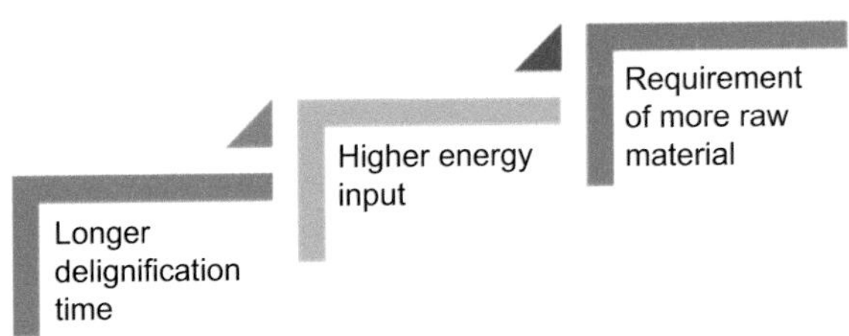

Figure 6.17 Drawbacks of the conventional oxidants as delignifying and bleaching agents.

6.3.8.1 Green chemistry innovation

To overcome these problems, Terrence Collins, from Carnegie Mellon University, developed TAML^{TM} activators to make the use of H_2O_2 more efficient to replace chlorine bleaches [37, 38]. TAML^{TM} activators are iron-based tetraamido-macrocyclic ligand activators (Fig. 6.18) that do not contain any toxic functional groups and work as catalysts for the conversion of H_2O_2 into hydroxyl radicals, which enables the oxidation/bleaching of the remaining lignin. Thus, H_2O_2 works as a powerful oxidizing agent. The high catalytic efficiency of the Fe-TAML activators allows H_2O_2 to break more of lignin in a shorter amount of time with higher selectivity. What is even more phenomenal is that this process generates only water and oxygen as the environmentally benign by-products, which makes this protocol much greener as compared to the traditional process using chlorine-based bleaching reagents.

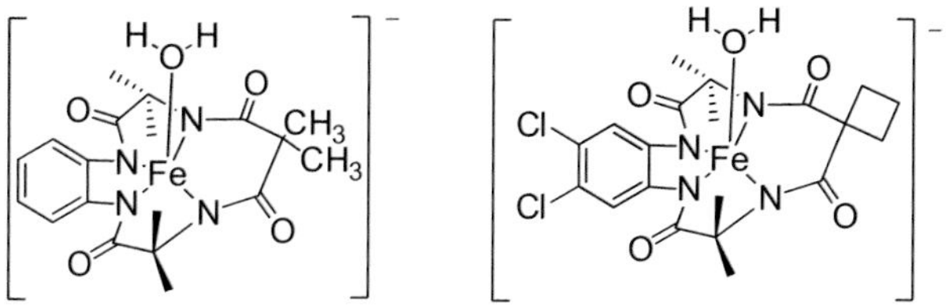

Figure 6.18 Fe-TAML activator complex.

This technology is patented. Apart from playing a key role in bleaching, TAML™ activators can also be used in laundry applications as they can prevent the transfer of dyes between fabrics and activate peroxides found in various bleaches. The various green chemistry principles involved are less hazardous chemical synthesis, use of renewable feedstocks, and use of catalysts.

6.3.9 Simple and Efficient Recycling of Rare Earth Elements from Consumer Materials Using Tailored Metal Complexes

Presidential Green Chemistry Challenge: 2017 Academic Award

Rare earth metal species are a group of chemical elements (La-Lu, Sc, and Y) existing in the earth's crust that keep the society running smoothly and driving toward a clean-tech-driven economy. In the United States, approximately 17,000 metric tons of rare earth oxides are used annually in diverse products, including wind turbines, catalysts, lighting phosphors, electric motors, batteries, and cell phones. Their extraordinary properties render them useful and irreplaceable in these modern technologies, right from energy to defense applications. Considering the way they have been overexploited, China has claimed that it is running out of them, with two-thirds of the country's supply already mined. Rare earths commonly co-occur geologically as mixtures of five to seven elements in ores that makes their primary mining, refining, and purification an extraordinarily energy-intensive and waste-generating process. Besides, the hard rock mining and refining processes need enormous quantities of water, acid, and organic solvents and produce large quantities of hydrofluoric acid, organics, and radionuclide wastes, which can include uranium and thorium and their decay products.

ammonium compounds are commonly found as disinfectants in households and industries and are even less toxic than CuO. In ACQ, these ammonium compounds serve as insecticides and specific quaternary ammonium salts are used for this purpose, such as dodecyldimethylammonium chloride (DDAC) and benzyl-C12-18-alkyldimethylammonium chloride (BAC) (Fig. 6.21).

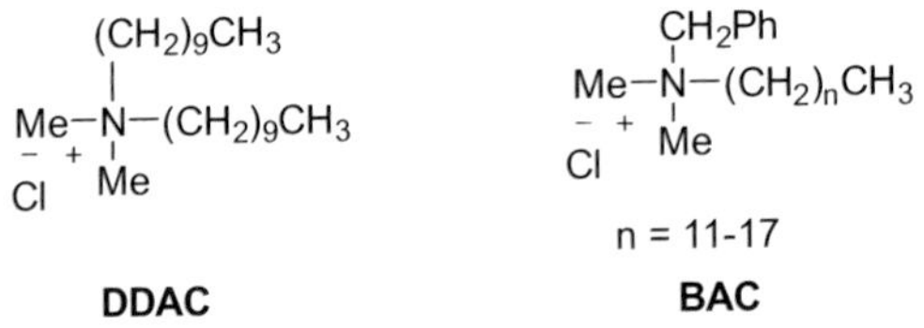

$(CH_2)_9CH_3$
Me–N–$(CH_2)_9CH_3$
Cl^- N^+ Me

DDAC

CH_2Ph
Me–N–$(CH_2)_nCH_3$
Cl^- N^+ Me

n = 11-17

BAC

Figure 6.21 Quaternary alkyl ammonium chlorides in ACQ.

To improve the formulation, CSI undertook various studies, such as compatibility with water repellents, foaming properties, mold prevention, and fasteners, and also developed the application technology. Similar to the CCA compounds, the chemicals in ACQ-treated wood are locked inside the wood due to chemical interaction with naturally occurring compounds present inside the wood. As discussed earlier, wood consists of polysaccharides and lignin. The ion-exchange process binds quaternary ammonium chlorides to the anionic sites on lignin and copper binds to wood by complexation with oxygen atoms in lignin and hemicellulose.

Replacement of CCA by ACQ has been one of the most dramatic achievements in pollution prevention as it has the potential to eliminate the use of huge amounts of arsenic and chromium each year. In addition, it eliminates the potential risks associated with the production, use, transportation, disposal in landfills, or incineration of CCA-treated wood. Furthermore, ACQ displays very high efficiency in controlling wood decay and is, thus, considered as the best substitute. As a result of the tremendous efforts made by CSI, it won the Presidential Green Chemistry Challenge Award in 2002.

6.4 Need for Industry-Academia Collaboration

The pharmaceutical industry lies at the core of medical innovation and has produced some brilliant life-saving blockbuster drugs till now. However, in face of the growing economic pressure and rising perplexity of societal and environmental problems, industries have realized the need to join hands with academic researchers. In fact, industry-academia collaborations have proven to be beneficial for both. Pharma industries are afflicted with critical problems, including declining research and development, low productivity, lack of innovation in the research pipeline, probability of the largest-selling drugs to go off-patent in the next few years, share losses, and reduced profit potential. They have realized that teaming up with academic researchers can only breathe new life into their translational research and help tackle the challenges of drug discovery and development. Working with academic researchers exposes the industrialists to technical thinking, innovative experimental strategies, and breakthrough research trends (Fig. 6.22). Not only industrialists, but also such kind of collaboration is critical for academia as they get a chance to create scientific knowledge, obtain industrial data, translate their basic discoveries into new therapeutics, and acquire access to resources and funding during times of dwindling grant support.

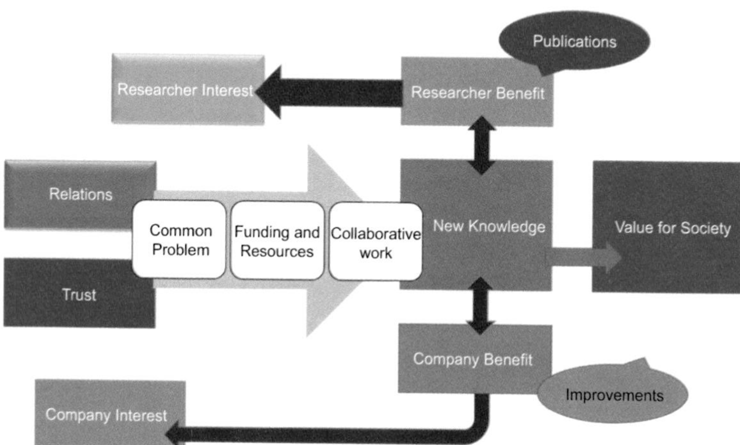

Figure 6.22 Benefits of industry-academia relationship.

There are numerous examples that clearly demonstrate how beneficial such partnerships have proven to be. For instance, among the earliest examples, the discovery of two crucial life-saving respiratory drugs corticosteroids (a class of drug that lowers inflammation in the body) and streptomycin (a class of drugs known as aminoglycoside antibiotics) were discovered with the joint efforts of industrial and academic researchers, which won them the Nobel Prize. Likewise, the discovery of many drugs, such as darunavir, emtriva, alimta, taxol, and vorinostat, are the result of academic intelligence coupled with industrial efforts [1]. Many of the close alliances have also given rise to innovations that have been recognized and awarded with Green Chemistry Challenge Awards. Through the ensuing examples, we have illustrated in brief some of these award-winning alliances on the development of greener routes.

6.4.1 An Efficient Biocatalytic Process to Manufacture Simvastatin

Professor Yi Tang, from the University of California, Los Angeles, collaborated with Codexis and developed an efficient biocatalytic pathway to manufacture simvastatin, which is a synthetic analogue of lovastatin—a natural fungal product used for the treatment of hypercholesterolemia and diabetic cardiomyopathy [40]. The traditional synthetic route required the conversion of lovastatin to simvastatin by the addition of a methyl group, which in turn required protection and subsequent deprotection of functionalities present in the lovastatin molecule. However, these were multistep processes and generated waste. Also the product yield was quite low (approximately 70%). To overcome these limitations, Professor Tang utilized the concept of "directed evolution," which enabled the creation of an engineered enzyme (cloned LovD, a natural acyltransferase produced by *Aspergillus terreus*, involved in synthesizing lovastatin) that could be employed for carrying out the initial synthesis of simvastatin. This methodology circumvented the need for tedious protection and deprotection steps required in the traditional synthesis, resulting in greater atom economy and minimizing waste generation (Scheme 6.19).

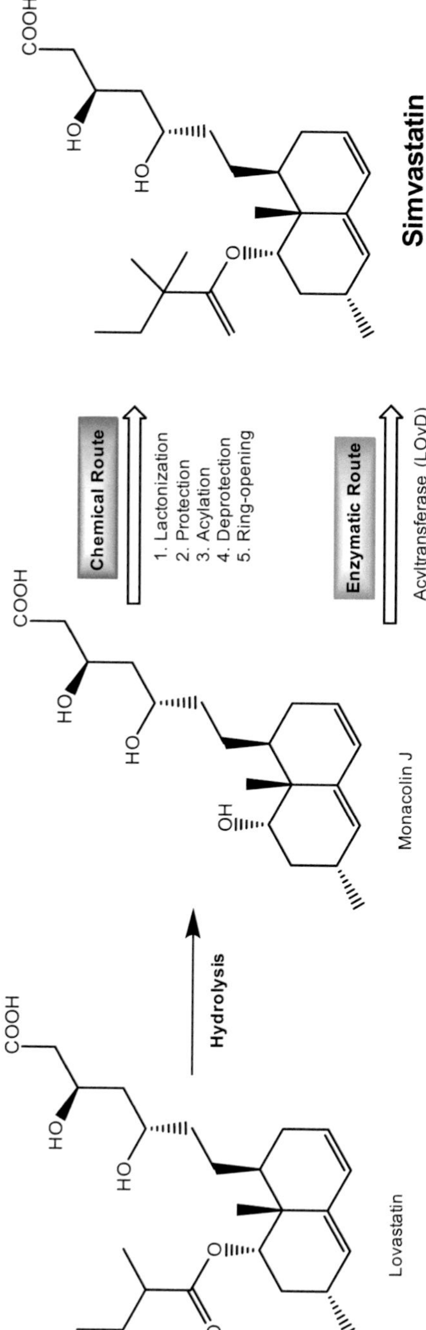

Scheme 6.19 Comparison of chemical and enzymatic routes to simvastatin synthesis.

6.4.2 Green Route for the Manufacture of Ranitidine

Another wondrous example, of ranitidine, highlights how important industry-academic relations are. Ranitidine HCl is world's largest-selling antihistamine drug, utilized widely for treating peptic ulcers, gastroesophageal reflux, and Zollinger–Ellison syndrome. The commercial production of this drug generates an obnoxious-smelling gas called "methyl mercaptan" as waste, which rapidly converts into the gaseous state at ambient temperature and pressure conditions, diffusing into the atmospheric air and contaminating it. Exposure to a high concentration of this gas may have serious repercussions. For instance, it may lead to fatal disorders, including central nervous system damage, malfunctioning of respiratory organs, and even death in certain cases.

To overcome the problems associated with the conventional route of ranitidine, which most of the industries had been following, Professor Mihir K. Chaudhuri, from IIT Guwahati and SMS Pharmaceuticals, Hyderabad, developed an environmentally benign catalytic protocol for converting the foul-smelling methyl mercaptan gas into the value-added dimethyl sulfoxide (DMSO, one of the starting materials used for the ranitidine manufacturing process) as shown in Scheme 6.20 [1]. The primary advantages of using DMSO is that it can be recycled back into the process, reducing the cost of ranitidine production by 40%. This protocol has been adopted increasingly by the industries in order to commercially manufacture this drug using compounds based on vanadium–titanium or titanium–phosphorous as catalysts. What is worth mentioning here is that it is again a striking result of the synergistic integration of the rich academic experience of Professor Chaudhuri and the industrial expertise of P. Ramesh Babu and T. V. Srihari of SMS Pharma and Suresh Babu of Rchem Industries.

6.5 Conclusion

Green chemistry is the future of all chemistry. This chapter showcases some greener routes for the synthesis of valuable compounds and also sheds light on some crucial real-world case studies that are broadly based on the US Presidential Green Chemistry Challenge

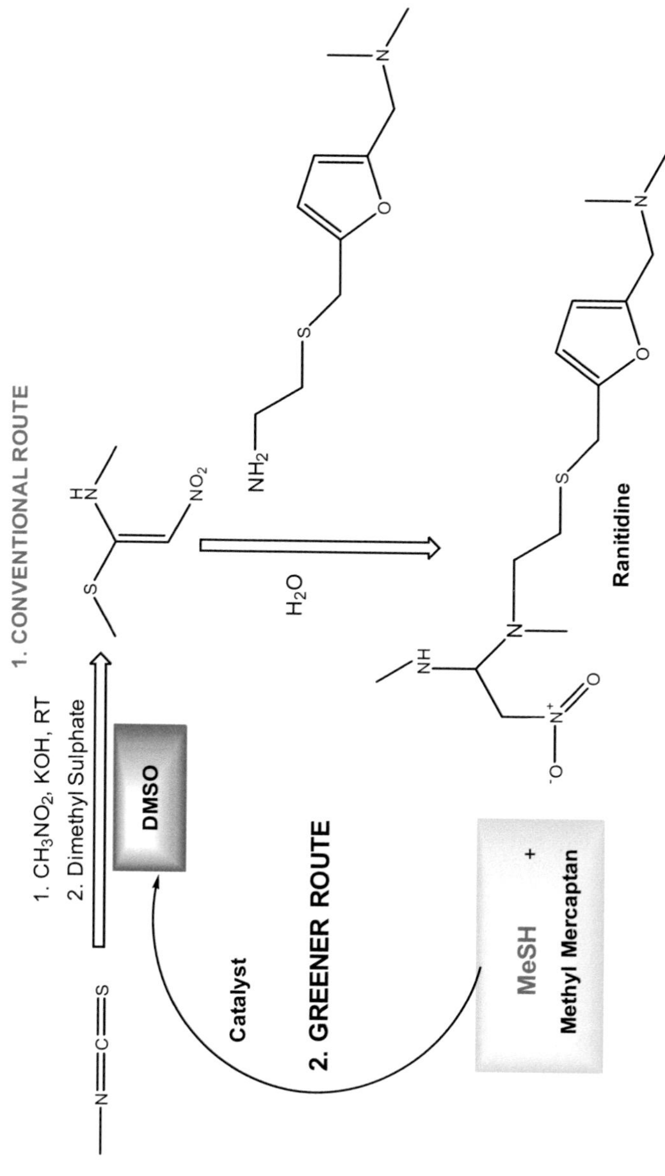

Scheme 6.20 Old versus greener routes for the synthesis of ranitidine drug.

Awards. These case studies provide a deep insight into how greening of chemical synthesis can be accomplished and what would be the large-scale benefits of doing so. It is certain that if the green chemistry principles outlined can be followed/implemented, one can contribute immensely toward a sustainable planet with improving human health as well as environment. For instance, the phasing out of deleterious organic solvents, substitution of homogeneous metal salts with recyclable heterogeneous catalysts, and replacement of conventional fossil fuel–based chemicals with renewable feedstocks have led to the greening of industries. The chapter concludes by discussing the need to fruitfully integrate the efforts of industrialists and academicians through close collaboration to green our planet.

6.6 Learning Outcomes

At the end of this chapter, students will be able to:

- Understand the benefits of greener protocols
- Compare conventional routes with alternative greener routes for the synthesis of valuable compounds
- Learn about the various real-world case studies on green chemistry that will enable them to understand how the practical implementation of green chemistry is beneficial in addressing many of the current global environmental and synthetic challenges
- Understand why the industry needs to join hands with academia for deriving maximum benefits
- Acquire knowledge about the Presidential Green Chemistry Challenge Awards and appreciate the work done by several companies toward sustainable development

6.7 Problems

1. How is the conventional method of antifoulant posing an environmental threat? How can this be overcome using a green methodology? Explain with example.
2. Explain how the cradle-to-cradle recycling concept is applied to carpets.

Chapter 7

Green Chemistry in Education, Practice, and Teaching

Reena Jain,[a] Anju Srivastava,[a] Manavi Yadav,[a,b] and Rakesh K. Sharma[b]

[a]*Department of Chemistry, Hindu College, University of Delhi, Delhi 110007, India*

[b]*Green Chemistry Network Centre, Department of Chemistry, University of Delhi, Delhi 110007, India*

reenajain_70@yahoo.co.in

Foster the development of the most innovative, relevant, and effective chemistry education in the world.

—ACS, 2014

7.1 Introduction

In 2015, the world came on a common platform to agree on a set of universal goals, the 17 sustainable development goals, that carry a vision for the planet as a world of good health, quality education, gender equality, safe drinking water, clean energy, and decent work

Green Chemistry for Beginners
Edited by Rakesh K. Sharma and Anju Srivastava
Copyright © 2021 Jenny Stanford Publishing Pte. Ltd.
ISBN 978-981-4316-96-5 (Hardcover), 978-1-003-18042-5 (eBook)
www.jennystanford.com

for all [1]. As we write this section for the book, we are passing through by far the biggest global crisis to date, the coronavirus disease 2019 (COVID-19) pandemic, and it is in these times that the pertinence of such universal goals is felt the most.

With not even 3700 days left, till the end of the year 2030, to achieve these goals, which will define our future into this century and beyond, large-scale innovations and solutions are required involving frontier science and technologies to transform our commitments into action.

Training the next generation of chemists on green chemistry practices is, therefore, needed more than ever before. Of course, ultimately the aim is to remove the word "green," implying, thereby, that the only way chemistry is taught and practiced is by using the approach and philosophy of green chemistry alone. It is not that green practices have happened only when the term "green chemistry" was introduced in 1990s. For example, even in the 1940s, catalytic methods such as iron-catalyzed Grignard additions were developed, though for reasons of iron being cheap and the reaction being a high-yield one. Green, besides referring to less or zero negative impact on the environment, also means low cost, waste minimization, and better efficiency. All of the latter inferences attract the attention of a far broader range of stakeholders, from industrialists, policy makers, and governments to scientists, researchers, educators, and students.

In a grand scheme of things, the ultimate result is important, that is, whether the term is used or not, the principles must be used. In the industry, the pathway taken to make an intended product is not as important as ensuring the end goal is met safely and effectively. So, for academicians and researchers, there is a huge opportunity to develop reactions at reduced costs or with reduced waste generation by employing any green chemistry principle, such as using renewable feedstock or a greener solvent or reducing the steps from the number of steps in the original pathway.

For these reasons, green chemistry must have a solid entry into a chemistry student's education so the development of green reactions becomes the only way of conducting reactions and practices.

This chapter deals with the efforts in the direction of strengthening the amalgamation of green into chemistry, so ultimately the word "green" can be removed from the term.

7.2 Green Chemistry in Classroom

A majority of students in today's times are sensitive about the growing environmental issues and are keenly interested in the sustainability of the world. However, their understanding of the causes of these issues may not be at par with their sensitivity or their keenness.

In the 1990s, the Pollution Prevention Act was passed, which shifted the focus of environment protection from controlling pollution to preventing pollution [2]. With this act was also brought in the concept of sustainability to address the environmental concerns and protect the future. Knowledge of and education on sustainability are so important that the United Nations declared the decade of 2005 through 2014 as the "Decade of Education for Sustainable Development." The year 2020 ushers in a decade, termed as the "Decade of Action," that calls for integrating the principles, values, and practices of sustainable development into all aspects of education and learning in order to address the economic, social, and environmental problems we face in the twenty-first century.

In this context, there is a huge need to teach chemistry from the perspective of contributing towards sustainable development, which would enable the students to design and develop environmentally benign products and processes. The unique contribution of our discipline to sustainability is green chemistry. Green chemistry is only 30 years old, but due to its inherent design, it has provided opportunities to address key weaknesses of the chemistry curriculum. All students of chemistry need to be well versed in environmental problems and potential solutions, and green chemistry gives them a unique opportunity to become scientists equipped in the right way to secure a healthy planet for future generations [3].

Green chemistry education will make it possible to build a workforce of educators, scientists, engineers, and innovators who will develop and practice sustainable solutions, whether it is in agriculture, medicine, energy alternatives, or luxury lifestyle products. To realize this mission, this workforce needs to be first trained in the philosophy, principles, and methodologies central to green chemistry.

The definition of green chemistry has been carefully worded and talks of several concepts in each word or phrase. The word "design" encompasses the criteria, principles, and methodologies of green chemistry. Similarly, the term "use or generation" necessitates the life-cycle assessment of any substance (cradle to cradle) and ensuring that green chemistry can be utilized at any stage of its life cycle. The word "hazardous" is used in a very broad sense, including specific and immediate hazards (corrosive, flammable, etc.) to not-so-distant hazards (carcinogenic, teratogenic, etc.) to far-reaching hazards (leading to ozone depletion, global warming etc.).

As an area of study, green chemistry is steadily making its way into the classroom and laboratory [4]. In the initial years of the introduction of the concept of green chemistry, the approach was to infuse the concept into the mainstream chemistry curriculum in order to best address the environment challenges. This idea faced several insurmountable barriers. While incorporation of the concept was entertained by curriculum experts, it was seen that topics on green chemistry were pushed to sidebars or into colored text boxes for optional reading in textbooks or confined to specialized courses. There were many reasons for this response to the subject, the primary reasons being an already overcrowded syllabus, limited education material on green chemistry that would help to teach the subject, and very few educators able to execute the goal. This trend continued for a substantial period. It was the last decade that saw an appreciable and increasing acceptance of green chemistry. This change has come with a lot of hard work and conscious efforts made, particularly in terms of strengthening the weak links, to overcome the barriers described.

Focus was drawn to empowering the educators by designing and developing quality education material highlighting the principles, tools, and strategies of green chemistry, which could replace old content and become a part of the curriculum [5, 6]. For classroom interaction, lecture modules were prepared in advance, supported by slides and handouts of case studies and success stories that were to be incorporated into the existing chemistry lectures.

The laboratory exercises were thoroughly checked for their reproducibility and supported with sufficient pre- and postlab information and worksheets. Older laboratory exercises were modified or extended to make them safer, friendlier, and thought provoking.

These efforts helped in providing green chemistry a strong foothold in the curriculum and garnering appreciation of the concepts associated with the 12 principles of green chemistry. Stakeholders began to realize and recognize the importance of this concept.

In parallel, a greater number of faculty were encouraged to join so capacity could be built up and a network set up among them, through sharing of resources, tools, and strategies. A common platform for sharing was utilized through the web that accelerated the dissemination and adoption of green chemistry.

In the last 20 years, a significant amount of education material has been collected and used primarily for two teaching methodologies for undergraduates and graduates—one in which the study material has been incorporated into the existing chemistry curriculum and the second wherein standalone courses have been designed. Both methodologies have their challenges and strengths. Therefore, both are being employed by academic institutions according to their circumstances. In general, the heartening news that emerges is that the interest in studying green chemistry in a structured manner has seen a noteworthy jump.

Professional societies, like the American Chemical Society (ACS) and the Royal Society of Chemistry London, in collaboration with the academic community, were the first in compiling in their respective countries educational material for green chemistry in a self-explanatory and interesting manner in the form of textbooks, case studies, and laboratory experiments [6]. Ample examples of green chemistry in action, supported by the theoretical background, drew the attention of teachers across countries. The popularity inspired many other professional societies, and soon German and Japanese societies prepared material to promote green chemistry education for their people. While the target audience of these materials is undergraduate and graduate chemistry students, senior secondary school students, faculty, and chemists working in industries also find the material most useful and informative. Other educational initiatives of these societies have been faculty training, school teacher training, symposia, professional workshops, and summer and winter schools. Nongovernment organizations (NGOs), like Beyond Benign, Wilmington, Massachusetts, US, spread green chemistry in a unique manner by conducting a lot of outreach programs to spread awareness about green chemistry, involving the students from schools, universities, and colleges.

The first college-level course in green chemistry was taught to graduate as well as advanced graduate students in the year 1992 by Professor Terry Collins at Carnegie Mellon University, US.Around the same time, the University of Nottingham, UK, started an undergraduate course in green chemistry. By the mid-1990s, the University of Oregon, US, had introduced green chemistry curriculum in chemistry laboratory teaching. In the last 15–20 years, green chemistry courses have started occupying place as stand-alone courses in colleges and universities in most parts of the world. European universities with programs at the master's level include the University of York, UK; Lund University, Sweden; Universidad Zaragoza, Spain; the University of Venice, Italy; the University of Patras, Greece; the University of Leuphana, Greece; the University of Strasbourg, France; and the University of Copenhagen, Denmark. While Oxford University is doing a graduate program, a research-based master's program in green chemistry is being conducted by Imperial College, the University of Bath, and the University of Lenceister, all in United Kingdom. The University of Monash, Australia, also conducts research programs in this field.

Today, we have well-established undergraduate and graduate courses in green chemistry. In the United States, some of the institutions that have launched graduate programs in green chemistry are the University of Toledo, Ohio; the University of Massachusetts Lowell; and the University of California, Berkeley. The green chemistry curriculum at the Yale-UNIDO University was developed jointly by the Center for Green Chemistry & Engineering at Yale and the United Nations Industrial Development Organization (UNIDO), under the dynamic leadership of Professor Paul T. Anastas [7].

At the Berkeley Center for Green Chemistry (BCGC), the University of California, Berkeley, US, a program titled "Greener Solutions," makes the students work in close partnership with various companies, nonprofits, and/or government agencies [7–9].

In all, green chemistry courses across nearly 40 colleges and universities of the United States have put in their efforts to popularize green chemistry among students and teachers. In Canada, undergraduate and postgraduate programs on green chemistry and in some cases interdisciplinary programs have been running successfully.

Over the years of efforts, a robust collection of educational materials has been developed that has enabled the permeation of green chemistry into the curriculum. Professional bodies, like the ACS Green Chemistry Institute (GCI), through conferences and workshops designed especially for the development of educational material, have been able to raise a formidable compilation of educational material in the form of an annotated bibliography, laboratory exercises, lecture modules, and videos [9]. Very early in this journey, a database of green chemistry education material (GEMs) for chemists was built that, among other materials, included chemical concepts useful in understanding green chemistry [8]. Today, there are many exercises that highlight the use of the 12 principles of green chemistry in a variety of scenarios. Some leading textbooks on introductory chemistry have also infused important green chemistry content.

Educational material has even been translated into Spanish and Portuguese. A step beyond is where non-chemistry-background students are also introduced to the relevance of green chemistry in the current times of societal challenges involving sustainability. This hugely challenging task is accomplished with the aid of external projects and case studies [10].

In India, green chemistry education initiatives began 20 years ago, with short-term courses being introduced in colleges of Delhi University. Around this time, the Department of Science and Technology, Government of India, constituted a task force in collaboration with experts such as Mihir K. Chaudhuri, S. Chandrasekaran, and B. C. Ranu. The task force invited Professor R. K. Sharma, coordinator Green Chemistry Network Centre (GCNC), Department of Chemistry, University of Delhi, India, to prepare a monograph of green chemistry experiments for undergraduate and postgraduate students [11]. This monograph includes interesting experiments that illustrate how green chemistry can be put into action with the incorporation of green chemistry principles and why the conventional synthetic techniques should not be followed/practiced. Also, green chemistry has acquired immense significance in the curriculum of Delhi University as three diverse papers (Skill Enhancement Course, Discipline Specific Elective, and Generic

Elective Courses) have been introduced in the undergraduate degree courses of Delhi University, which also include the laboratory component. Similarly, the University of Rajasthan, India, is running a master's program in green chemistry and the syllabus of this course is designed by Professor Sharma.

Further, it is worth mentioning here that in order to spread the significance of green chemistry in India, GCNC was set up at the Department of Chemistry, University of Delhi, India, in the year 2003 under the recommendation of a panel of world leaders headed by Professor Anastas, which has played a key role in the promotion of green chemistry education at the level of schools, colleges, and universities. The green initiatives by this center have been truly beneficial in spreading awareness on the need to adopt green chemistry, revolutionizing the educational sector of India. Under the able leadership of Professor Sharma, every year GCNC has been hosting a myriad of activities, including conferences, workshops, and symposia, at both national and international levels on crucial issues related to green chemistry. GCNC has also been instrumental in research initiatives, particularly in developing sustainable and cost-effective solutions for combating environmental and socioeconomic challenges.

In Hendrix College, Arkansas, US, the entire curriculum at the Chemistry Department is green-sensitive. It is mandatory for the general chemistry students to participate in the Green-SWAT (Green Soil-Water Analysis at Toad Suck) programs while the organic chemistry students work with the Toad Suck Institute for Green Organic Chemistry [8]. They have a zero-effluent laboratory in which the waste and effluents are minimized and solvents are reused.

7.3 Green Chemistry in a Teaching Laboratory

Laboratories for undergraduates and postgraduates have been traditionally using toxic, corrosive, and carcinogenic substances on a routine basis. To develop practical skills, students are trained to handle chemicals, apparatus, and instruments under normal lab settings. The usual time spent by a student in a chemistry laboratory at the undergraduate and postgraduate levels is more than 12 hours per week. Inside these laboratories, the experiments conducted are age-old, with no relevance in current times or links with the modern-time chemistry practiced in industries and academic research laboratories. These experiments include preparing various organic and inorganic compounds that involves various harmful chemicals and toxic solvents and release of noxious fumes. Qualitative analyses, in particular, consist of concentrated acids, thionyl chloride, phosphorous pentachloride, hydrogen sulfide, benzoyl chloride, lead(II), arsenic(III), mercury(II), cadmium(II), and many more. The reactions involved in qualitative analyses evolve large amounts of toxic fumes, which cannot be prevented as most of the laboratories have been built many years ago and are not well ventilated or well lit. This poses further risk of undue exposure to undesirable fumes and odors, thereby affecting students, laboratory staff, and teachers. Some other examples are nitration reactions that involve fuming nitric acid and sulfuric acid and distilling phenol and various other liquid organic compounds, especially in cases where peroxides are formed. Students are thus given a word of caution to handle such substances with care. They perceive laboratories to be synonymous with risk and hazard, which develops in them an aversion toward the subject and a perception of labs being risky and hazardous.

Academic laboratories also generate a lot of solid and liquid wastes, which are difficult to manage, especially the liquid waste, which is directly thrown into the sink, which eventually leads to water contamination. Experiments involve macroscale preparations of several compounds that require longer reaction times and high-temperature conditions. Curricula emphasize on yields of the desired products and are indifferent toward the side-products generated, which are ultimately dumped as waste. Therefore, students remain unaware of the amount of waste generated in the overall process. In most cases, the product is not employed as a starting material for other experiments and remains as a waste in the record files of the students.

The cost of waste disposal, chemicals, and construction makes the new laboratories too expensive to be maintained within the budgets of the academic institutions [10, 12]. Here is where the approach of green chemistry provides a tremendous opportunity to remove all the shortcomings of the old and new laboratories used in undergraduate courses. Green chemistry laboratories use alternate starting materials, solvents, reagents, and reaction media that reduce risks and hazards and are inherently safe and environment friendly. Need for fuming hoods is minimized, and recycling and reuse of materials cuts down the costs associated with running the laboratories. All these enable the students to develop a natural liking for the subject and enthuses them to develop sustainable solutions for a better world. Through these experiments, the students learn to think critically and compare the effects of conventional and green experiments on the planet and become sensitive to how to embed green chemistry principles and metrics into their routine chemistry experiments.

Even before the introduction of green chemistry, a laboratory manual entitled *Zero Laboratory Effluent Laboratory Manual* was published by Professor Tom Morton in 1973 [13]. Arising out of recommendations in the green chemistry program at the University of Oregon, US, a manual entitled *Green Organic Chemistry: Strategies, Tools, and Laboratory Experiments* was published in 2004. In India, the first monograph on green chemistry experiments was published in the year 2013.

In 2011, BCGC at the University of California, Berkeley, US, launched its first interdisciplinary graduate-level course in green chemistry, along with experiments incorporating green chemistry principles. Rice University, Texas, US, also began an undergraduate laboratory in green chemistry. At the organization called Beyond Benign, Wilmington, Massachusetts, US, teaching of green chemistry in laboratory classes featured waste reduction, economic benefits, and hazard reduction. This greening of laboratory experiments not only allowed students to perform chemistry outside the traditional lab settings but also equipped them with the tools to assess and improve chemical processes.

To get a more effective impact of the benefits accompanying a green chemical reaction, students need to perform conventional methods alongside corresponding greener alternatives. Instead of being satisfied with the successes from hazardous traditional methods, all teachers must train the future chemists to inspect methods more closely and appreciate risk reduction and waste prevention through green chemistry (Fig. 7.1).

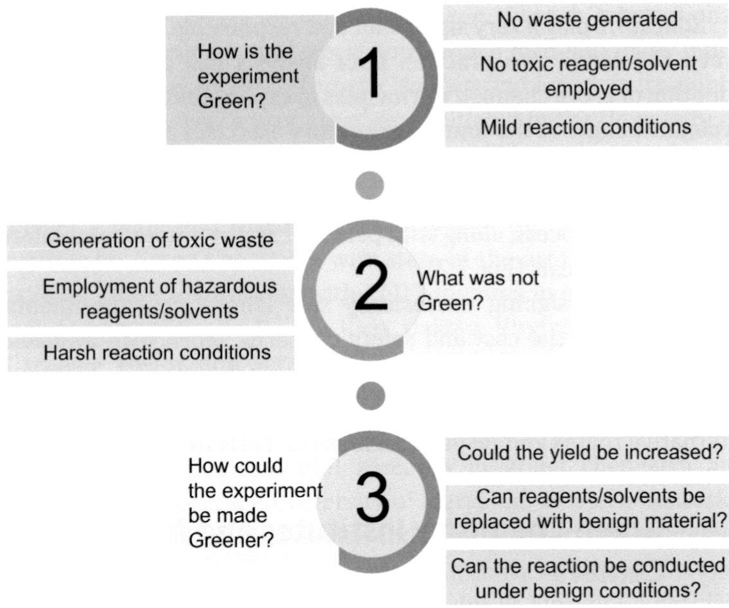

Figure 7.1 Questions that students must be able to answer after performing any experiment.

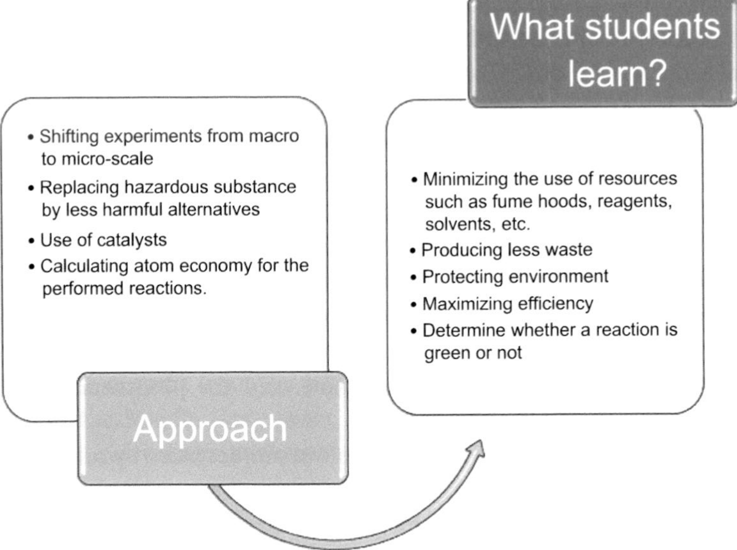

Figure 7.2 Approaches to emphasize the greenness of a chemical process.

in Green Chemistry, and Denmark's Centre for Green Chemistry and Sustainability.

The ACS GCI has always been the driving force behind the promotion of green chemistry. The ACS GCI, in collaboration with Green Chemistry and Commerce Collaboration, launched a portal via which one can visualize the growing network of organizations involved in green chemistry. BCGC is a central node in the network, connecting industry and academia and bridging the fields of chemistry, public health, and environmental sciences.

With time, more and more universities and network centers, not only in the United States, but all across the world, started promoting and popularizing green chemistry.

With the kind of awareness that has been generated for green chemistry, certain NGOs, such as BIZ, are also working in collaboration with business and environmental leaders to advance healthy materials and a safer chemicals economy.

In Australia, Monash University, the University of Queensland, and the University of Sydney offer PhD programs in green chemistry and green nanotechnology, while in Canada, Centre in Green Chemistry and Catalysis (CGCC) is promoting green chemistry research.

The European Union is also coming up in popularizing green chemistry. Sustainable Industrial Chemistry, the European doctoral program on sustainable industrial chemistry, is a three-year international program run successfully by the European Union. It is offered by a consortium of 33 partner institutions with strong industrial links. Ecole Polytechnique Federale de Lausanne, Switzerland; Institute of Science and Technology (IUST), Universidad Complutense de Madrid, Spain; and Gothenburg/Chalmers University of Technology, Spain, conduct research in green chemistry whereas Portuguese Science and Technology Foundation, Portugal, and Delft University of Technology, Netherlands, offer PhD program in related areas.

In the United Kingdom, the University of Oxford, the University of York, the University of Leicester, Durham University, and Queen's University are popular hubs for research activities in green chemistry. The Green Chemistry Centre of Excellence, University of York, UK, is now a world-leading academic facility for pioneering pure and applied green and sustainable chemical research. To spread awareness among the nonscientific audience, the University of

Nottingham, UK, even started funding a "Public Awareness Scientist" position at the university. The research scholars of the university also interact regularly with the young students in secondary schools to give them an understanding of green chemistry.

The National Environment Research Institute, an integral part of the National University of Singapore, Singapore, focuses on the development of integrated sustainability solutions for the environment. It collaborates with NGOs, industries, government agencies, and the academia to consistently address real-world issues. A PhD program in green chemistry is offered by the University of Hong Kong, Hong Kong. The Centre for Green Chemistry and Catalysis in Chinese Academy of Sciences, China, is conducting research programs.

Green initiatives under the leadership of the father of green chemistry led to the establishment of the GCNC at the Department of Chemistry, University of Delhi, India. Under the guidance of Professor Sharma, GCNC coordinator, this center has helped build a strong network for the exchange of ideas and expertise between academicians, chemists, engineers, and industries. Besides promoting research, it also prepares and disseminates educational materials for schools, colleges, and universities. This center also designs training programs, workshops, seminars, symposiums, career counseling, and conferences to promote green and sustainable chemistry and discuss its advancement at the national and international levels. As a result of the hard work put in by the team members of GCNC, green chemistry is offered as an elective paper under the choice-based credit system, followed in all colleges and universities across India.

The global market for green chemistry is predicted to grow exponentially to US$98.5 billion by 2020. *Source*: ACS.

7.5 Important Journals and Websites

Appreciating the relevance of green chemistry, the Royal Society of Chemistry, London, UK, started publishing a journal *Green Chemistry* and provided a forum for the publication of new and innovative research on the development of sustainable and alternate green technologies. The journal was started in 1999, by the efforts of James Clark, from the University of York, London, UK. With more and more research being conducted in green chemistry, several journals with a high-impact factor can now be accessed easily. Table 7.1 depicts some of the major journals on green chemistry and sustainable chemistry.

Table 7.1 List of major journals

No.	Journal name	Publication
1.	Green Chemistry	The Royal Society of Chemistry
2.	ACS Sustainable Chemistry and Engineering	American Chemical Society
3.	ChemSusChem	Wiley
4.	Current Opinion in Green and Sustainable Chemistry	Elsevier
5.	Green Chemistry Letters and Reviews	Taylor and Francis
6.	Green and Sustainable Chemistry	Frontiers in Chemistry
7.	Green and Sustainable Chemistry	Scientific Research Publishing
8.	Trends in Green Chemistry	iMedPub LTD
9.	Journal of Chemical Education	American Chemical Society
10.	International Journal of Green Chemistry and Bioprocess	Universal Research Publications
11.	Asian Journal of Green Chemistry	Sami Publishing Company
12.	Current Green Chemistry	Bentham Science
13.	Current Research in Green and Sustainable Chemistry	Elsevier

Due to the efforts put in by the pioneers in green chemistry, a lot of study material is now available on the GEMS site, whereas to find individual teachers, one must log on to the Green Chemistry Education Network site. Some more relevant websites for green chemistry are enlisted in Table 7.2.

Table 7.2 List of important websites

No.	Organization/Institute	Website
1.	American Chemical Society Green Chemistry Institute	https://www.acs.org/content/acs/en/greenchemistry.html
2.	Green Chemistry United States Environmental Protection Agency	https://www.epa.gov/greenchemistry
3.	Beyond Benign, Green Chemistry Education	https://www.beyondbenign.org/
4.	Exploring the Role of Green Chemistry at Harvard	https://green.harvard.edu/news/exploring-role-green-chemistry-research-university
5.	Center for Green Chemistry & Green Engineering at Yale	https://greenchemistry.yale.edu/education
6.	Green Chemistry Education Network, University of Oregon	http://cmetim.ning.com/
7.	Berkeley Center for Green Chemistry University of California, Berkeley	https://bcgc.berkeley.edu/
8.	Institute for Green Science, Carnegie, Mellon University	https://www.cmu.edu/mcs/research/areas/greenchem-environment/index.html
9.	Green Chemistry Excellence Centre, University of York	https://www.york.ac.uk/chemistry/research/green/
10.	Greening Across the Chemistry Curriculum, University of Scranton	https://www.scranton.edu/faculty/cannm/green-chemistry/english/index.shtml
11.	Green Chemistry Assistant, St. Olaf College	https://www.stolaf.edu/apps/chemistry/gca/
12.	Green Chemistry, University of Oregon	http://greenchem.uoregon.edu/
13.	iSUSTAIN Green Chemistry Index v2.0	https://www.isustain.com/
14.	Interuniversity National Consortium "Chemistry for the Environment"	http://www.incaweb.org/
15.	Warner Babcock Institute for Green Chemistry	https://www.warnerbabcock.com/
16.	Green Chemistry Network Centre, University of Delhi, Delhi, India	http://greenchem.du.ac.in

4. Explain the role of a student while performing any chemical reaction.
5. List any five green chemistry centers.
6. What can educators do to bring green chemistry into lab practice?

References

1. https://www.un.org/sustainabledevelopment/sustainable-development-goals/
2. Anastas, P. T. and Kirchhoff, M. M. (2002). Origins, current status, and future challenges of green chemistry. *Acc. Chem. Res.*, **35**(9), pp. 686–694.
3. Andraos, J. and Dicks, A. P. (2012). Green chemistry teaching in higher education: a review of effective practices. *Chem. Educ. Res. Pract.*, **13**(2), pp. 69–79.
4. Anastas P. T, Beach, E. S. (2009). *Green Chemistry Education,* ACS Symposium Series, Changing the Course of Chemistry.
5. Hjeresen, D. L., Boese, J. M. and Schutt, D. L. (2000). Green chemistry and education. *J. Chem. Educ.*, **77**(12), pp. 1543.
6. Haack, J. A. and Hutchinson, J. E. (2016). Green chemistry education: 25 years of progress and 25 years ahead. *ACS Sustainable Chem. Eng.*, **4**(11), pp. 5889–5896.
7. https://www.acs.org/content/acs/en/greenchemistry/students-educators/academicprograms.html
8. http://advancinggreenchemistry.org/benchmarking/education/green-chemistry-is-emerging-in-academic-institutions-all-over-the-world/
9. https://www.migreenchemistry.org/toolbox/directory/
10. Kennedy, S. A. (2016). Design of a dynamic undergraduate green chemistry course. *J. Chem. Educ.*, **93**(4), pp. 645–649.
11. http://greenchem.du.ac.in/aboutus.html
12. Zuin, V. G. and Mammino, L. (2015). *Worldwide Trends in Green Chemistry Education* (Royal Society of Chemistry, London).
13. https://www.hendrix.edu/uploadedFiles/Departments_and_Programs/Chemistry/Green_Chemistry/ACS%20talk%20and%20lab.pdf
14. http://cires.colorado.edu/env_prog/chemrawn/

Chapter 8

Green Chemistry: Vision for the Future

Pooja Rana,[a] Sriparna Dutta,[a] Anju Srivastava,[b] and Rakesh K. Sharma[a]
[a]*Green Chemistry Network Centre, Department of Chemistry, University of Delhi, Delhi 110007, India*
[b]*Department of Chemistry, Hindu College, University of Delhi, Delhi 110007, India*
pooja.rana.kmc@gmail.com

> *A key path to a sustainable future is if the chemists invent better technology.*
>
> —John C. Warner

8.1 Introduction

In the preceding chapters, we have seen how green chemistry has improved the existing processes or made them completely sustainable, changing the age-old perceptions of industrialists, ultimately leading to a win-win situation for all the stakeholders. The impetus for the same was provided in the form of the

Green Chemistry for Beginners
Edited by Rakesh K. Sharma and Anju Srivastava
Copyright © 2021 Jenny Stanford Publishing Pte. Ltd.
ISBN 978-981-4316-96-5 (Hardcover), 978-1-003-18042-5 (eBook)
www.jennystanford.com

- Lack of research plans that can enable and expedite technology transfer within academia, government organizations, and industries
- Lack of patent extension grant for greener method optimization

All these factors if taken into account can facilitate the adoption of environmentally benign processes.

Education challenges

The significance of green chemistry education has been recognized since its birth. Education is the forerunner of implementation, and hence it is important to educate our society, especially students, who are also our future, for developing and using products in an environmentally benign manner. We need to equip students with tools necessary to support and promote global sustainability. However, some of the key challenges that prevent us from achieving the goals of sustainability are:

- Lack of appropriate educational materials, tools, and resources for teaching green chemistry
- Effective dissemination strategies
- Lack of emphasis on enabling students to recognize hazard/toxicity
- Lack of fundamental emphasis on making students understand the need to replace "yield" with "atom economy"

At the fundamental level, they should be introduced to the basic tenets of green chemistry. Not just for students, but green chemistry is important for all the stakeholders, including policy makers, business people, and the general public. It teaches problem-solving skills and instills critical thinking with valuable skills to innovate, being principle based.

"The astounding accomplishments of Green Chemistry and Green Engineering thus far are pale in comparison to the power and the potential of the field in the future," says Anastas. With the effective proposition and implementation of green chemistry solutions and intervention of appropriate stakeholders—policy makers, legislators, environmentalists, chemists, and government agencies—it is anticipated that we will be able to eradicate all issues and render the planet earth greener and safer.

8.3 Future Directions: Focus of the Future Researchers

The focus of the chemical sector now and in the future should be conserving the efficacy of function while simultaneously cutting down on various hazards, such as physical (explosivity, flammability, corrosivity, etc.), toxicological/health (e.g., carcinogenicity and endocrine disruption), and universal (e.g., greenhouse gas emission, aquatic toxicity, and challenges of ozone layer restoration). Catastrophic results can occur due to negligence in risk management and exposure control mechanisms.

The following attributes should be given due consideration while developing any process/product.

8.3.1 Nondepleting Nature

A thoughtful design is required within the integrated system context for metamorphosing from nonrenewable to renewable chemistry by considering the deleterious consequences that could result from either land transformation or utilization of water resources or food production [16]. A paradigm shift toward renewable chemistry and benign processes can be achieved by adopting the circular processes (Fig. 8.1) rather than following the linear approach. The concept of circular economy was given due attention in Chapter 2. In circular processes, waste can be rendered as a valuable resource in the chemical sector that eliminates the challenges and repercussions of disposal and treatment, for example, adopting the abundant CO_2 in the petroleum industry for the production of polyurethanes, which not only act as a good substitute for propylene oxide but also resolve the problem of global warming [16–19].

8.3.2 Nontoxic Nature

The goal of designing nontoxic chemicals can only be accomplished through collaborative partnerships between different related fields, like chemistry, toxicology, and genomics. For this, one should have the expertise and understanding of the primary molecular mechanism of absorption, distribution, and metabolization in and excretion from the body and what would be the influence of various sustainable synthesis, including low energy requirements, catalytic amount, superior selectivity, diminished consumption, and superior separation, also allowing the use of less hazardous materials. We have already discussed the basics of catalysis in Chapter 3 and have seen how green chemistry can be coupled with nanotechnology for designing green catalysts. The heterogenization approach has proven to be beneficial in accomplishing the key goals of green catalysis: (a) activity, (b) durability, and (c) recyclability on a single platform. For doing so, the homogeneous metal complexes are immobilized (tethered) onto different solid support materials, such as silica, titania, graphene oxide, carbon nanotubes, zeolites, clays, and ferrite nanoparticles. The solid support helps in holding the active sites through strong coordination during a catalytic process. In this way, the best attributes of both homogeneous (high activity, selectivity, etc.) and heterogeneous catalysts (facile separation and excellent reusability) can be integrated. Once the solid support is nanosized, tremendous enhancement in catalytic activity is observed due to an overall increase in the surface-area-to-volume ratio [28]. Several ecofriendly, recyclable, and competent catalysts have been designed using this strategy.

However, in view of urgent environmental and economic needs, the focus of the future researchers should be on utilizing renewable and abundant raw materials as precursors for designing catalysts and substitute the unsustainable use of expensive, toxic, and fossil resources as catalysts. Additionally, the designed catalyst should be synthesized in a sustainable manner in the minimum number of steps possible, which is usually ignored during the conventional synthesis of a catalyst. Also, to mitigate issues related to climate change, we need to shift our emphasis to engineering a more sustainable world through green catalysis. From this perspective, efforts must be directed toward valorization of unavoidable waste in order to achieve a circular economy. All the progress in developing sustainable catalysis that has been accomplished so far remains mostly at the laboratory level; in the future, we must try new tactics to assimilate lab and industry outcomes to scale up the use of green catalytic material and processes [29].

8.4.4 Catalyst-Free Reactions in Organic Synthesis

From the previous section, we could fundamentally infer that catalysts possess outstanding advantages, such as the ability to reduce waste, cost, and labor during the process of making and breaking of bonds. However, the toxicity that could arise because of metal contamination, longer reaction time, expensive nature, and limited selectivity prevails in most of the cases may come in the way of large-scale industrial application. To overcome these disadvantages, attempts are being directed toward the development and design of catalyst-free protocols to render syntheses greener and more sustainable. It is envisioned that significant catalyst-free reactions and processes will be developed, designed, and utilized by researchers and chemical industries in near future [30].

8.4.5 Energy-Efficient Synthesis

"Programming" a process in a cascade manner can lead to a reduction in the amount of solvent or reagent and minimize or eradicate unwanted by-products, streamlining the synthesis of natural products. As we discussed in Chapter 5, alternative renewable driving sources of energy, such as microwave irradiation, ultrasonic irradiation, photochemical, and electrochemical reactions, can be employed to accomplish the goals of a greener sustainable synthesis. However, there is an urgent need to develop and utilize more value-added renewable energy-efficient methods for the future cyclic processes. The main aim should be achieving an energy-efficient process that will have the capability to fulfil the need of today's as well as tomorrow's generation without harming the biosystem and can be practically retained in the industry on a large scale [3, 23, 31, 32].

In the future, we should not rely on the use of nonrenewable fuel sources (e.g., crude oil, gasoline, and coal). We should move toward improved synthetic methods based on nonfossil carbon sources and renewable energy resources that would have a positive impact on the environmental footprint. Scientists have shown interest in renewable carbon sources as well as the "decarbonized" energy of hydrogen technologies and electrochemical processes. A circular economy can be efficiently propagated by utilizing the abundant raw and expertise will be required to combat the challenges and embrace the promises. Some of the current challenges confronting the researchers are the probability of the results being different when miniaturized and the lack of true commercial standardization, as exists with larger-scale, undesired variability (with the reduced quantity there may be the possibility of getting interference from variables) [36].

8.6 Biomimetic: Green Chemistry Solution

Whoever knew that the solution to most of our critical challenges would lie in the hands of Mother Nature? Delightfully, nature provides an enormous scope to tune and model reactions by applying similar principles. Biomimetics, or biomimicry, is the science and art of emulating nature's best biological ideas by understanding the principles of the underlying mechanisms to improve human life by successful implementation in science, engineering, and medicine [37]. The term "biomimetics," originating from the Greek words *bios*, meaning life, and *mimesis*, meaning to imitate, was introduced by Schmitt in the year 1957. He proclaimed it as a cornerstone for biology and technology. Since its inception, this field has shown its existence in our day-to-day life with or without our knowledge and has always evolved along with human civilization, from blades and hatchets instigated by dental structures to the strongest cutting-edge carbon nanomaterials. In 1903, The Wright brothers developed and designed a mechanical airplane by utilizing the aerodynamic principle of the wings of eagles. The fundamentals of biomimicry are also reflected in the work of Leonardo da Vinci. He also designed a "flying machine" inspired by a bird. Figure 8.5 highlights some of the champion adaptors that provide potent new biomimetic chemistry-based solutions [38].

Biomimetic chemistry particularly relies on deriving information from biology and infusing it with chemistry. A number of biocatalytic chemical modifications exploiting the concept of biomimetics have been achieved successfully in a sustainable manner, and these transformations have been found to be fast, stereospecific, and single-step. In biomimetic transformations, the target molecule is synthesized by mimicking the enzymes present in the natural system

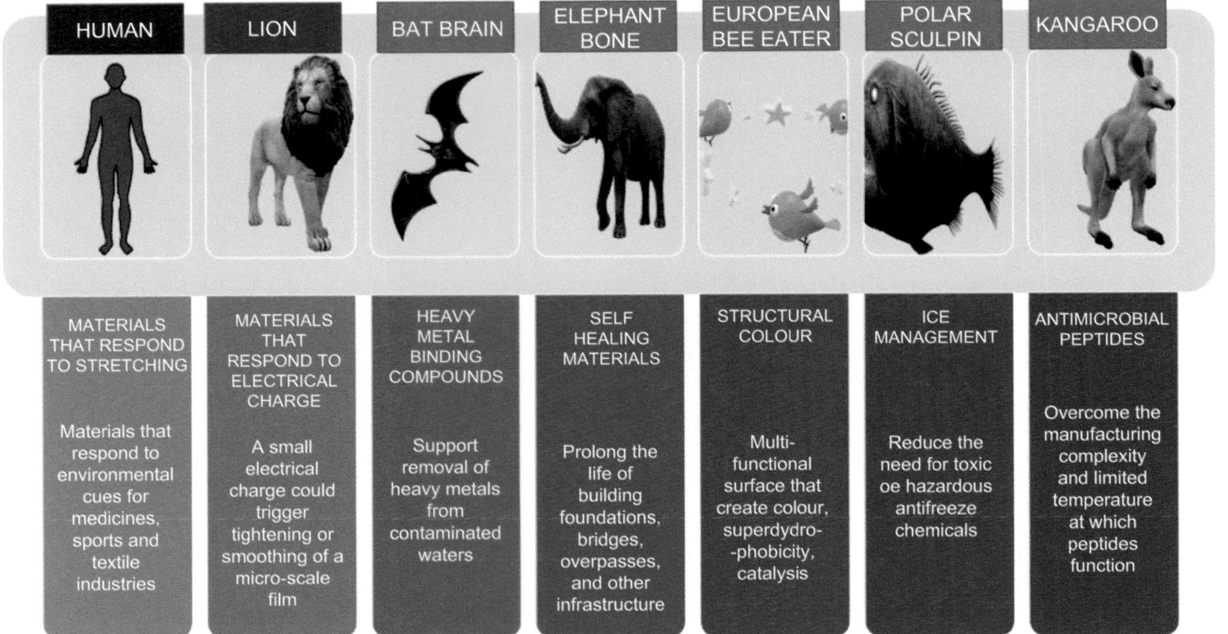

Figure 8.5 Seven champion adaptors that can inform new biomimetic chemistry-based solutions.

to carry out a transformation. So, fundamentally, it is the synthesis of molecules that imitate enzymes and help in the development of artificial enzymes. The beauty of enzymatic transformations is that they do not require any protection and deprotection of functional groups, which leads to a reduction in waste as the number of steps are reduced. For example, aldol reaction based on enzymatic chemistry fits in easily with nature's chemistry and has played a significant role in helping chemists to create synthetic versions of natural compounds. Noble Prize winner John Cornforth said, "Nature, it seems, is an organic chemist having some predilection for the aldol and related condensation." In a similar manner, human beings can apply the knowledge built up by the time-tested genius of 10–30 million species by identifying, developing, and implementing this knowledge, which can offer important lessons for moving toward living sustainably on this planet. Figure 8.6 projects the advantages of biomimetic chemistry [39].

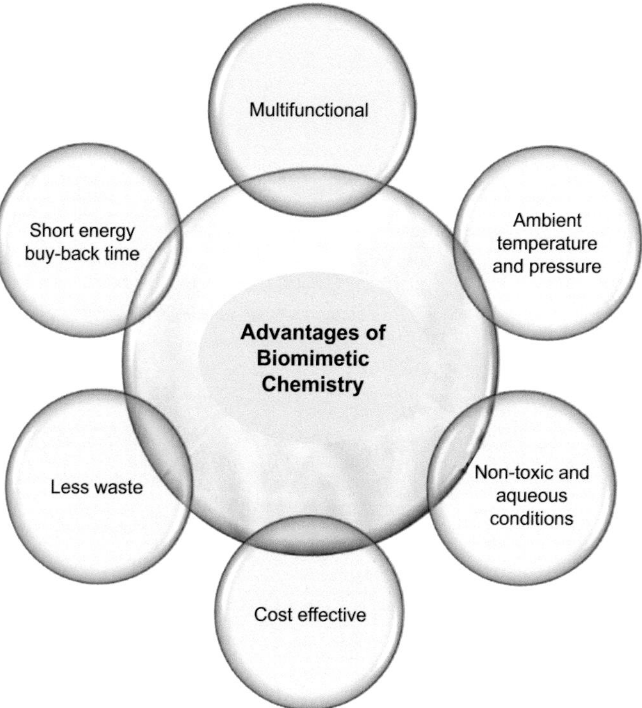

Figure 8.6 Advantages of biomimetic chemistry.

Since the field of biomimetics is highly rewarding, researchers are putting in efforts to imitate the ideas, concepts, and principles of various biological processes occurring in nature in the laboratory. Professor Li from Oregon State University was inspired by the alluring ability of mussels to adhere to rocks by utilizing a special protein. The group successfully employed this wonderful idea of nature for developing a formaldehyde-free soy-based adhesive protein by modifying amino acids. This interesting innovation was conferred the Presidential Green Chemistry Challenge Award in the year 2007. Not just catalysis, but the principles of nature have been utilized in various fields, including computing, structural engineering, and vehicle design. In Table 8.2, we have outlined the various biomimetic strategies that have been employed in the diverse field of chemistry [37, 40, 41].

Table 8.2 Biomimetic strategies in chemistry

Diverse fields of chemistry inspired by biomimicry	Inspiration	Approach
Materials chemistry	Silicate-based bionanocomposites	Nacre synthesized by inorganic layers of aragonite crystals
	Clay-based bionanocomposites	Chitosan–montmorillonite bionanocomposites
Catalysis	Enzymes	Monomeric chromium epoxidation catalysts
		Tethered dichromium epoxidation catalysts
		Mn porphyrin oxidation catalysts
Surfaces	Gecko feet	Polymer-based dry adhesives
		Carbon nanotube–based dry adhesives
Light-harnessing systems	Photosynthesis	Dendrimers, C_3N_4
Organic transformations	Biosynthesis	Synthesis of usnic acid
		Synthesis of progesterone

Researchers have successfully utilized the lotus effect—the hydrophobic effect shown by the leaves of the lotus flower—to create self-cleaning windows and windshields, hard disks, and magnetic tapes.

Currently, US-based researchers are employing the biomimicry taxonomy as an applied record (Fig. 8.7) [40].

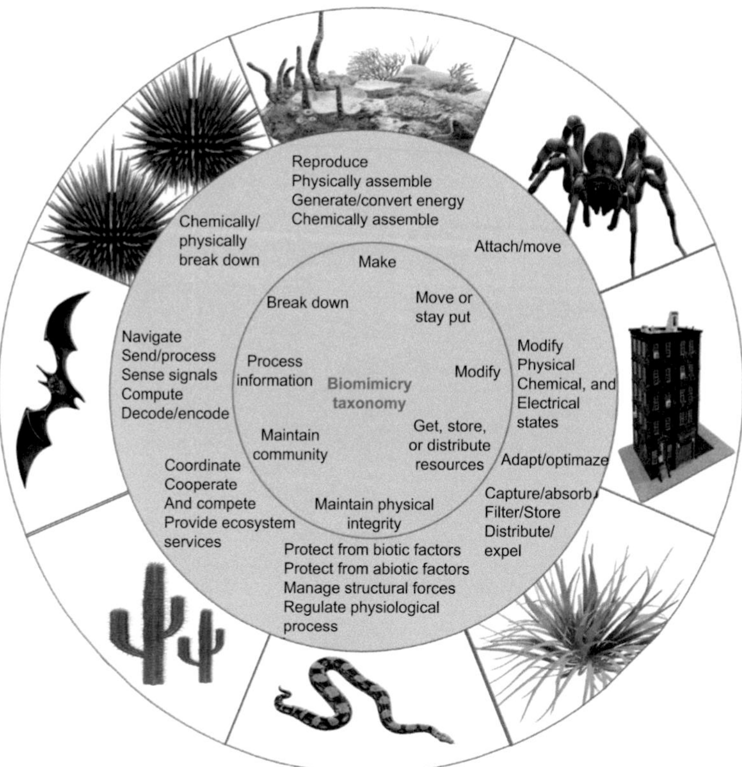

Figure 8.7 Categorized representation of biomimicry taxonomy.

However, it is rather surprising to note that the potential of biomimetics has not been fully exploited in the area of green chemistry. Since nature is the oldest, the wisest, and by necessity the best green chemist, as highlighted by Mark Dorfman, a biomimicry chemist, tremendous benefits could be derived if a green chemist would try to utilize the concept of biomimetics when trying to design greener, cheaper, more efficient, and high-performance materials. One of the most critical challenges facing the researchers working in the area of biomimetics is that although biomimetic tools assist in generating ideas and applying these ideas from biomimetics to other promising applications, they is rarely actionable advice offered on the fundamental understanding of the biological ideas into manufacturable prototypes. Also, the practical challenges lie in scaling up at a reasonable cost and efficacy.

A symbiotic relationship is required between biomimetics and industry to design nature-inspired processes to deliver convenience in various fields, such as chemistry, biology, architecture, engineering, medicine, and biomedical engineering. For us humans to coexist with nature, there is a great demand that various aspects of biomimetics, such as economic, environmental, and social aspects, be brought into our day-to-day life. Various steps have been taken for advancement in this field. One such achievement is the development of the website Asknature.org, by Janine M. Benyus, US, to enforce the practice of biomimicry to unravel countless obstacles and challenges, create a substantial impact, and generate plentiful value to fulfil the needs of the upcoming generation. In a similar way, Benyus and others initiated a social enterprise named Biomimicry 3.8 for the exchange of ideas related to biomimicry and provide an opportunity to network with interdisciplinary investigators, researchers, scientists, engineers, businessmen, participants, and investors [42].

> The Ronald Breslow Award is every year granted for any remarkable achievement in the field of biomimetic chemistry. The award consists of a reward of U$5000, a certificate, and travel overhead.

8.7 Continuous Flow Technology

Over the past decades, versatile, durable, chemical, and thermal-resistant inexpensive lab wares/flasks have been utilized to carry out reactions on the milligram to gram scale. However, academic and industrial researchers are now directing their attention toward the viable application of continuous operation of chemical reaction in pipes and tubes for better mixing, better heating, scalability, ease of manipulation of pressure and temperature, new opportunities offered in the field of heterogeneous catalysis, multistep synthesis, etc. This is called the continuous flow (CF) chemistry, which can be elucidated as a sequence of chemical processes conducted in continuous flowing streams within microreactors [43]. Currently, the "digital economy" in the industries has driven the utilization of the CF method as a superior alternative to batch processing, which helps reduce the manufacturing cost and achieve the agenda of sustainability. The CF method has been fruitfully employed in large-scale industries in the gas phase for the production of small molecules, such as ammonia in the Haber–Bosch and Fischer–Tropsch processes for the generation of synthetic fuels. Subsequently, the CF approach is ready to lend a hand in improving safety measures, lowering the reaction time, reducing waste generation, improving reproducibility, and increasing energy efficiency to align with the goals of green chemistry and engineering [44, 45]. The real and impactful advantages of CF have also been realized in heterogeneous catalysis and photochemical reactions that cannot be carried out in conventional round bottom flasks (RBs) and batch processes. All the advantages associated with the CF method have been highlighted in Fig. 8.8.

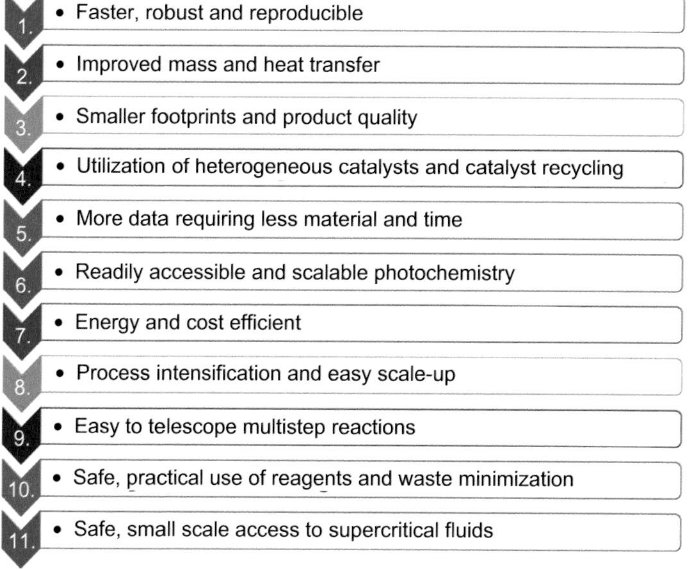

Figure 8.8 Numerous advantages of CF technology.

The commonly known CF reactors are plug flow reactors and column reactors, whereas more sophisticated chemical processes require specifically designed CF reactors (such as photoreactors and electrochemical reactors) to fulfil the need.

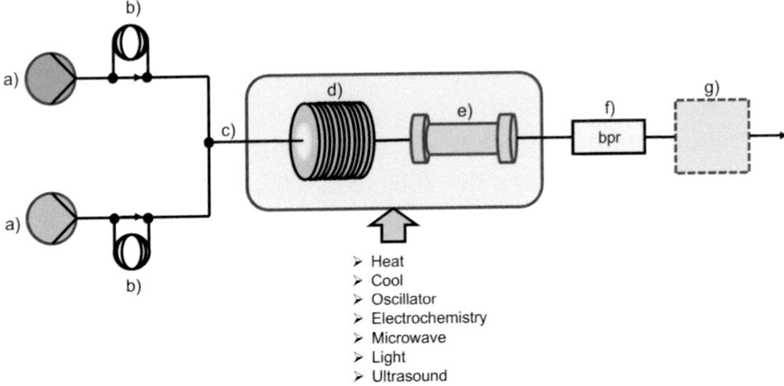

Figure 8.9 Flow chemistry setup.

A typical flow chemistry setup has been projected in Fig. 8.9, and the function of each component is demonstrated in Fig. 8.10 [55].

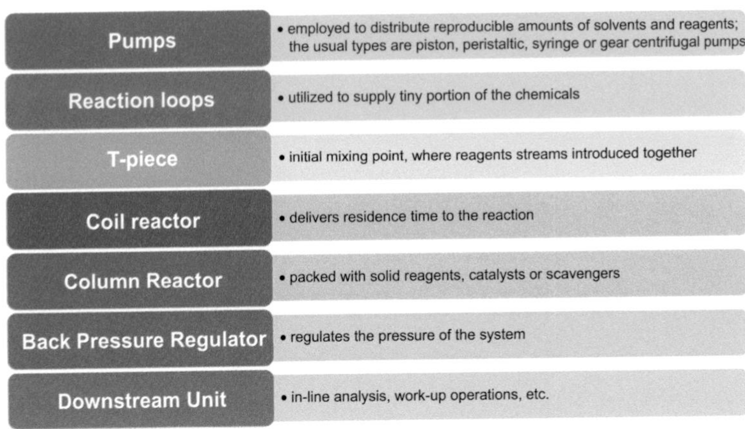

Figure 8.10 Functions of the continuous flow reactor's components.

However, for the wider applicability of CF in the future, numerous challenges need to be overcome, which are discussed next [46, 47].

- Selectivity and atom economy: The CF process should be selective and atom efficient to suppress the formation of side-products, the intention being to eliminate complex and resource-intensive workup procedures.

- More robust catalysts and reactions: The benefits of integrating heterogeneous catalysis with CF have been demonstrated by their prevalence in the production of commodity chemicals. Yet, there are a few limitations associated with the use of heterogeneous catalysts in CF synthesis that hamper their large-scale applicability. One of the most critical problems encountered is leaching of the catalyst into the liquid phase, which necessitates the shift in the focus of the researchers toward this direction.
- Solvents: Precipitation of the product or by-product may cause blockages by interrupting the flow of the reaction mixture. It is, therefore, important to carefully select and use solvents that will prevent this issue.
- Blockages: Researchers still need to focus on the challenges of handling the solid particles in the reaction mixture, which might cause blockages.

In addition to these, some other challenges are encountered arising because of the insolubility of the precursors, switching to other solvents (which is primarily witnessed during multistep synthesis), and faulty design of the CF reactor and processes. The barriers associated with the reactor design can be removed with the aid of green chemistry and engineering by economical use of space, operation, and time; efficient mass and heat transfer; and upscaling to deploy a more environmentally friendly approach [43, 44, 48].

8.8 Combinatorial Chemical Technology

The traditional approach toward obtaining a novel molecule possessing specific properties is synthesizing different compounds one at a time and investigating their properties also one at a time. This is a highly laborious, time-consuming, as well as waste-generating process, and only after successful investigation is one able to identify the lead compound. The introduction of combinatorial chemical technology has revolutionized the approach by making possible the fabrication of a diverse library of compounds. Figure 8.11 illustrates the fundamental features of a traditional methodology versus those of a combinatorial methodology for the development of a drug. Combinatorial chemical technology is a smart scientific technology used for the synthesis and characterization of structurally diverse compounds and for automated screening of their biological, pharmaceutical, and industrial utility on a small scale through reaction matrices [49, 50].

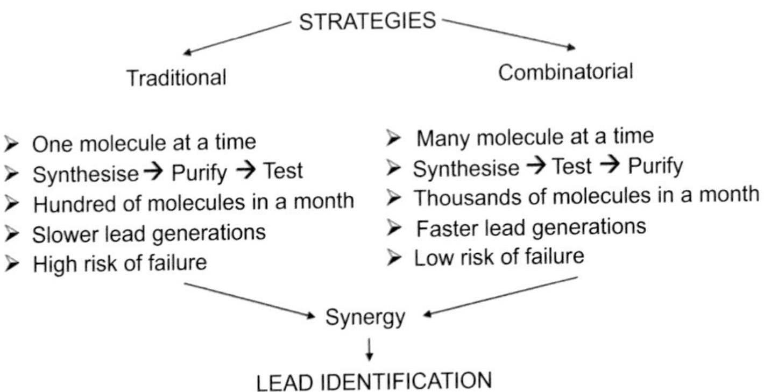

Figure 8.11 Principal features of traditional versus combinatorial strategies for the development of drugs.

This technique enables the synthesis of several combinations of $X_n \times Y_n$ within the single reaction rather than carrying out multiple reactions of the $X \times Y$ type.

Traditional reaction: $X + Y \rightarrow XY$

Combinatorial chemistry: $X_{1-n} + Y_{1-n} \rightarrow X_{1-n}\, Y_{1-n}$, where n is the number of reagents utilized

Previously, the technology was specifically showing profound effect in the pharmaceutical sector for the synthesis of peptide and oligonucleotide libraries involving amide chemistry. Scientists are still eager to fully uncover the potential of combinatorial chemical technology and enhance its "phase trafficking" abilities for integrating synthesis with purification. The ability to synthesize, purify, chemically analyze, and biologically test all the structures in the library is the hidden goal of the combinatorial chemical technology. By applying the technology, pharmaceutical companies have been able to develop biologically active lead compounds and their derivatives, which are synthesized and analyzed for their biological efficacy without causing any adverse impact on the environment, and the biodegradable nature of the products is also taken into account.

The combinatorial technology is so efficient that it can be successfully employed for generating a sounding library of 10^5 or more compounds [50]. Despite these successes, the following complexities are associated with this technique:

- Possible complexity associated with the synthesized compounds because of their amidic and lipophilic nature, high molecular weight, etc.
- Lack of quality control related to the generated compound.
- Optimum mixture size to establish higher-quality biological results with a high number of compounds.
- Size of library to achieve success in screen.

Here is a list of the various types of combinatorial chemical technologies and their characteristics:

- Solid-phase combinatorial chemistry
    - Large excess of reagents allowed
    - Multiple synthesis allowed
    - Easy workup and isolation
    - Mix and split possible
- Solution-phase combinatorial chemistry
    - Use of all organic reactions allowed
    - No chemistry assessment
    - No linker/cleavage chemistry
    - Unlimited product quantities

Various researchers are adopting solid-phase combinatorial chemistry over the solution phase in order to synthesize diverse compounds prepared in parallel arrays for optimization due to their advantages, such as preparation in large amounts, facile quality control, avoidance of "tagging" or "deconvolution" methods, and ability to format flexibly according to the needs of each screen [50].

To exploit the full potential of combinatorial chemistry, there is a need to fulfil the gaps that persist in the knowledge about the technology. As we discussed earlier, combinatorial chemistry has been successfully utilized for amide chemistry. However, there is a growing quest to practice it for nonamide chemistry, explore a range of linkers (easy to cut down in the reaction mixture), and discover various solid supports to attain a more rapidly and cheaply synthesized focused library of diverse structurally active compounds [51].

8.9 Green Chemistry and Sustainability

In recent years, academicians and industrialists have focused on resolving challenges that persist in sustainable development by adopting the 12 tenets of green chemistry. Undeniably, green chemistry is an important tool for achieving sustainability and has provided a new direction to budding researchers for developing technologies that can meet the needs of the present generation without compromising on the resources of the future generation. This is also what sustainability fundamentally preaches. As we have seen, green chemistry shows the competence to efficiently utilize raw material and eradicate the utilization of toxic materials or production of toxic products when designing chemical products in multidisciplinary fields. In fact, practicing the 12 tenets of green chemistry and green engineering in our day-to-day life will lead us toward the path of sustainability [52]. So far, using green chemistry, we have been able to:

- Rule out the use of harmful chlorofluorocarbons in refrigerators, responsible for ozone layer depletion
- Design more efficient solar cells
- Design and develop life-saving drugs, including Taxol (chemotherapy drug) and ibuprofen (painkiller), in a greener and cheaper way
- Use safer diphenyl carbonate for the synthesis of polycarbonate as an alternative to phosgene and methylene chloride
- Design biodegradable plastics
- Develop biofuels
- Develop photocatalytic technologies for pollution abatement

There are so many more examples of what we have been able to accomplish with the aid of green chemistry that have been discussed in Chapter 6.

We should admit that for a green and sustainable future, the conventional chemical synthesis and processes need to be altered to prevent the adverse effects of global climatic changes. The current need is to educate the young generation about the hazardous effects of toxic chemicals and processes and encourage them to look for greener and sustainable alternatives to overcome the long-lasting hazardous impacts in economic and ecological way [1, 53]. In addition, we need to fill the education gap by spreading awareness about toxicity, harm, environmental impact, persistence, policies, and laws governing chemicals, and this should be taught in the curriculum of core chemistry. Databases and software are required that can correctly predict the persistence and toxicity of various chemicals and suggest greener alternatives of the required chemicals to reduce chemical waste, toxicity, and energy. Moreover, various green chemistry programs should be launched to enhance the knowledge of sustainable chemistry at academic as well as industrial levels to foster the practice of green chemistry. An overhaul of chemical policies is also essential. Figure 8.12 provides a chart that highlights various parameters to achieve the various goals of sustainable development by adopting the green energy strategies and technologies [54].

Also, safety should be assessed in three steps before any innovation is commercialized:

1. Standardize chemical safety tests: Assessing liver toxicity, ozone depletion, endocrine disruption, etc., is important.
2. Test finished products: In this test, the end product should be tested for toxicity because testing only starting materials (even if they are found to be benign) could be misleading as their combination could bring about more harmful end products.
3. Publicize test results: The toxicity test score should be printed on the product for better understanding of the toxicity and risk associated with the concerned product.

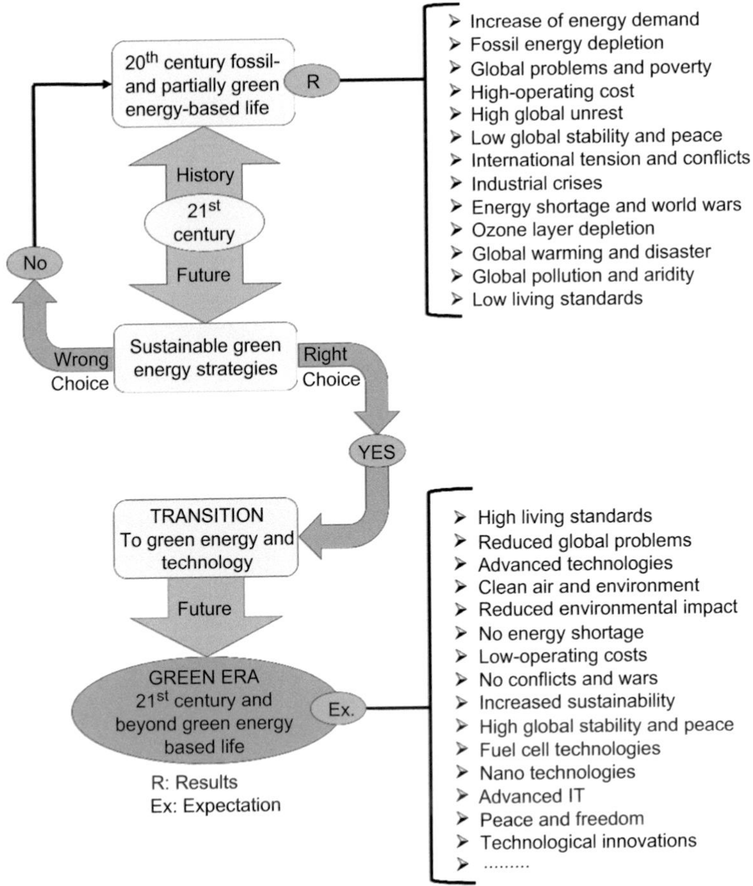

Figure 8.12 Achievement of sustainable development by opting for various greener energy strategies and technologies.

8.10 Conclusion

The future of this planet earth is certainly not dark, and it is green chemistry that has made us believe so. With the innumerable achievements recognized in the form of the US Presidential Green Chemistry Challenge Awards, green chemistry has shown its potential to stay in the mainstream and has changed the mindset of researchers, especially those working in the area of chemical sciences. The new trend now is to practice chemistry in a safer way and design all the synthesis methods by incorporating the 12 principles, also called the basic tenets, of green chemistry outlined by pioneers Paul T. Anastas and John C. Warner. Industries have enormously benefitted by adopting green chemistry. Conventional multistep syntheses of drug manufacturing have been replaced by one-pot cascade processes. New CF reactors are slowly showing the capability to replace the conventional batch reactors. More emphasis is now being placed on increasing atom economy and using environmentally friendly reagents. All these changes have helped in reducing the waste generated by these industries. Also, stringent protocols are being followed for waste disposal. Through green chemistry, we are also being able to find promising solutions for combating major environmental issues, including increasing levels of CO_2 and other pollutants in the atmosphere. The latest green technologies, like combinatorial chemistry, biomimetics, and miniaturization, that have been described in brief in this chapter are being exploited largely. Yet a number of challenges remain, for instance, we are still dealing with pollution caused by microplastics. Delightfully, different parts of the globe have joined hands to work for the mission of accomplishing sustainability, as evident from the 2030 Agenda of UN Sustainable Development Goals. However, meeting all these goals will require fruitful collaborations between scientific communities, including industrial and academic researchers, policy makers, and other stakeholders. Only then can effective implementation of green chemistry solutions be achieved. All the efforts that have been directed so far are just the beginning; the urgency of the need to implement green chemistry needs to be spread far and wide so that we are able to overcome various education and implementation challenges summarized in this chapter and are able to afford to drop the prefix "green" in the near future as chemistry itself would be quintessentially green anyway.

8.11 Learning Outcomes

At the end of this chapter, students will be able to:

- Understand various opportunities and barriers in the field of green chemistry

- Unlock the full potential of promising green technologies, such as biomimetics, combinatorial green chemistry, and continuous flow
- Design greener protocols by employing sustainable reagents and routes such as utilization of various green solvents, like water and DMSO, and green catalysts in various organic transformations; dry-media or solventless fabrication; catalyst-free reactions for diverse synthesis; and energy-efficient synthesis
- Wisely choose raw materials like feedstock and energy resources in a cyclic process invoking the concept of circular economy
- Develop system thinking for designing and developing various cyclic chemical processes

8.12 Problems

1. What do you understand by the term "combinatorial chemistry"?
2. Discuss the barriers and challenges to the adoption of green chemistry.
3. Explain the concept of miniaturization with appropriate examples.
4. How can you make a protocol "green" by using various selective reagents? Describe in brief.
5. What do you understand by the term "multifunctional reagents"?
6. Explain the various challenges prevailing in continuous flow technology.

References

1. Sheldon, R. A. (2017). The E factor 25 years on: the rise of green chemistry and sustainability. *Green Chem.*, **19**, pp. 18–43.
2. Zimmerman, J. B., Anastas, P. T., Erythropel, H. C. and Leitner, W. (2020). Designing for a green chemistry future. *Science*, **367**, pp. 397–400.

3. Anastas, P. T. and Warner, J. C. (1998). *Green Chemistry: Theory and Practice*, eds. Anastas, P. T. and Warner, J. C., Principles of green chemistry (Oxford University Press, New York) pp. 29–56.

4. Erythropel, H. C., Zimmerman, J. B., de Winter, T. M., Petitjean, L., Melnikov, F., Lam, C. H., Lounsbury, A. W., Mellor, K. E., Janković, N. Z. and Tu, Q. (2018). The Green ChemisTREE: 20 years after taking root with the 12 principles. *Green Chem.*, **20**, pp. 1929–1961.

5. Anastas, P. T. and Zimmerman, J. B. (2003). Peer reviewed: design through the 12 principles of green engineering. *Environ. Sci. Technol.*, **37**, pp. 94A–101A.

6. Anderson, P. W. (1972). More is different. *Science*, **177**, pp. 393–396.

7. Li, C.-J. and Anastas, P. T. (2012). Green Chemistry: present and future. *Chem. Soc. Rev.*, **41**, pp. 1413–1414.

8. Kümmerer, K. (2009). Antibiotics in the aquatic environment–a review–part I. *Chemosphere*, **75**, pp. 417–434.

9. Fedoroff, N. V., Battisti, D. S., Beachy, R. N., Cooper, P. J., Fischhoff, D. A., Hodges, C., Knauf, V. C., Lobell, D., Mazur, B. J. and Molden, D. (2010). Radically rethinking agriculture for the 21st century. *Science*, **327**, pp. 833–834.

10. Jones, K. C. and De Voogt, P. (1999). Persistent organic pollutants (POPs): state of the science. *Environ. Pollut.*, **100**, pp. 209–221.

11. Anastas, P. T. (2019). Beyond reductionist thinking in chemistry for sustainability. *Trends Chem.*, **1**, pp. 145–148.

12. Matlin, S. A., Mehta, G., Hopf, H. and Krief, A. (2016). One-world chemistry and systems thinking. *Nat. Chem.*, **8**, p. 393.

13. Anastas, P. and Eghbali, N. (2010). Green chemistry: principles and practice. *Chem. Soc. Rev.*, **39**, pp. 301–312.

14. Liu, J., Mooney, H., Hull, V., Davis, S. J., Gaskell, J., Hertel, T., Lubchenco, J., Seto, K. C., Gleick, P. and Kremen, C. (2015). Systems integration for global sustainability. *Science*, **347**, p. 1258832.

15. Anastas, P. T. and Kirchhoff, M. M. (2002). Origins, current status, and future challenges of green chemistry. *Acc. Chem. Res.*, **35**, pp. 686–694.

16. Tuck, C. O., Pérez, E., Horváth, I. T., Sheldon, R. A. and Poliakoff, M. (2012). Valorization of biomass: deriving more value from waste. *Science*, **337**, pp. 695–699.

17. Langanke, J., Wolf, A., Hofmann, J., Böhm, K., Subhani, M., Müller, T., Leitner, W. and Gürtler, C. (2014). Carbon dioxide (CO_2) as sustainable feedstock for polyurethane production. *Green Chem.*, **16**, pp. 1865–1870.

18. von der Assen, N. and Bardow, A. (2014). Life cycle assessment of polyols for polyurethane production using CO_2 as feedstock: insights from an industrial case study. *Green Chem.*, **16**, pp. 3272–3280.
19. Zimmerman, J. and Anastas, P. (2005). *Sustainability Science and Engineering: Defining Principles*, eds. Zimmerman, J. and Anastas, P., When is a waste not a waste (Elsevier B.V.) pp. 201–221.
20. Melnikov, F., Kostal, J., Voutchkova-Kostal, A., Zimmerman, J. B. and Anastas, P. T. (2016). Assessment of predictive models for estimating the acute aquatic toxicity of organic chemicals. *Green Chem.*, **18**, pp. 4432–4445.
21. Geyer, R., Jambeck, J. R. and Law, K. L. (2017). Production, use, and fate of all plastics ever made. *Sci. Adv.*, **3**, p. 1700782.
22. Makone, S. and Niwadange, S. N. (2016). Green chemistry alternatives for sustainable development in organic synthesis. *Green Chem.*, **3**, pp. 113–115
23. Lancaster, M. (2016). *Green Chemistry: An Introductory Text*, 3rd Ed. (Royal Society of Chemistry, UK).
24. Anastas, P. T. and Beach, E. S. (2007). Green chemistry: the emergence of a transformative framework. *Green Chem. Lett. Rev.*, **1**, pp. 9–24.
25. Sheldon, R. A. (2005). Green solvents for sustainable organic synthesis: state of the art. *Green Chem.*, **7**, pp. 267–278.
26. Capello, C., Fischer, U. and Hungerbühler, K. (2007). What is a green solvent? A comprehensive framework for the environmental assessment of solvents. *Green Chem.*, **9**, pp. 927–934.
27. Nagendrappa, G. (2002). Organic synthesis under solvent-free condition: an environmentally benign procedure—I. *Resonance*, **7**, pp. 59–68.
28. Rana, P., Gaur, R., Gupta, R., Arora, G., Jayashree, A. and Sharma, R. K. (2019). Cross-dehydrogenative C (sp^3)–C (sp^3) coupling *via* C–H activation using magnetically retrievable ruthenium-based photoredox nanocatalyst under aerobic conditions. *Chem. Commun.*, **55**, pp. 7402–7405.
29. Suib, S. L. (2013). *New and Future Developments in Catalysis: Catalysis for Remediation and Environmental Concerns* (Elsevier B.V.).
30. Brahmachari, G. and Banerjee, B. (2015). Catalyst-free organic synthesis at room temperature in aqueous and non-aqueous media: an emerging field of green chemistry practice and sustainability. *Curr. Green Chem.*, **2**, pp. 274–305.

31. Stankiewicz, A., Clark, J. H., De La Hoz, A., Fan, J., Chain, R. M., Hessel, V., Gascon, J., Van Gerven, T., Hristov, J. and Larachi, F. (2016). *Alternative Energy Sources for Green Chemistry* (Royal Society of Chemistry, UK).
32. Clark, J. H. and Macquarrie, D. J. (2008). *Handbook of Green Chemistry and Technology* (John Wiley & Sons, US).
33. Wong, H. and Cernak, T. (2018). Reaction miniaturization in eco-friendly solvents *Curr. Opin. Green Sustainable Chem.*, **11**, pp. 91–98.
34. de la Guardia, M. and Garrigues, S. (2012). *Handbook of Green Analytical Chemistry* (John Wiley & Sons, US).
35. Kricka, L. J. (1998). Miniaturization of analytical systems. *Clin. Chem.*, **44**, pp. 2008–2014.
36. Chater, R. and McPhail, D. (2004). Laboratory teaching of SIMS to university undergraduates. *Appl. Surf. Sci.*, **231**, pp. 141–145.
37. Lehn, J.-M. and Benyus, J. (2012). *Bioinspiration and Biomimicry in Chemistry: Reverse-Engineering Nature* (John Wiley & Sons, US).
38. Bar-Cohen, Y. (2006). Biomimetics—using nature to inspire human innovation. *Bioinspiration Biomimetics*, **1**, pp. 1.
39. Heathcock, C. H. (1982). *Acyclic Stereoselection via the Aldol Condensation* (ACS Publications, US).
40. Hwang, J., Jeong, Y., Park, J. M., Lee, K. H., Hong, J. W. and Choi, J. (2015). Biomimetics: forecasting the future of science, engineering, and medicine. *Int. J. Nanomed.*, **10**, p. 5701.
41. Bello, O., Adegoke, K. and Oyewole, R. (2013). Biomimetic materials in our world: a review. *J. Appl. Chem.*, **5**, pp. 22–35.
42. Peters, T. (2011). Nature as measure: the biomimicry guild. *Archit. Des.*, **81**, pp. 44–47.
43. Vaccaro, L., Lanari, D., Marrocchi, A. and Strappaveccia, G. (2014). Flow approaches towards sustainability. *Green Chem.*, **16**, pp. 3680–3704.
44. Newman, S. G. and Jensen, K. F. (2013). The role of flow in green chemistry and engineering. *Green Chem.* **15**, pp. 1456–1472.
45. Baumann, M., Moody, T. S., Smyth, M. and Wharry, S. (2020). A perspective on continuous flow chemistry in the pharmaceutical industry. *Org. Process Res. Dev.*, https://doi.org/10.1021/acs.oprd.9b00524.
46. Zhang, Y. and Jiang, H.-R. (2016). A review on continuous-flow microfluidic PCR in droplets: advances, challenges and future. *Anal. Chim. Acta*, **914**, pp. 7–16.

47. Akwi, F. M. and Watts, P. (2018). Continuous flow chemistry: where are we now? Recent applications, challenges and limitations. *Chem. Commun.*, **54**, pp. 13894–13928.

48. Hellgardta, K. and Hii, K. (2018). *Advanced Green Chemistry, Part 1: Greener Organic Reactions and Processes*, eds. Horváth, I. T. and Malacria, M., Continuous flow technologies in the development of "green" organic reactions and processes (World Scientific, Singapore).

49. Brown, D. (1997). Future pathways for combinatorial chemistry. *Mol. Diversity*, **2**, pp. 217–222.

50. Mishra, A., Gupta, A., Singh, A. and Bansal, P. (2010). Combinatorial chemistry and its application: a review. *Int. J. Chem. Anal. Sci.*, **1**, pp. 100–105.

51. Seneci, P., Fassina, G., Frecer, V. and Miertus, S. (2014). The effects of combinatorial chemistry and technologies on drug discovery and biotechnology: a mini review. *Nova Biotechnol. Chim.*, **13**, pp. 87–108.

52. Varma, R. S. (2016). Greener and sustainable trends in synthesis of organics and nanomaterials. *ACS Sustainable Chem. Eng.*, **4**(11), pp. 5866–5878.

53. Varma, R. S. (2014). *Greener and Sustainable Chemistry* (Multidisciplinary Digital Publishing Institute, Switzerland).

54. Midilli, A., Dincer, I. and Ay, M. (2006). Green energy strategies for sustainable development. *Energy Policy*, **34**, pp. 3623–3633.

55. Flow Chemistry (URL: https://www.organic-chemistry.org/topics/flowchemistry.shtm).

Index

absorption 180–182, 289
absorption apparatus 96
accident 5, 7, 19, 20, 121, 163
  chemical 7, 12
  industrial 8
  nuclear 6
acetaldehyde 53, 211, 258
acetone 66, 67, 137, 139, 168
acetonitrile 124, 136, 142, 167, 197, 198, 298
acetylene 53, 93, 211
ACQ *see* alkaline copper quaternary
active ingredient 63, 67, 86, 123, 240
active pharmaceutical ingredient (API) 38, 120
adhesives 2, 60, 61, 120, 303
adsorption 67, 190, 193
advanced oxidative process 187
agent
  antifouling 258
  bleaching 243
  carbonylating 94
  foam-blowing 4
  infectious 113
  oxidative decomposing 189
  reducing 127, 196
alcohol 17, 92–94, 108, 112, 120, 126, 139, 143, 144, 169, 177, 186
aldehyde 85, 92, 138, 149–152, 171, 209, 221, 237
aldol condensation 152, 153, 293
alkaline copper quaternary (ACQ) 250, 251
alkene 92–94, 96, 126, 128, 138, 142, 170, 179, 184, 220, 221
alkylation 108, 178
alkyne 30, 92, 93, 142
alumina 95, 98, 99, 107, 171, 178
ammonia 16, 19, 81, 95, 96, 210, 236, 292, 306
analgesics 38, 239
anhydride 93, 172
aniline 85, 169
anion 93, 135–137, 299
  cyanide 85
  nucleophilic 100
  organic 135
  tetraalkylammonium salt 197
antifoulants 226, 228, 229, 257
API *see* active pharmaceutical ingredient
arsenic 41, 97, 249–251, 271
asymmetric induction 146
atmosphere 4, 147, 175, 198, 235, 292, 314
atom economy 12, 13, 25–29, 38, 75, 84, 240, 241, 284, 288, 291, 307, 314
autoxidation 220, 221
azomethine 153, 293

bacteria 207, 219, 226, 246, 248
ball milling 293
Barbier–Grignard-type reactions 128
barnacles 226, 229
Barton reaction 185
Baylis–Hillman reaction 26
benzaldehyde 87, 109, 144, 151–153, 158, 172
benzene 12, 13, 15, 17–19, 26, 85, 121–123, 177, 184, 207–209, 241, 245

benzoin 85–87, 114
bioaccumulation 8, 230, 242
bioadhesives 61
biocatalyst 112, 114, 297
biochemical oxygen demand 37
biodegradability 141, 143, 145, 248, 249, 292
biodiesel 58, 59, 83, 143, 144
biodiesel synthesis 58, 143, 144
biomass 55, 57, 59, 143, 146, 152, 156, 248, 297
biomimetics 300–305, 314, 315
biomimicry 300, 303, 305
bioplastics 60
biosolvents 123, 147
bisphenol A (BPA) 14, 45, 102
block copolymer 238, 239
bond 27, 81, 110, 140, 142, 161, 162, 220, 295
  carbon 134, 138, 140–142, 147, 152
  chemical 124
  covalent 285
  hydrolysable 47
BPA *see* bisphenol A
bubble 175–177
Bunsen burner 162

cadmium 48, 231, 232, 271
caffeine 16, 122, 132, 237
cancer 45, 224, 225, 231, 235, 250
carbon 84, 91, 95, 99, 129, 134, 138, 140–142, 147, 152, 212, 215, 220, 221
carbon dioxide 14, 85, 169, 258
carbon monoxide 14, 93, 94, 98, 129, 149
carbon tetrachloride 41, 123, 124, 133
carpet 21, 224–226, 257
catalysis 79–84, 86, 88–90, 92, 94, 96, 98, 100, 102, 104, 106, 112–114, 180, 181, 284, 285, 303
  asymmetric 103, 104
  chemical 88
  enzymatic 223
  green 294
  homogeneous 91, 93, 94, 113
  hypochlorite decomposition 18
  role of 89, 114
  sustainable 294
catalyst 15–18, 23, 67, 80–85, 89–91, 93–96, 102–105, 108–115, 125, 137, 138, 148–151, 154, 155, 162, 181, 188–190, 193, 207, 243, 244, 290, 291, 293–295
  asymmetric 104, 114
  chiral 104
  complex 144
  copper 210, 211
  copper-zinc oxide 152
  fluorous 150
  fluorous-modified 150
  fluorous-soluble 148
  green 86, 90, 114, 291, 315
  heterogeneous 67, 82, 91, 94, 95, 105, 114, 182, 193, 257, 306, 308
  homogeneous 91, 102–104, 154
  iodide-promoted cobalt 93
  iridium-based 93
  metal 86, 91, 125, 129, 137, 151, 219
  palladium Smopex 67, 68
  peroxidase mimetic 111
  phase transfer 100, 101, 114, 169
  rhodium 64, 91, 92, 149, 209
  robust 308
  triarylphosphine 92
  tungsten 102
  use of 18, 67, 81, 213, 241, 244, 273, 291
  water-soluble 129
  Wilkinson's 92, 93
  Zeigler-Natta 96

zeolite 18
catalytic converter 97–99
catalytic method 82, 84, 89, 264
catalytic oxidation 84, 96, 211
catalytic process 80, 154, 294
catalytic reaction 80–82, 89
catechol 207, 208, 258
cavitation effect 174, 176, 177, 200
CCA *see* chromated copper arsenate
cellulose 56, 57, 59, 242
CFC *see* chlorofluorocarbon
chemical process 10–12, 19, 20, 22, 23, 25, 56, 57, 61, 86, 88, 119, 120, 122, 272–274, 306, 307
chemical reaction 13, 16, 23, 25, 28, 29, 97, 98, 112, 114, 1255, 161, 164, 166, 174, 177, 192, 195, 272
chemical synthesis 61, 115, 156, 164, 199, 206, 213, 244, 257, 284, 285, 299
chemical transformation 25, 162, 177, 199
electron-initiated 182
ultrasound-mediated 177
chemistry 1, 2, 10, 123, 173–175, 177, 179, 264, 265, 269, 270, 272, 274, 276–278, 281, 300, 302, 303, 305, 306
amide 309, 311
analytical 119
computational 113
environmental 195
enzymatic 302
fundamental 21
inorganic 293
introductory 269
linker/cleavage 310
medicinal 195
miniaturized 298
modern-time 271
nonamide 311
physical 192
renewable 289
supramolecular 293
chemists 1, 2, 164, 167, 264, 267, 269, 272, 277, 284, 288, 291
analytical 70
medical 280
pharmaceutical 63, 280
synthetic 86, 162
chlorofluorocarbon (CFC) 3, 4, 14, 235, 236, 311
chromated copper arsenate (CCA) 250, 251
coating 2, 60, 120
cobalt 91–93, 110, 149
coffee 16, 122, 132, 237
column reactor 307
combinatorial chemistry 309–311, 314, 315
combustion 51, 235
compound 4, 16, 18, 206, 208, 228–230, 251, 255, 257, 301, 308–310
active 311
active methylene 212
alkyl aluminum 96
ammonium 251
antimicrobial 248
aromatic 208
azo 198
carbonyl 112, 127, 128, 139, 183
chlorine-containing 3
hydrophilic 180
hydrophobic 180
inorganic 81, 271
multifunctional 89
natural 302
nucleophilic 93
oily 224
organochlorine 242, 243, 246
organometallic 137
pharmaceutical 38

polyether 139
synthesized 310
toxic 19, 52
water-insoluble 129
conduction band 182, 188–190
contaminants 179, 231
contamination 42, 69, 187, 190
corrosion 199, 214
corrosivity 41, 289
cosmetics 46, 60

DDT *see* dichlorodiphenyltrichloroethane
defects 103, 106, 114
degradation 19, 72, 73, 75, 179, 180, 189–191, 220, 286, 290
environmental 154
microbial 73
thermal 149
depolymerization 225, 242
derivative 12, 63, 74, 94, 197, 310
acetophenone 152
chemical 17
detergent 2, 18, 73, 238
diabetic cardiomyopathy 253
dichlorodiphenyltrichloroethane (DDT) 3, 246, 247
Diels–Alder reaction 14, 16, 126, 134, 140, 170, 171, 178
dimethyl formamide (DMF) 123, 140–142, 147, 167
dimethyl sulfoxide (DMSO) 36, 101, 110, 123, 140, 141, 255, 256, 298, 315
dioxin 4, 50, 225, 242, 243
disaster 3, 5, 313
discovery 193, 206, 239, 248, 253, 287
disease 2, 3, 45, 53, 55, 221, 248
disinfectant 4, 251
disodium iminodiacetate (DSIDA) 210, 211, 258
disorder 232, 255
distillation 16, 96, 98
azeotropic 66
DMF *see* dimethyl formamide
DMSO *see* dimethyl sulfoxide
drug 3, 36–38, 42, 63, 64, 75, 86, 114, 206, 239, 240, 253, 255
antidepressant 21, 38, 123
antihistamine 255
anti-inflammatory 42, 86
chemotherapy 311
life-saving 65, 311
sitagliptin 75
thalidomide-based 3
DSIDA *see* disodium iminodiacetate
dye 17, 83, 101, 111, 181, 187, 189, 190, 230, 231, 244
dye degradation 187, 188, 190, 191

electrocatalysis 193, 195, 196, 200
electrochemical synthesis 193, 195, 197–199
electrochemistry 174, 192, 193, 195, 198–200, 307
emission 25, 50, 52, 59, 97, 113, 124
automotive 97
carbon 163
environmental 122
sulfur 59
ultrasonic 175
enantiomer 3, 15, 103
energy 19, 21, 24, 57–59, 80, 82–84, 161–163, 165, 166, 188, 199, 200, 287, 290, 291, 295, 297, 312
activation 80, 89
bandgap 188
chemical 57
clean 20, 55, 83, 156, 163, 263
decarbonized 295
fossil fuel 219
green 313
kinetic 177

mechanical 293
photochemical 162
photon 188
renewable 291
solar 181, 218
sustainable 53, 163
energy consumption 20, 120, 162, 194, 198, 199, 225
environment 2, 3, 5, 6, 8–11, 24, 25, 34, 42–45, 47, 48, 72, 73, 75, 156, 230, 231, 246, 247, 274, 285, 290
environmental problem 4, 212, 226, 252, 265
enzyme 26, 81, 86, 112, 114, 223, 232, 258, 285, 300
artificial 302
engineered 253
lipase 223
natural 112
zinc-containing 232
epoxidation 16, 110, 126, 137, 138, 303
ester 58, 64, 92, 93, 120, 170, 220
esterification reaction 133, 146, 170
ethanol 30, 67, 112, 123, 139, 140, 146, 155, 211
ethylene 85, 93, 211, 225
e-waste 47–49
explosion 5, 12, 28, 121, 122, 167
exposure 28–30, 121, 184, 224, 235, 250, 255, 271, 299
chronic 250
disastrous 5
incessant 42
long-term 224, 232, 235
extraction 67, 98, 122, 125, 132, 143, 293, 297

Faraday's laws 192
fat 58, 220, 222, 238
saturated 220
unsaturated 220
vegetable 60
feedstock 11, 152, 207, 215, 219, 286, 315
biobased 20
chemical 56
degradable 292
renewable 17, 19, 57, 143, 155, 213, 257, 264, 280, 285, 291
fermentation 30, 56, 112, 146, 215, 216, 219, 249
fiber 46, 215, 224, 225
nylon 226
polysaccharide 242
textile 21
filtration 16, 98, 111, 187, 223
Fischer–Tropsch process 306
fluid 129–131, 175–177, 292
fluorous biphasic system 148–150, 157
fluorous ponytails 148, 157
food chain 3, 42, 46, 53, 247
forest 8, 57, 143, 163
fossil fuel 57, 58, 132, 206, 218, 226, 257, 296
fouling 191, 226, 228
Friedel-crafts reaction 84
fuel 57, 58, 89, 120
biobased 280
engine 286
synthetic 306
traditional 162
fumes 163, 271
noxious 271
obnoxious 43, 69
toxic 225, 271

garbage 34, 44, 72
gas emissions 57, 70, 147, 163, 289
gas 5, 36, 97, 99, 129, 131, 132, 177, 237, 255, 299
explosive 199

flue 225
greenhouse 52, 83, 237
methane 50
methyl sulfide 36
obnoxious-smelling 255
poisonous 5, 14, 50
glucose 112, 207, 208, 215, 216, 218
fermentation of 26, 216
nontoxic 17
glycerol 58, 76, 143, 144, 157, 220
goals 20, 21, 54, 55, 66, 73, 156, 162, 163, 264, 266, 289, 295, 297, 298
global 55
hidden 309
independent 287
key 293, 294
noble 22
primary 292
universal 20, 263, 264
greener route 115, 206, 208, 210, 219, 241, 253, 255–257
greenhouse effect 51, 121
green protocol 172, 292
green technologies 74, 162, 278, 284, 314, 315
Grignard-type coupling reactions 128

Haber–Bosch process 95, 306
harpin 248, 249
hazards 3, 28–30, 40, 41, 121, 122, 266, 271, 272, 285, 288, 289
environmental 122, 215
far-reaching 266
global 286
health 120, 122, 173, 192
molecular 286
not-so-distant 266
heterogeneous catalysis 94, 95, 97–101, 113, 306, 308
HFCS *see* high-fructose corn syrup
highest occupied molecular orbital (HOMO) 181, 182, 193
high-fructose corn syrup (HFCS) 112, 113
Hofmann elimination reaction 168
holes 181, 182, 188, 189
HOMO *see* highest occupied molecular orbital
homogeneous sonochemistry 176
hormones 38, 45
HYDECAT *see* hypochlorite decomposition catalysis
hydrocarbons 3, 16, 47, 98, 99, 143, 192
aliphatic 139, 237
chlorinated 4, 120
low-molecular-weight 120
unsaturated 93
hydrogenation 63, 64, 84, 92, 127, 128, 137, 219–221, 223
hydrolysis 27, 64, 139, 168, 254
hydrophobic effect 126, 167, 184, 304
hydroquinone 85, 207, 208
hypochlorite decomposition catalysis (HYDECAT) 18, 19

ibuprofen 36, 42, 86, 239–241, 259, 311
industry 37, 38, 43, 83, 120, 121, 156, 191, 192, 206, 213, 251, 252, 255, 257, 258, 274, 276, 277, 280, 284, 305, 306, 314
agrochemical 112
beverage 113
biotechnology 66
chemical 2, 3, 22, 24, 65, 80, 83, 85, 120, 293, 295
cosmetic 293
dairy 180
food 132
garment 258
large-scale 306
petroleum 289

pharmaceutical 24, 37, 42, 51, 63, 74, 75, 185, 186, 197, 213, 251, 298
polymer 82
pulp 241
textile 301
innovation 10, 33, 54, 55, 61, 68, 156, 231, 241, 252, 253, 303, 312
disruptive 286
incredible 213
large-scale 264
medical 251
inorganic salt 22, 84, 85, 137, 139, 143
insecticides 246, 250, 251
interesterification 222, 223
chemical 222, 258
enzymatic 219, 223, 224
ionic liquid 15, 123, 135–137, 157, 198, 292
ions 136, 137, 192, 195, 196
isolation 110, 154, 248, 310
isomerization 18, 104, 112, 180, 184, 210, 222
isomer 184, 218, 221

jewelry 131, 174

ketone 84, 85, 102, 152, 169, 170, 221, 237
kidneys 121, 232
Koch reaction 94

landfills 34, 38, 47, 48, 50, 52, 70, 72, 215, 225, 241, 251
laws 8, 49, 192, 312
environmental 8
federal 40
national 34
regulatory 6
LCA *see* life cycle assessment
life cycle assessment (LCA) 22, 24, 25, 122, 266
ligand 110, 125, 142, 148, 150
lignin 59, 242, 243, 251
liquid-crystal display 48
liver 49, 121, 224, 250
lovastatin 253, 254
lowest unoccupied molecular orbital (LUMO) 181, 182, 193
lubricant 2, 17, 120, 207
LUMO *see* lowest unoccupied molecular orbital

magnetic nanoparticle (MNP) 110–112
maximum concentration guideline levels 8
mechanism 80, 164, 194, 200, 220, 232, 239, 248, 258, 300
conduction 165
early-defense 248
environmental 290
exposure control 289
molecular 289
natural defense 248
photocatalytic redox 188
polarization 165
semiconductor-based photocatalytic 188
medicine 3, 37, 38, 113, 246, 265, 280, 300, 301, 305
melting point 95, 135–137, 152, 220–222
menthol 104
mercury 48, 53, 211, 271
metal 41, 49, 51, 52, 65, 84, 97, 106, 128, 195, 231, 233
abundant 285
heavy 41, 48, 52, 56, 57, 231, 237, 246, 250, 301
palladium 67
rare 285
rare earth 245
toxic 49, 52
transition 93, 144

methanol 14, 16, 43, 58, 67, 92, 93, 108, 126, 139, 152
microorganism 19, 47, 226, 230
microplastics 44, 46, 314
microwave irradiation (MW irradiation) 165–172, 284, 295
Minamata disease 53
miniaturization 68–70, 75, 297–299, 314, 315
MNP *see* magnetic nanoparticle
model 54, 65, 285
  circular economic 36, 54, 56
  integrated 71
  linear 36
  linear economic 54, 74
monomers 101, 215, 216, 225, 226
Monsanto process 93
Mother Nature 125, 180, 300
MSW *see* municipal solid waste
municipal solid waste (MSW) 34, 50, 51
MW irradiation *see* microwave irradiation
MW technology 165, 166, 173

nanocatalysis 104, 105, 107, 109, 111
nanocatalyst 105, 106, 112, 114
nematodes 246, 248, 249
nucleophile 193–196
nylon 17, 18, 46, 185, 207, 225, 226

ocean 34, 44, 55, 163
oil 4, 37, 52, 56, 60, 79, 147, 216, 219–223, 235, 238, 239
  brown 153
  crude 5, 59, 215, 224, 295
  edible 121
  heavy 237
  motor 4
  plant 59
  vegetable 59
  waste cooking 58, 59
olefins 109, 127, 149, 150, 183, 184
organic compound 16, 22, 41, 124, 125, 139, 167, 194, 235, 242, 245, 271
organic reaction 109, 125, 127, 139, 143, 193, 293, 310
organic synthesis 124, 125, 127–129, 131, 133, 135, 137, 139, 141, 143, 145, 147, 149, 157, 293, 295
organisms 42, 103, 124, 215, 228, 247, 248
Ostwald process 96
oxidant 111, 126, 140, 142, 209, 243
oxidation 16, 84, 85, 98, 142, 145, 150, 169, 177, 188, 189, 192–194, 211
oxidation reaction 126, 127, 134, 137, 138, 140, 142, 145, 150, 169, 177, 209
oxide 97, 98, 152, 186
  graphene 294
  propylene 90, 289
  rare earth 244
  tributyl tin 226
ozone depletion 6, 14, 266, 311, 312
ozone layer 4, 15, 121, 207, 235, 236

paint 2, 16, 21, 46, 60, 174, 226, 229, 230, 280
palladium 67, 84, 98, 107, 211, 219
Paterno–Büchi reaction 183, 184
pathways 9, 12, 42, 62, 178, 206, 235, 264, 280, 290, 292
  biocatalytic 253
  chemical 10
  electrochemical 193

linear 285
redesigned 74
shikimic acid 207, 208
solvent-free 293
surface run-off 42
synthetic 63, 75, 121, 213, 258, 284
PEG *see* polyethylene glycol
pesticides 1, 3, 17, 41, 226, 246–249
pests 3, 246, 248, 249
pharmaceutical company 3, 63, 121, 206, 309
phase transfer catalyst (PTC) 100–102, 114
phenol 12, 13, 18, 85, 110, 171, 207, 271
phone 47, 49, 68, 164, 244
phosgene 102, 212, 311
photocatalyst 181, 186–188, 190, 191
photocatalytic reaction 188, 191
photochemistry 180, 181, 185, 186, 191, 306
photodimerization 184
photodissociation 236
photoexcitation 182, 186, 188
photons 181, 186, 188, 189, 285
photosynthesis 57, 147, 180, 182, 248
pigments 101, 230, 231, 258, 280
planet 4, 20, 50, 55, 71, 257, 263, 265, 272, 292, 302
plant 5, 57, 60, 66, 147, 182, 248, 249
chemical 5
hydroformylation 92
marine 226
organic 57
pesticide manufacturing 5
rosaceous 248
sewage treatment 46
wastewater treatment 72
plasticizers 17, 207, 224, 225
plastics 2, 19, 21, 34, 35, 44–47, 51, 60, 61, 65, 74, 214, 215, 230
biobased 60
biodegradable 46, 75, 311
compostable 60
conventional 60
petroleum-based 45, 219
pollutants 2, 6, 8–10, 34, 180, 314
acidic 9
inorganic 231
ocean 45
organic 179, 187–189
pollution 20, 22, 38, 44, 46, 50, 55, 57, 82, 89, 265, 311, 314
polyethylene glycol (PEG) 100, 139, 140, 155, 292
polymerization reaction 133, 216
polymer 15, 18, 59–61, 75, 101, 107, 108, 145, 211, 215, 225, 238
biobased 59
biodegradable 19
high-molecular-weight 216
petroleum-based 21
soluble 292
synthetic 224
thermoplastic 215
polystyrene 45, 208, 215
polyurethane 17, 21, 211, 289
polyvinyl chloride (PVC) 48, 51, 53, 224, 225
potassium cyanide 178, 200
potassium permanganate 177, 209
potato 60, 61, 111
poverty 55, 83, 156, 163, 206
pressure 93, 95, 129–132, 162, 166, 175, 176, 237, 249, 250, 252, 302, 305, 307
acoustic 176
blood 221
critical 130
environmental 60
hydrostatic 176

negative 175
supercritical 237
vapor 120
pressure-treated wood (PTW) 250
principles 10, 11, 29, 35, 69, 75, 264–266, 287, 288, 291–293, 297, 298, 300, 303
aerodynamic 300
basic 20, 164, 181, 200
fundamental 88
green chemistry 11–13, 15, 17, 19–21, 206, 213, 214, 257–259, 264, 267, 269, 272, 274, 284, 287
protein 112, 147, 232, 237, 248, 303
protocol 71, 112, 113, 150, 153, 186, 198, 241, 243, 255, 315
catalyst-free 295
catalytic 255
electrochemical 199
photochemical 186
solvent-free 154
stringent 314
PTC *see* phase transfer catalyst
PTW *see* pressure-treated wood
purification 97, 119, 120, 190, 244, 285, 309
PVC *see* polyvinyl chloride

quaternary ammonium 100, 107, 108, 250

radiation 171, 180, 191
electromagnetic 180
infrared 237
MW 164, 166, 167, 173
solar 191
ultrasonic 111
ultraviolet 4, 121, 235
ranitidine 36, 255, 256
raw material 11, 12, 54, 59–61, 65, 70, 75, 82, 89, 218, 224, 311, 315
reactant 12–14, 23, 25, 27, 80, 81, 98, 100, 101, 125, 126, 150–154, 156, 166, 167, 170, 171, 181, 193, 293
gaseous 95
nonpolar 126
organic 129, 137
solid 152
reaction conditions 89, 100, 101, 110, 120, 128, 142, 165, 168, 169, 171, 198, 207
reaction medium 15, 105, 110, 133, 137, 140, 141, 149, 157, 166, 194, 272
reaction miniaturization 297–299
reaction rate 89, 101, 120, 125, 126, 131, 140, 180, 191, 192
reactor 82, 150, 307, 308, 314
batch 314
chemical 162
continuous 298
electrochemical 307
plug flow 307
ultrasonic 180
recycling 48–50, 52, 54, 55, 64–66, 70, 72, 195, 199, 224, 226, 245, 280, 281, 292
reduction 64, 65, 69, 70, 127, 128, 186, 188, 192, 194, 196, 215, 219, 295, 299, 302
reduction reaction 127, 139, 144, 146
renewable resource 46, 59, 60, 142, 214, 219, 285, 296
rhodium 64, 67, 91, 98, 137
river 4, 5, 35, 46, 53, 85

SCF *see* supercritical fluid
SDGs *see* Sustainable Development Goals
ship 35, 164, 226
shock wave 176
silica 18, 95, 96, 98, 99, 107, 294
sitagliptin 63, 64, 86

soap 59, 73, 238
sodium hydroxide 12, 41, 84, 242
soil 42, 47, 49, 50, 72, 247
solubility 131, 137–139, 149, 150, 197, 238, 239, 290
solvent 23, 64–66, 100, 101, 119–121, 123–125, 131, 132, 137, 139, 140, 142, 148, 151, 154–157, 272, 273, 297–299, 307, 308
  aprotic 101
  biobased 143, 155
  biomass-based 143
  biomass-derived 143, 157
  carbon-neutral 147
  chlorinated 15, 16, 132
  classical 122, 133
  dipolar 141
  discarded 154
  eutectic 292
  flammable 28
  fluorous 150
  fluorous biphasic 148, 155
  green 122–124, 149, 156, 292, 299, 315
  halogenated 178
  harsh 169
  nongreen 140
  nonpolar 238
  nonvolatile 139
  organic 16, 89, 126–128, 136, 139, 142, 143, 149, 157, 235, 237, 244
  polar 238
  traditional 141
solvent-free synthesis 151, 153, 154, 293
sonochemistry 17, 174, 176, 177, 180
species 3, 44, 93, 95, 105, 154, 179, 302
  aquatic 19
  bald eagle 247
  dibutyltin 230
  inorganic 231
  marine 34, 230
  organometallic 179
  reactive 185
  reactive carbanion 128
spectrum 61, 164, 180, 187, 230, 247
starch 57, 59, 147, 215, 216
starting material 19, 22, 25, 36, 38, 62, 69, 207–209, 215, 222, 240, 242, 271, 272
Strecker synthesis 210, 258
styrene 45, 208, 209, 258
substitution reaction 27, 139, 178, 220, 242
substrate 64, 82, 90, 112, 125, 154, 193, 195, 206
sugar 30, 57, 61, 86, 147, 182
supercritical CO_2 16, 21, 122, 131, 132, 236, 237
supercritical fluid (SCF) 16, 123, 129–132, 155, 157, 198, 237, 292, 306
surfactant 17, 18, 235, 238, 239
sustainable chemistry 10, 21, 29, 213, 274, 277, 278, 312
sustainable development 6, 20, 21, 36, 54, 55, 205, 257, 265, 284, 286, 311, 313
Sustainable Development Goals (SDGs) 20, 36, 55, 56, 74, 83, 156, 162, 163, 263, 280, 312, 314
sustainable solutions 9, 265, 272
Suzuki–Miyaura reaction 108
Suzuki reaction 29
system 40, 66, 130, 175–177, 190, 287, 307
  automated 66
  automated solvent recycling 66
  biphasic 149, 150
  choice-based credit 277
  exhaust 97
  heterogeneous catalytic 198

human reproductive 42
immune 228, 242
nervous 121
reactive 166
sewage 121
weak immunity 192

technique 37, 52, 61, 76, 83, 113, 154, 162, 164, 242, 309, 310
analytical 69
catalyst recovery 67
fluid/solvent extraction 297
food processing 219
inorganic analysis 69
miniaturized extraction 297
miniaturized-sample preparation 69
solvent microextraction 297
sustainable 64
synthetic 269
traditional 74
waste management 34, 74
tetrahydrofuran (THF) 123, 139, 145, 155, 179
textiles 60, 65, 172, 280, 296
THF *see* tetrahydrofuran
thiamine 85–87
toluene 15, 108, 121, 123, 124, 139, 149, 150, 178, 200, 235, 298
toxic by-products 50, 52, 74
toxic chemicals 9, 43, 69, 312
toxicity 2, 3, 11, 40, 41, 141, 143, 222, 224, 228, 230–232, 234, 247–250, 285, 288, 290, 312
acute 132, 228
aquatic 289
chronic 8, 122
liver 312
reproductive 230
toxins 219, 231
triglycerides 144, 219–223

Ullmann cross-coupling reaction 110
ultrasonic irradiation 177–179, 200, 291, 295
ultrasound 17, 173–179, 293, 307
urethanes 211, 212, 258

van der Waals forces 238
virus 248, 249
viscosity 16, 18, 131, 136, 154, 176, 237
VOC *see* volatile organic compound
volatile organic compound (VOC) 16, 41, 120, 121, 141, 144, 156, 235

Wacker Chemie oxidation process 211
waste 11–13, 17, 18, 21–23, 25, 34–38, 40–44, 50–66, 68–72, 74–76, 82–86, 162, 163, 210, 240, 241, 249, 270, 271, 273, 274, 284, 285, 297–299, 302
academic 43
agricultural 57
animal 57
chemical 8, 35, 41, 51–53, 312
conventional 71
corrosive acid 41
food 58, 60
gaseous 34, 50
hazardous 19, 34
hospital 38
ignitable 41
industrial 57
liquid 34, 43, 50, 69, 271
nonhazardous 38, 41
noxious 86
organic 51, 52, 55
pharmaceutical 37, 38, 40–42
plastic 34, 44, 45, 47, 56, 215, 297
reactive 41
recycle paper 50
shredder 52

solid 34, 38, 40, 47, 50, 51, 218, 224
solvent 63
toxic 34, 41, 86, 273, 280, 285
urban 143
zero 61
waste generation 34, 36–39, 41, 43, 45, 47, 49, 52, 54, 62, 64, 66, 70, 71, 73, 74, 296
waste management 13, 35, 36, 52, 55, 64, 71, 74
waste materials 22, 36–38, 52, 56, 57, 60, 76, 187
waste minimization 9, 37, 44, 66, 72, 89, 172, 264, 280, 281, 297, 306
waste products 23, 28, 62, 225, 280
waste reduction 59, 63, 64, 69, 71, 272, 293, 298
waste source 58, 74, 156
water pollution 8, 38, 52, 291
water purification 181, 195
Wittig reaction 26, 27
wood 51, 57, 242, 249–251

xylene 169, 208, 235

zeolites 18, 95, 107, 167, 171, 293, 294
Zollinger–Ellison syndrome 255